Wireless Identification and Sensing Systems for Harsh and Severe Environments

IEEE Press
445 Hoes Lane
Piscataway, NJ 08854

Wireless Identification and Sensing Systems for Harsh and Severe Environments

Edited by

Smail Tedjini
Grenoble INP/LCIS
University Grenoble Alpes
Valence
France

Valentina Palazzi
Department of Engineering
University of Perugia
Perugia
Italy

IEEE Press
WILEY

Published by John Wiley & Sons, Inc., Hoboken, New Jersey.
Published simultaneously in Canada.

Library of Congress Cataloging-in-Publication Data is applied for

Hardback ISBN 9781394169078

Cover Design: Wiley
Cover Images: © Longhua Liao/Getty Images, © Weiquan Lin/Getty Images

Set in 9.5/12.5pt STIXTwoText by Straive, Pondicherry, India

To our families and to all the colleagues and mentors who inspired our research.

Foreword

When Smail and Valentina contacted me, asking to write a foreword for a new book which they were putting together as co-editors, I was thrilled. I have known Smail for nearly 20 years: first, through his pioneering publications on radio frequency identification (RFID), the field in which we both started working around the same time, and then we finally met in person at one of IEEE Antennas and Propagation Symposiums in USA and kept in touch ever since. As for Valentina, I have known her publications on chipless and harmonic tags for years, and finally had the pleasure to meet her in person last year in Valence, France. Both Smail and Valentina are world-class researchers, and they managed to put together a team of other high-caliber researchers in RFID, who wrote individual chapters for this great book.

RFID is a wireless technology that has been around for a while and significantly evolved. It now extends beyond pure identification (originally it was conceived as sort of a wireless barcode) and encompasses low power wireless devices and systems with various sensing, networking, security, and localization capabilities. Hence, the generic term "RFID" is often used loosely to denote a variety of active and passive RF technologies: actively transmitting or passively backscattering, with energy-harvesting or battery-powered tags, etc. The focus of this particular book is passive RFID and its operation in severe environments.

While many active RF or wireless technologies can be used for sensing and communication, passive RFID is especially attractive and presents a very compelling low-cost, reliable and scalable alternative to solutions based on various active wireless protocols. It is well positioned as a core technology for enhancing everyday physical objects with sensors and communication abilities and connecting them to the Internet of Things (IoT). Passive RFID deployments are growing in size and number as they help to solve numerous business use cases across a variety of industries and supply chains. An estimated 40 billion passive RFID tags (also called transponders) have been sold in 2023 to a wide range of industries and businesses including consumer goods tracking, apparel tagging, security, access, healthcare, construction, airlines, etc., with majority of that volume being RAIN and NFC tags.

Passive RFID can be classified into chipless and chip-based. Chipless RFID encompasses pure chipless tags (composed of multiple RF resonators), surface acoustic wave (SAW) tags (which may include a special non-semiconductor SAW chip), and harmonic tags (diodes are the only semiconductor devices used in those). Chip-based RFID tags use complex semiconductor ICs, containing more transistors than microprocessors used in PCs in mid-80's. Passive RFID can work in various frequency bands such as LF (Low Frequency), HF (High Frequency, also known as NFC), UHF (Ultra High Frequency, also known as RAIN), or microwave. Chip-based and chipless tags of different kinds described above have varying degrees of sophistication, and their design is often a tradeoff between functionality and cost. All passive RFID systems may be deployed in both far field and near field scenarios.

Basic RFID principles are well known by now, but many challenges lie in the practical implementation. This is exactly the area which this book targets: how various passive RFID systems (chip-based and chipless) work in real environments: in presence of metal objects and moisture, in propagation environments that are not governed by Friis free-space equation, in applications where tags can be implanted in human body or placed on food items, in situations where extreme radiation may occur, etc. The collection of chapters in this book represents a sampling of various active research topics in the field of passive RFID implementation in real environments, covering all of those cases above.

I am very excited to see this book in print and can't wait to see what you, dear reader, will think about it. You opened this book because you are interested in RFID. I envy you, because I have already read it, and you still have this experience ahead of you. RFID technology will continue to develop and pose many interesting challenges for researchers and engineers in both academia and industry, such as yourself. I hope that this book will inspire you and will lead you to some new ideas. Who knows, maybe you will become a contributor to the next book on RFID?

Pavel Nikitin
May 2024,
Seattle, WA

Contents

List of Contributors

Sara Amendola
RADIO6ENSE S.r.L.
Roma
Italy

Thierry Aubert
Université de Lorraine, CNRS, IJL
Nancy
France

Rahul Bhattacharyya
Auto-ID Labs, Massachusetts Institute of Technology
Boston, MA
USA

Francesca Benassi
DEI "Guglielmo Marconi"
University of Bologna
Bologna
Italy

Martí Boada
Department of Electronic, Electric and Automatic Engineering
Universitat Rovira i Virgili
Tarragona
Spain

Michele Borgese
Research and Development Department, SIAE Microelettronica
Milan
Italy

Mathieu Le Breton
Géolithe Company
Crolles
France

ISTerre, Université Grenoble Alpes
Grenoble
France

Mathieu Cassel
ENS, Université de Lyon
Lyon
France

Fanny Coffigniez
Institut d'Électronique et des Systèmes, Univ Montpellier
Montpellier
France

Filippo Costa
Dipartimento Ingegneria dell'Informazione, Università di Pisa
Pisa
Italy

Alessandra Costanzo
DEI "Guglielmo Marconi"
University of Bologna
Bologna
Italy

Francesco Alessio Dicandia
Istituto di Elettronica e di Ingegneria dell'Informazione e delle Telecomunicazioni, Consiglio Nazionale delle Ricerche (CNR)
Torino
Italy

Yvan Duroc
Univ Lyon, Université Claude Bernard Lyon 1, INSA Lyon, Ecole Centrale de Lyon, CNRS
Villeurbanne
France

Omar Elmazria
Université de Lorraine, CNRS, IJL
Nancy
France

Cécile Floer
Université de Lorraine, CNRS, IJL
Nancy
France

Simone Genovesi
Dipartimento Ingegneria dell'Informazione, Università di Pisa
Pisa
Italy

David Girbau
Department of Electronic, Electric and Automatic Engineering
Universitat Rovira i Virgili
Tarragona
Spain

Jasmin Grosinger
Institute of Microwave and Photonic Engineering, Graz University of Technology
Graz
Austria

Sami Hage-Ali
Université de Lorraine, CNRS, IJL
Nancy
France

Amirhossein Karami-Horestani
CIMITEC, Departament d'Enginyeria Electrònica, Universitat Autònoma de Barcelona
Bellaterra
Spain

Antonio Lázaro
Department of Electronic, Electric and Automatic Engineering
Universitat Rovira i Virgili
Tarragona
Spain

Giuliano Manara
Dipartimento Ingegneria dell'Informazione, Università di Pisa
Italy

Gaetano Marrocco
Tor Vergata University of Roma and RADIO6ENSE S.r.L.
Roma
Italy

Ferran Martín
CIMITEC, Departament d'Enginyeria Electrònica, Universitat Autònoma de Barcelona
Bellaterra
Spain

Diego Masotti
DEI "Guglielmo Marconi"
University of Bologna
Bologna
Italy

Alicja Michalowska-Forsyth
Institute of Electronics, Graz University of Technology
Graz
Austria

Carolina Miozzi
RADIO6ENSE S.r.L.
Roma
Italy

Manuel Monedero
NXP Semiconductors GmbH
Munich
Germany

Cecilia Occhiuzzi
Tor Vergata University of Roma and RADIO6ENSE S.r.L.
Roma
Italy

Giacomo Paolini
DEI "Guglielmo Marconi", University of Bologna
Bologna
Italy

Valentina Palazzi
Department of Engineering
University of Perugia
Perugia
Italy

Ferran Paredes
CIMITEC, Departament d'Enginyeria Electrònica, Universitat Autònoma de Barcelona
Bellaterra
Spain

Camille Ramade
Tageos
Montpellier
France

Benjamin Saggin
IES, Univ Montpellier, CNRS
Montpellier
France

Brice Sorli
IES, Univ Montpellier, CNRS
Montpellier
France

Robert Staraj
Université Côte d'Azur, LEAT
Laboratoire Electronique Antennes et Télécommunications
Sophia Antipolis
France

Smail Tedjini
LCIS, University of Grenoble-Alpes
50 Rue Barthelemy de Laffemas
Valence
France

Philippe Le Thuc
Université Côte d'Azur, LEAT
Laboratoire Electronique Antennes
et Télécommunications
Sophia Antipolis
France

Arnaud Vena
IES, Univ Montpellier, CNRS
Montpellier
France

Ramón Villarino
Department of Electronic, Electric
and Automatic Engineering
Universitat Rovira i Virgili
Tarragona
Spain

Konstantinos Zannas
LCIS, University of Grenoble-Alpes
50 Rue Barthelemy de Laffemas
Valence
France

About the Editors

Smail Tedjini, Life Fellow IEEE, URSI Fellow; Doctor d'Etat in Physics 1985 Grenoble University; senior researcher of CNRS 1986–1993; University full professor since 1993, now with Grenoble Institute of Technology. His main topics for both research and teaching, concern Applied Electromagnetism, RF, Optoelectronics, RFID, and Wireless Systems. He served as coordinator and staff member in numerous academic programs both for education and research. He was coordinator for PhD program, master and bachelor programs for the Universities at Grenoble, some of these programs are under collaboration with international universities from Europe, Canada, Brazil, Vietnam, Egypt, and Maghreb. He served as the director of Esisar, Department of Grenoble-INP. He is involved in academic research supervision since 1982. His main topics in research are applied electromagnetism, modeling of devices and circuits at both RF and optoelectronic domains. Current research focuses on wireless systems with special attention to RFID technology and its applications. He is the founder in 1996 and past director of the LCIS Lab. Now, he is within ORSYS group that he leaded until 2014 and founded in 2002. He supervised tens of research contracts with public administrations and industries. He supervised 50 PhD, co-authored 350 publications, 3 books, 7 patents, 10 book chapters, and book: Wiley "Non-linearities in Passive RFID Systems" and served as reviewer/opponent for tens of PhD worldwide.

TPC-Cochair of IEEE-RFID2011 TPC Member for several editions of IEEE-RFID, TPC member or cochair for RFID-TA since 2010. General Chair IEEE RFID-TA 2012. Organizer of IEEE-RFID-TA 2012 in France. Organizer/co-organizer of RFID special sessions in numerous conferences in particular: IEEE-MTTs, IEEE-APs, IEEE-RFID-TA, EuMW, EuCAP, URSI-GASS, URSI-ATRASC, local conferences in France, Brazil, Vietnam, Tunisia. Organizer of RFID summer school in France, Brazil, Vietnam, and Maghreb. Organizer of (3) Student Design Competitions at IEEE-IMS. Editor of special issue on RFID at IEEE-microwave magazine, and Annales des telecom. Vice-chair IEEE MTT-TC

24 (2017–2019). Chair IEEE MTT-TC-26 (2020–2021), since 2024 he is the chair of French Chapter of IEEE-Antenna and Propagation. Since October 2022 he is nominated as Emeritus professor at University Grenoble Alpes. Since 2024 he is Distinguish Lecturer of the IEEE Council on RFID.

Valentina Palazzi received the MS degree in electrical engineering and the PhD degree in industrial and information engineering from the University of Perugia, Italy, in 2014 and 2018, respectively. During her PhD, she was visiting PhD student at the Tyndall National Institute, Cork, Ireland, at the Centre Tecnològic de Telecomunicacions de Catalunya, Barcelona, Spain, and at the Georgia Institute of Technology, Atlanta, GA, USA. Currently, she is a researcher with the High Frequency Electronics Laboratory at the University of Perugia, Perugia, Italy. She has been the scientific coordinator for the University of Perugia of the European projects CHARM - Challenging environments tolerant Smart systems for IoT and AI (G.A. n. 876362) and OPEVA - Optimization of Electric Vehicle autonomy (G.A. n. 101097267).

She wrote more than 80 scientific papers (both journal and conference proceedings), and a book chapter, and she holds three patents. Her current research interests include the design of RF components, wireless sensors, radar front ends, wireless power transfer technologies, additive manufacturing, and conformal electronics.

Dr. Palazzi is an elected member of the IEEE Microwave Theory and Technology Society (MTT-S) Administrative Committee from 2024 to 2026. She is a member of IEEE MTT-S Technical Committees 16 “Microwave and Millimeter Wave Packaging Interconnect and Integration,” 17 “Microwave Materials and Processing Technologies,” and 26 “RFID, Wireless Sensor and IoT.” She was the Chair of the Technical Committee 26 from 2022 to 2023. She has been an Early Career Representative of the URSI Commission D “Electronics and Photonics” of the International Union of Radio Science (URSI) from 2021 to 2026. She won a number of international prizes, including the IEEE Graduate Fellowship in 2017, 2017 MTT-S Prize–Italy Chapter Central and Southern Italy, and the International Union of Radio Science (URSI) Young Scientist Best Paper Award conferred at the 2019 URSI Italian National Meeting. Since November 2024, she has served as an Associate Editor for IEEE Microwave and Wireless Technology Letters.

Preface

In the continuously evolving digital landscape, wireless systems are experiencing considerable developments. Emerging concepts such as digital twins, Industrial IoT, telemedicine, precision agriculture, and fail-operational systems are changing the approach toward the design of electronics and telecommunication systems, opening the way to new markets and applications that were unimaginable only a few decades ago. The urgence for pervasive real-time monitoring systems is pushing technology beyond its current limits, raising new challenges for both academic and industrial R&D labs.

In this context, wireless sensors and RFID technologies can offer very effective solutions provided that their design and implementation take into account the characteristics and constraints imposed by the envisioned applications. For both wireless sensors and RFID devices the use of radiofrequency (RF) technologies is pivotal not only for enabling the data transfer but also for sensing and energy scavenging. While the simplicity and the high performance of RF devices in free space and line-of-sight (LOS) are well known, the situation is quite different for real applications that can include the presence of lossy materials, non-LOS communication, metallic elements, and so forth. Therefore, in most applications the real environment is very different compared to free space and introduces severe propagation conditions from an electromagnetic point of view. Additionally, harsh environmental conditions, such as high temperatures or humidity, can challenge electronics operation, calling for new materials and design approaches.

This book is aimed at both doctoral students and engineers developing R&D projects about wireless and RFID technologies. It provides a unique source of examples of successful wireless system solutions leveraging RFID technologies and similar energy-efficient wireless approaches, enabling the implementation of advanced concepts such as Internet of Things (IoT) in severe and harsh environments. Indeed, while several literatures on wireless technologies and RFID systems are available and can be considered for the design of basic systems mostly in controlled environments (i.e. anechoic chambers) and academic labs, only a few focus on the actual implementation of advanced and effective solutions in real environments. In real applications wireless sensors and RFIDs have to communicate their information by means of RF signals which propagate in heterogeneous and complex environments. The presence of metal, liquids, biological tissues, plants, and so forth cause significant degradation, distortion or even cancellation of the RF signals, not to mention the detuning effects on the antenna and on the other RF signal components that are generated by the presence of heterogeneous objects in the application environment. Real environment exhibits severe behavior for electromagnetic signals and RF devices. The same characteristics are encountered in harsh environment where high temperatures, strong radiation, and corrosive materials are present. Last but not least, wireless systems that operate in real applications have to comply with the constraints imposed by standards and regulations which are rarely considered in controlled environments.

The book covers the recent advances in wireless and RFID systems where severe electromagnetic behavior and harsh conditions are taken into consideration, while complying with RF standard and regulations. So, this book provides the reader with the design rules and methodologies to obtain satisfactory performance and possibly avoid the typical oversights and mistakes that can be made when first approaching to this topic.

The book is organized in 12 independent chapters grouped in 3 sections. The first section is dedicated to RFID design approaches to implement passive wireless sensors able to operate in harsh and severe environments. This section includes 6 chapters concerning different use cases, which are briefly described herein below:

Chapter 1 entitled "UHF RFID Identification and Sensing for the Industrial Internet of Things (I-IoT)." This chapter analyzes the challenges for the successful integration of RFID sensor networks within the industrial and consumer IoT contexts. The challenges in RFID sensor networks encompass robust communication links and accurate sensing measurements. Industrial applications confront additional difficulties due to wireless powering in presence of lossy mediums and metallic objects, compounded by high temperatures and sensor placement on fast-moving mechanical parts. Temperature impacts sensing ICs, reducing communication ranges, and affecting performance. External probes connected via analog front ends experience stronger temperature sensitivity, causing nonlinear drift. Monitoring high-speed components introduces further issues, including backscattering link robustness and data transmission/decoding issues, exacerbated by electromagnetic noise from high-power motors. Ensuring data integrity across the IoT platform necessitates comprehensive security assessments. Adopting new-generation of microchips with sensing capabilities and low-power encryption algorithms for data security requires careful evaluation, considering implications for power consumption and data rates. All these challenges are carefully described in the chapter, and possible countermeasures are highlighted.

Chapter 2 entitled "RFID Sensing In Power-Plant Generators and Power Transformers." The harsh environment in this chapter is predominately represented by the dense metallic parts which need to be monitored wirelessly. Such environments hinder the antenna performance. In addition, fast rotating parts cause mechanical stress on potential sensors, and the high electric/magnetic field needs to be taken into consideration when selecting the antenna. To tackle the aforementioned harsh environment impact, there are specific choices to be made: the antenna structure should offer good performance when positioned on metallic surfaces, the weight of a potential sensor should be minimized to avoid adding extra strain to the rotating system and loop antenna types must be avoided since they can be current generated by the movement inside of a high magnetic field. Overall, solutions are shown, which are dictated by the expected environment of operation.

Chapter 3 entitled "Design of Passive UHF RFID Sensors Meeting Food Industry Regulations." RFID tags transformed into sensors sensitive to certain parameters of their environment must be produced in large quantities and at low cost. These requirements can affect measurement reliability, especially in harsh environment conditions. The solutions presented in this chapter aim to avoid this lack of reliability and could therefore be considered for a variety of applications. So by using a simple RFID antenna and a suitable RFID chip, the described systems are easy to duplicate and can be used in groups. Food products are delivered as batches and are separated at the very end in the delivery to the consumer. Although monitoring until the purchase is of interest, it is essential during the first few hours/days of transport, because if the repeated exchanges between the various logistics partners and the exponential growth of micro-organisms. Furthermore, unlike many RFID systems, the described solutions are independent of external databases. This is an advantage, especially in crowded electromagnetic or severe environments.

Chapter 4 entitled "Challenges of Using RFID for Outdoor Environmental Monitoring." The deployment of RFID tags outdoors brings a new spectrum of technical difficulties that must be overcome. The presence of water rain, dew, snow, or frost on tags usually decreases signal strength and modify phase difference of arrival, by coupling with the tag antenna. In the far field, propagation through non-air mediums such as soil, water, snowpack, snowy terrains, or vegetation increases loss, phase delay and multipath interferences. These conditions make it challenging to reliably use passive tags outdoors for localization, sensing but also identification. In this chapter all these challeges are accurately described, with the aim to provide an estimate of the performance loss.

Chapter 5 entitled "Harmonic Transponders for Tracking and Sensing." This chapter is dedicated to a special category of backscatter radios, called harmonic transponders. Harmonic transponders can operate in environments characterized by strong reflections and metal parts, which are particularly severe for signal propagation, as the contribution from the tags can be easily separated from the rest of the reflections. Additionally, due to their simple circuitry (they are usually based on single diodes), they are particularly robust, which makes them good candidates to operate in environments characterized by harsh conditions (such as low or high temperature, high humidity, and so forth), thereby expanding the possible fields of application of wireless sensors.

Chapter 6 entitled "Passive Wireless Sensors in Radiation Environments." This chapter examines the radiation environment and its impact on passive/batteryless wireless sensors using the ultrahigh frequency (UHF) radiofrequency identification (RFID), specifically the ramifications of radiation on electronic devices. The chapter explores the use of RFID sensors in radiation environments, including potential radiation effects on RFID tags and methods for protecting these tags against radiation damage. In particular, the discussion links the fundamentals of radiation effects in CMOS circuits with the architectural characteristics and operation features of the RFID tag chip frontends, going down to the circuit block level. Although RFID sensor tags are not yet widely used in radiation environments, there is promising potential demonstrated by existing sensor tag prototypes. These prototypes have successfully addressed major issues, such as tight power constraints resulting from batteryless operation and harsh radiation environments caused by ionizing radiation. Signal-pattern-based sensor tag systems have proven particularly successful in areas such as wireless power transfer, setup independence, and robustness in moderate static multipath environments. Additionally, using scaled CMOS for RFID chips and their passive operation is a promising trend that offers a certain degree of radiation hardness.

The second section entitled chipless includes four chapters, dedicated to wireless sensors based on both surface acoustic wave (SAW) and RF resonators, as reported hereafter:

Chapter 7 entitled "SAW Devices Combining RFID and Sensor Functionalities for Harsh Environments." This chapter is dedicated to provide an introduction to the surface acoustic wave (SAW) sensors. In our modern society, due to the development of the Internet of Things (IoT) or the industry 4.0, the continuous monitoring of physical data has become a real need. This is still a challenge in harsh environments such as high-temperature environments or difficult to access environments. Thanks to an electro-mechanical conversion, the surface acoustic wave (SAW) technology enables to overcome the limits of conventional sensors. Indeed,

SAW sensors are passive systems and thus there is no embedded electronic and no need to regularly replace the battery. Thanks to their small size they can be placed in small and tight spots or even work as abandoned sensors (buried sensors for example). SAW sensors can also be remotely interrogated and can thus offer exciting perspectives for remote monitoring and control of moving parts.

Chapter 8 entitled "Wireless Sensing for Harsh and Severe Environments Based on SAW Sensors." Surface acoustic wave (SAW) sensors are able to measure various physical parameters like temperature, pressure, and stress without an external power supply. This type of sensors can withstand high temperatures (up to 600 °C), voltages up to 530 kV, currents up to 20 kA, as well harsh chemical environments (heavy fuel oil, SF6 gas, or mineral oil). Therefore, sensors based on SAW technology are used in specific harsh environments like ship engine room, medium- or high-voltage electrical equipment either air- or gas-insulated switchgears or oil-filled circuit-breakers. These environments contain many metallic parts where the sensors can be placed, which in some scenarios involve movement, as well as the presence of chemical. The proximity of metallic parts in motion or chemical in different states, with a temperature and age-dependent complex permittivity, in the vicinity of the antenna associated with the SAW sensor can modify its input impedance leading to a significant measurement error. Actual SAW transducers are described in this chapter, and their performance are toroughly discussed.

Chapter 9 entitled "Microwave Encoders for Motion Control and Chipless RFID Applications." This chapter presents microwave encoder systems for motion control. They represent an interesting option over optical (encoder) systems based on a similar principle, due to the fact that microwaves are more tolerant to the effects of dirtiness and pollution, as compared to optical radiation. Namely, if the apertures in optical encoders are clogged, the system cannot exhibit its functionality, contrary to microwave encoders, more robust against the effects of pollution and dirtiness. Such harsh and polluted environments are encountered in many industrial systems, and, for this reason, microwave encoders are especially suited in such applications.

Chapter 10 entitled "Chipless RFID Technology for Operations in Harsh Environments." There are two main relevant aspects of this chapter that deals with the benefit of using chipless RFID sensors. On the one hand, the chapter addresses the different wireless sensor paradigms, namely surface acoustic wave sensors, near-field ones as well as far-field radio frequency backscattering. Moreover, it puts in evidence the advantages with respect to solutions that resort to wired connection to recover the data at the sensor or solutions that require batteries. On the other hand, several application domains are addressed (space, oil and gas, automotive, industrial tools monitoring) and great attention is paid to the employed materials.

The third section is entitled systems. It includes two chapters and is dedicated to the impact of specific environments on system based on wireless and contactless devices. Chapter contents are briefly described herein after:

Chapter 11 entitled "Energy-Autonomous Wireless Architectures for Predictive Maintenance in Harsh Closed Applications". In the application described in this chapter, the sensor nodes are placed all around an electromagnetically harsh environment full of metal parts, and they are energized remotely by RF sources operating in the Industrial, Scientific and Medical (ISM) 2.4 GHz band; these sources act as "illuminators," and a multitude of wireless batteryless sensor nodes are placed in the key points of the bonnet, in contact with parts of the car that need to be monitored. These spaces are generally made of metallic parts, creating an environment that is hostile to the electromagnetic fields. Suitable solutions are described and their performance is discussed.

Chapter 12 entitled "Implanted NFC Tags: Study of Energy Harvesting and Reading by Means of Smartphones." This chapter describe an IoT solution for implanted electronics based on near-field communication (NFC). Information gathering from implanted electronics is hampered by a number of factors, including poor coupling due to the vastly different sizes of the reader and the tag antennas; the unknown position of the implanted device causing misalignment; the constrained quality factor of the communication bandwidth; the detuning of the antennas; and, most importantly, the effects of the body that increase the losses, degrading the quality factor of the antenna. The implanted electronics as well as its antenna must be properly protected to ensure proper operation for a sufficient amount of time. To avoid infections, these protections must also be biocompatible. Consequently, the electrical properties of the antenna are modified. Maximum power transfer to load design should be taken into consideration to effectively power the implanted electronics since the power source comes from the smartphone. Consequently, it is crucial to know the NFC IC input impedance. Hence, considering this type of design, the modeling of the IC presents another issue because of its non-linear behavior. Finally, the presence of ferrites in the smartphone, which allows its antenna to be electrically isolated from the battery and other metallic parts, affects the separation between the mobile and the implant, since they are directly linked to the coupling coefficient.

Smail Tedjini and Valentina Palazzi

Section 1

RFID

1

UHF RFID Identification and Sensing for the Industrial Internet of Things (I-IoT)

Carolina Miozzi, Sara Amendola, Cecilia Occhiuzzi, and Gaetano Marrocco

Tor Vergata University of Roma and RADIO6ENSE S.r.L., Roma, Italy

1.1 Introduction

Internet of Things (IoT) is currently considered the paradigm for the next applications in gaming, leisure, and home automation [1] and is expected to also foster changes in modern medicine, manufacturing, energy, agriculture, and transportation, especially thanks to synergy with the upcoming Fifth/Sixth Generation (5G-6G) wireless communication systems [2]. Indeed, the ability to wirelessly interconnect humans, objects, and machines promises unprecedented technical and economic opportunities in many industrial branches [3]. The capabilities to control complex industrial systems and predict events are enabled by properly processing data gathered directly from surrounding environments and machines in operation, with a positive impact on production, safety, and overall efficiency. This particular implementation of IoT, denoted as *Industrial IoT (I-IoT)*, refers to highly reconfigurable wireless sensor networks that affect the costs, energy, and procedures of the manufacturing process. By adding novel sensors to existing machinery without significantly modifying current architectures, their functionalities (retrofit) will be upgraded to achieve, over the long term, a fully automated, networked, and data-oriented industry [4].

In the last decade, energy-autonomous wireless sensors for ambient monitoring, personal tracking, [6] and manufacturing control have greatly benefited from the well-assessed Radio Frequency Identification (RFID) standard known as EPC

Wireless Identification and Sensing Systems for Harsh and Severe Environments,
First Edition. Edited by Smail Tedjini and Valentina Palazzi.

Published 2025 by John Wiley & Sons, Inc.

C1G2, which nowadays enables sensing functionalities [7], [8, 9] in addition to basic identification. RFID-based sensors require relatively limited maintenance compared to other wireless technologies such as ZigBee, Bluetooth, or WiFi [10]. The required energy is provided by external interrogators that can power and interact with a multiplicity of sensors at a time, thus enabling a single-to-multipoint link with a remarkable reduction in overall wiring. In addition to passive RFID analog tags that take advantage of interactions between the antenna and the environment to indirectly collect sensing information [11], digital sensor tags equipped with dedicated sensors are becoming attractive for real applications [12]. The latter incorporates ICs provided with internal sensors (generally low-medium temperature sensors) and analog/digital front end to connect a variety of external low-power devices. Therefore, several physical parameters can be monitored: high temperature (>80 °C), humidity, and pressure are among the most required in the industrial field. The relatively low cost and maintenance of such tags, their versatility in use and application, and not least, the possibility to exploit conventional RFID readers and standard protocols are expected to boost the massive integration of RFID platforms in industrial scenarios, from automotive to hot/cold manufacturing, from logistics to predictive maintenance. Early studies envisage UHF RFID as a pillar for the implementation of *physical* or *perception layer* of the monitoring platforms [9], especially when active devices are not acceptable in terms of size (e.g. for integration within objects), costs, and operating conditions (severe thermal and mechanical stresses) or in general, when battery replacement is not easy. Recent use cases have proven the effectiveness of RFID sensor networks in the food sector [13], pharma [14], chemical industry [15], and smart grids [16].

The large amount of data generated from these physical sensing layers can be exploited to build a digital representation of plants, machinery, and products, the so-called *digital twins,* in which the physical operation process is virtually judged, analyzed, predicted, and optimized and then transferred back to the physical world to execute the optimized solutions [17]. For this purpose, RFID sensor networks must be integrated within cloud-based infrastructures, in which big data can be efficiently processed according to *data-driven* approaches. Taking advantage of the ability of the modern reader to transmit the data collected through many different communication protocols, for example, MODBUS [18], TCP/IP [19], OPC-UA [20], or MQTT [21], interoperability between platforms, technologies, and monitoring layers is achieved, with benefits in terms of effectiveness of action, usability, and cost optimization.

To establish an efficient monitoring platform through RFID sensors, a robust communication link is required, regardless of the operative conditions of the networks. In fact, in industrial scenarios high temperatures, mechanical solicitations, electromagnetic interference, and the massive presence of lossy and scattering objects such as water and metal could sensibly impact the communication capabilities of sensors, especially in case battery-less configurations are adopted [9]. Operability in a harsh environment has been partially investigated for HF tags subjected to freezing cycles and exposure to gamma irradiation [22] and UHF tags subjected to high-temperature cycles (20–180 °C) combined with immersions in water [23]. Similarly, the effects of weather conditions and exposure to washing cycles have been investigated in [24] and [25], respectively. The results suggest that RFID tags can withstand high-temperature cycling, but the simultaneous presence of water may jeopardize their reliability and sensibly reduce the read distance [26]. Scattering and shielding objects often impose coverage optimization through the selection and positioning of *ad hoc* reader antennas, such as high-directive arrays or omnidirectional monopoles. Additional difficulties arise in the case of moving machinery and objects or when the asset to be monitored is enclosed in shielding cavities. In such cases, the reader antennas need to be placed in close proximity to the tags, that is directly integrated within the shielding.

This chapter provides an overview of the applications and challenges related to RFID in industrial scenarios. Architectures and components are described in Section II followed by examples of real deployments of RFID system for item-level (Section III) and plants (Section IV) monitoring. The main challenges and possible mitigation and countermeasures are finally discussed in Section V.

1.2 I-IoT Ecosystem: Architectures and Components

RFID sensor tags can be deployed in industrial scenarios according to two approaches (see Table 1.1): (i) tags embedded into objects during the production and (ii) tags attached over objects, equipment, machinery components, and connecting infrastructures. In both cases, the tagged item can be stationary or moving with respect to the interrogating system. The two architectures are naturally complementary and can coexist. The detection of moving objects

Table 1.1 Architectural options for RFID sensor network for Industry 4.0.

		Tag Application to the item	
		Embedded	**Attached**
Tag vs reader	**Fixed**	Pipelines tanks	Plants Complex Systems
	In movement	Items on conveyor belts, carousels	Rotating components

through a fixed-reader infrastructure is the most common RFID architecture, for example, the tracking of luggages and items along known trajectories of conveyor systems and crossing through gates. By integrating an RFID sensor tag within a product that undergoes consecutive manufacturing steps along a predetermined path, we can gather information about the sequence of the imposed mechanical, thermal, and chemical treatments. Thus, continuous item-level qualification can be achieved, thus revolutionizing traditional industrial practices based only on sporadic sample checks. Thanks to the one-to-many interaction of the UHF RFID protocol, a plethora of batteryless sensors can be displaced over and within machinery and simultaneously interrogated by one or multiple reader system integrated within the factory data management system. Sensors can be attached over moving/rotating machinery as well as integrated within cabinets and pipelines. A factory as a whole can hence be monitored and controlled without heaving procedures, hardware, and software tools.

The I-IoT ecosystem (Figure 1.1) can be organized in three parts, data generation, data transmission, and data analytics, which will be described next.

1.2.1 Data Generation

Using the taxonomy of I-IoT architectures, the RFID label and sensors are the building blocks of the *Perception Layer,* which is in charge of generating data from the physical world and transmitting them wirelessly to the first node of the network.

The stealthy nature of RFID technology allows it to implement the perception layer to an unprecedented level, bringing the digital intelligence inside the objects, as battery-less labels can seamlessly fit where no other sensing device could.

Off-the-shelf UHF RFID tags can be classified as: (i) conventional logistics tags for identification only and (ii) advanced sensing tags integrating ICs with more sophisticated electronics, such as embedded sensors, an analog-to-digital converter (ADC), power management units, and sensors front end. Both types can be adopted in industrial scenarios as data generators. Tags of the former type can be successfully adopted as sensors, provided that the process/phenomena to be sensed is capable to affect their communication performance with the reader. Using the *antenna-as-a-sensor* approach [27], even conventional RFID passive labels can therefore indirectly collect sensing information. Relevant examples are the monitoring of air quality through a polymer-doped tag sensitive to humidity [28], and the evaluation of fruit ripeness, as described in Sect. 1.4.3.

New-generation RFID ICs are instead provided with on-board ADCs for external sensor interface, embedded sensing capabilities, integrated amplifiers for signal conditioning, and programmable functionalities (e.g. rolling averaging) to preprocess the sensing data. These ICs offer RFID technology to considerably expand the I-IoT application boundaries. However, advanced features are achieved

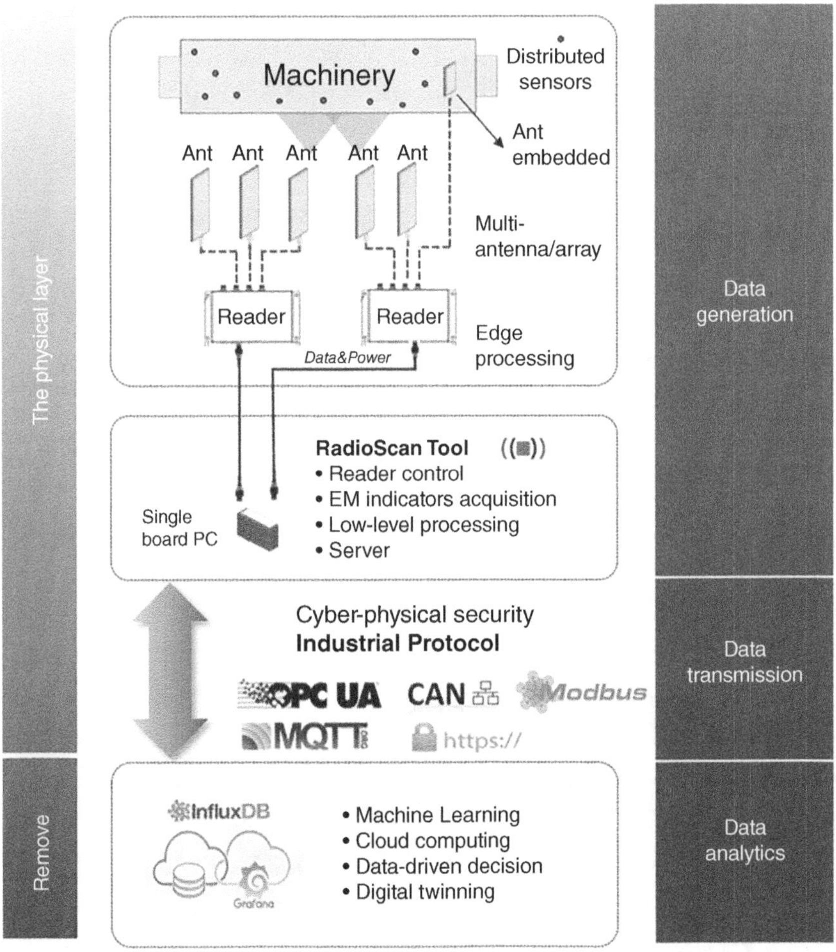

Figure 1.1 Logical layers of an RFID-based sensor network for Industrial Internet of Things.

at the expense of bulkier form factors, higher costs, and greater activation powers to supply the integrated electronics. Commercially available tag-mounting sensing-oriented ICs can be in the form of demo boards, general-purpose inlays, and ad hoc transponders for specific applications. Demo boards (some examples in Figure 1.2) are mainly intended to demonstrate the feasibility of sensing solutions based on RFID technology and to facilitate the implementation of pilot projects. Figure 1.3 shows COTS demo boards connected to an external PT1000 NTC probe to monitor the temperature inside the metal shielded cavities of rolling stock electrical equipment during testing and certification procedures.

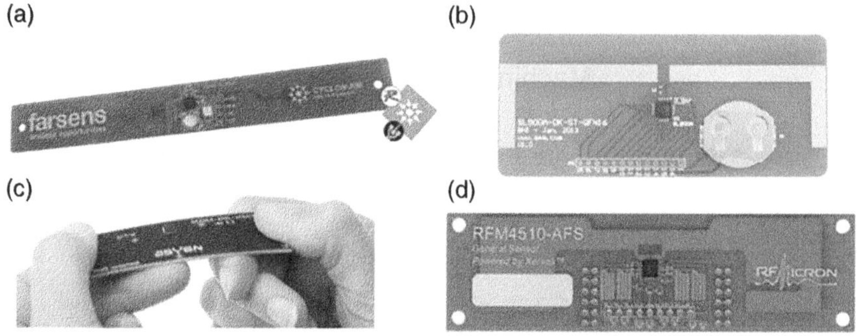

Figure 1.2 Demo boards from sensor-oriented IC's manufacturer. (a) Cyclon RM by Farsens, (b) SL900A by AMS, (c) Demo board by Asygn, (d) RFM4510 by RFMicron. *Source:* Shenzhen Ruichi IoT Technology Co., Ltd., http://www.farsensiot.com/#footer, last accessed January 18, 2024; Xabier Eguiluz et al. 2017/MDPI/CC BY 4.0; ASYGN, https://asygn.com/, last accessed January 18, 2024; Neumüller Elektronik GmbH.

Figure 1.3 COTS demo board applied on electrical equipment of rolling stocks provided with external probes (PT1000 and NTC) to monitor the temperature inside the metal shielded cavities.

Recently, many RFID smart labels manufacturers have begun to expand their product portfolio with ready-to-use sensor tags that can be broadly classified for metal and nonmetal applications (Figure 1.4). The availability of these components significantly contributes to quickly deliver a proof-of-concept of an IoT solution in a real scenario.

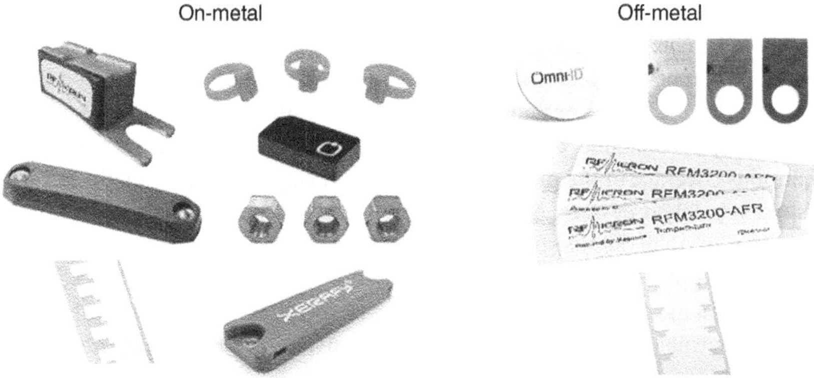

Figure 1.4 Commercial RFID UHF tags with sensing capabilities suitable for on-metal and off-metal applications. *Source:* PQSense Technology Co., Ltd. https://pqsense.com.tw/, last accessed 18, January 2024; Confidex, https://confidexpartners.com/, last accessed January 18, 2024; Xerafy Singapore Pte Ltd., https://xerafy.com/, last accessed January 18, 2024; www.pqsense.eu; http://www.pqsense.eu/; Azxon, https://www.axzon.com/, last accessed January 18, 2024.

However, in many unconventional environments, the adoption of COTS layout is not viable, especially when the tag is conceived for application into lossy mediums (e.g. rubber, carbon-loaded composites, concrete) or integration inside metal cavities.

Therefore, ad hoc tag solutions must be designed (Figure 1.5), by tailoring the shape of the radiator to the target item and selecting the proper materials to optimize communication and ensure durability against harsh operating conditions (e.g. high temperature, pressure, or mechanical stress). For instance, sensor tags for integration within rubber should have substrates permeable to the polymeric

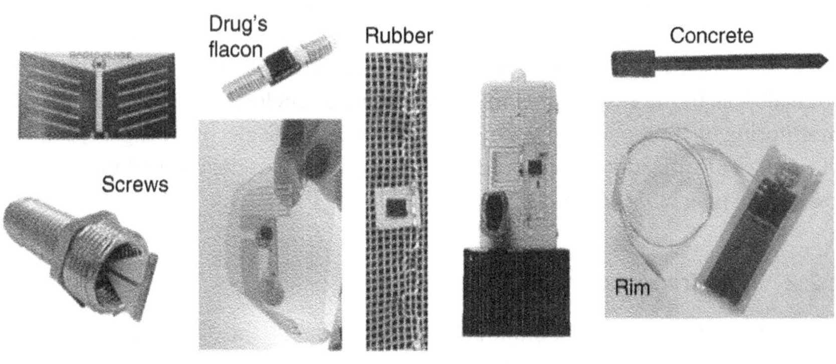

Figure 1.5 Overview of custom RFID sensors designed to solve the need for unconventional scenarios and tag placements.

compounds during the process to make the resulting device seamlessly part of the object. On the contrary, inside metallic or extremely lost objects, such as screws or concrete, when possible, the radiating part of the sensor should be distanced from the sensing one. Batteries or photovoltaic panels can support both communication and sensing by enabling data logging operations even in the case where the continuous coverage of the radio frequency cannot be guaranteed. Finally, flexible substrates, patched devices, or helical layouts can be considered to add sensing capabilities to small objects subjected to strong mechanical solicitations.

Figure 1.6 shows miniaturized cubic-like commercial and custom RFID sensors optimized for measuring the temperature over metal surfaces. A typical use is real-time thermal mapping of the engine compartment of passenger vehicles in service (Figure 1.6).

Besides sensors, RFID readers and interrogating antennas are part of the *Perception Layer*. Nowadays, different classes of UHF RFID readers are commercially available (Figure 1.7), encompassing solutions for compact portable devices for integration on mobile platforms (robots and UAVs) up to fixed reading points. Often rugged devices are required to withstand harsh industrial operative conditions (rain, powders, and mechanical solicitations). In the case of moving objects, reading speed is another critical performance parameter.

The monitoring of extended areas - such as sensors distributed on large-scale machinery or multiple equipments on a production site - poses additional coverage issues.

RFID readers natively provided with multiple ports for interrogation antennas or multiplexed architectures are adopted to significantly enlarge the overall footprint of the reading region or clusterize the different plant sectors. Figure 1.8 shows an interrogating system designed to maximize the linear extension of the reading region (along the main dimension of the machinery) by deploying multiple arrays. Communications metrics in terms of power-on-chip and read-rate of hundreds of sensors within such a region are also reported.

Finally, different challenges arise for interrogating systems when the measurement site is partially shielded up to electromagnetically inaccessible from the "outside," as in the case of sensors placed in metal enclosures. Examples include monitoring of components (bearings, rotors, etc.) inside electrical engine, control of rubber vulcanization inside molds [29] (Figure 1.9a) or the thermal mapping of coaxial cavities [30] (Figure 1.9b).

1.2.2 Data Transmission

Due to the intrinsically random nature of RFID events, the data generated at the physical layer can be sparse and inaccurate, especially in harsh environments and under variable boundary conditions. Despite preliminary data filtering and

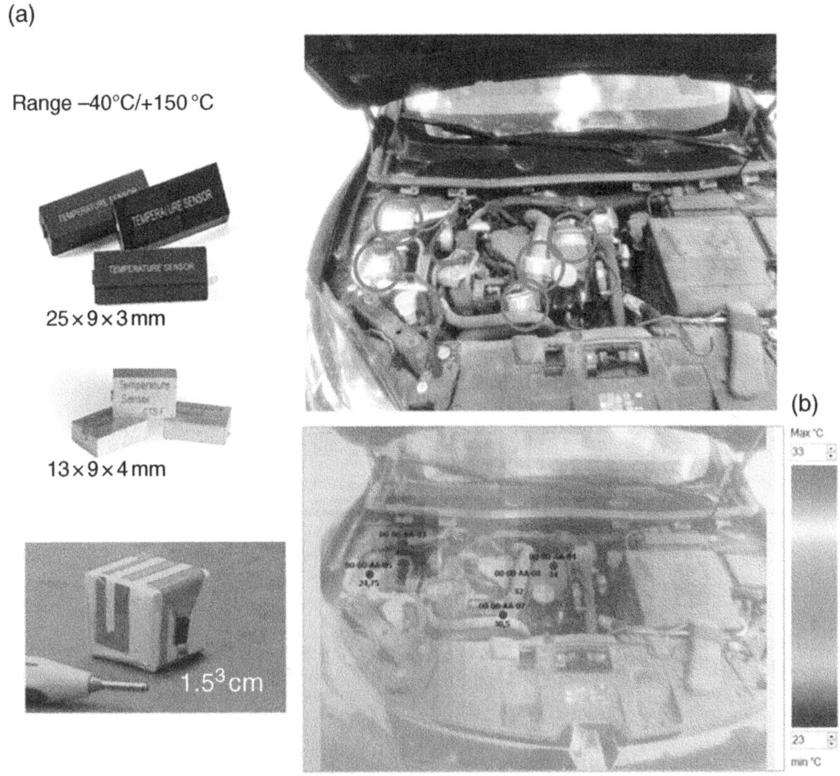

Figure 1.6 (a) Miniaturized RFID temperature sensors optimized for metal mounting and for measuring the contact temperature of the metal surface. (b) Thermal map of a car's hood in motion. The engine is clearly distinguishable from the area hosting the cooling systems.

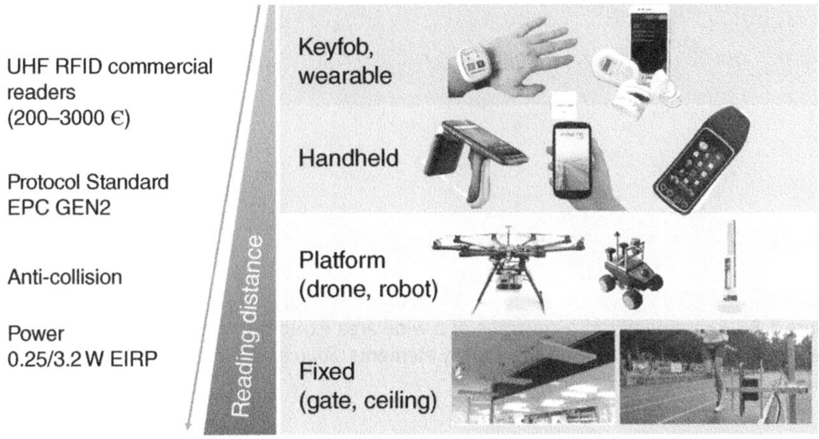

Figure 1.7 Options for an RFID reader in the UHF band. *Source:* Jiri B/Wikimedia commons/CC BY-SA 3.0.

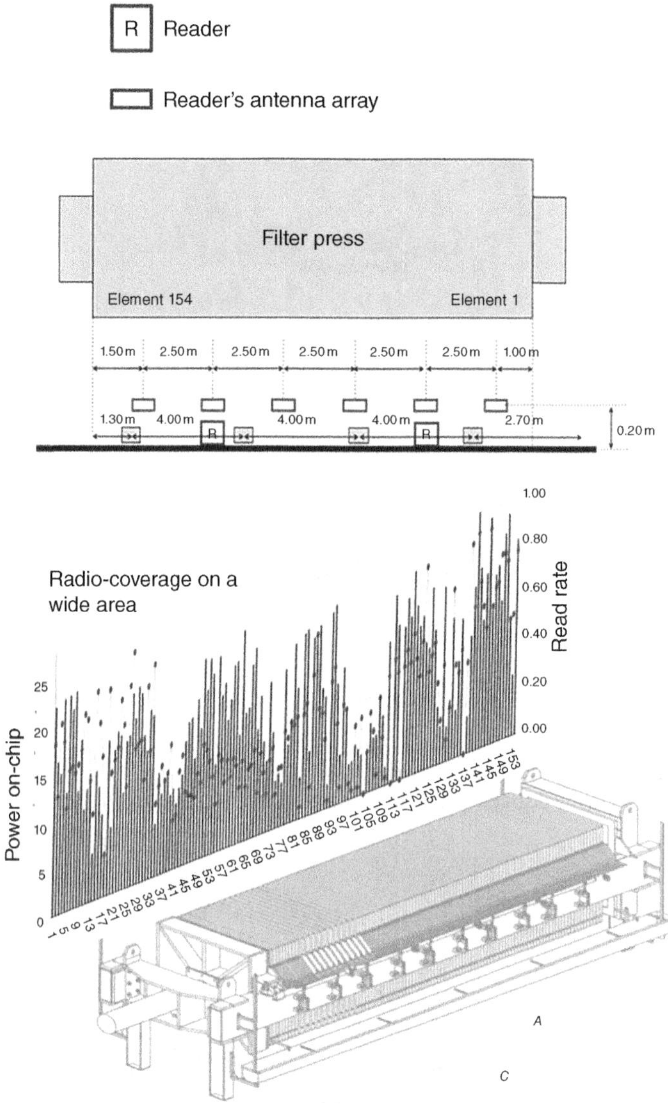

Figure 1.8 Example of radio-coverage of a wide area exploiting a multi-antenna interrogating system consisting of six array elements. *Source:* [32] /with permission of IEEE.

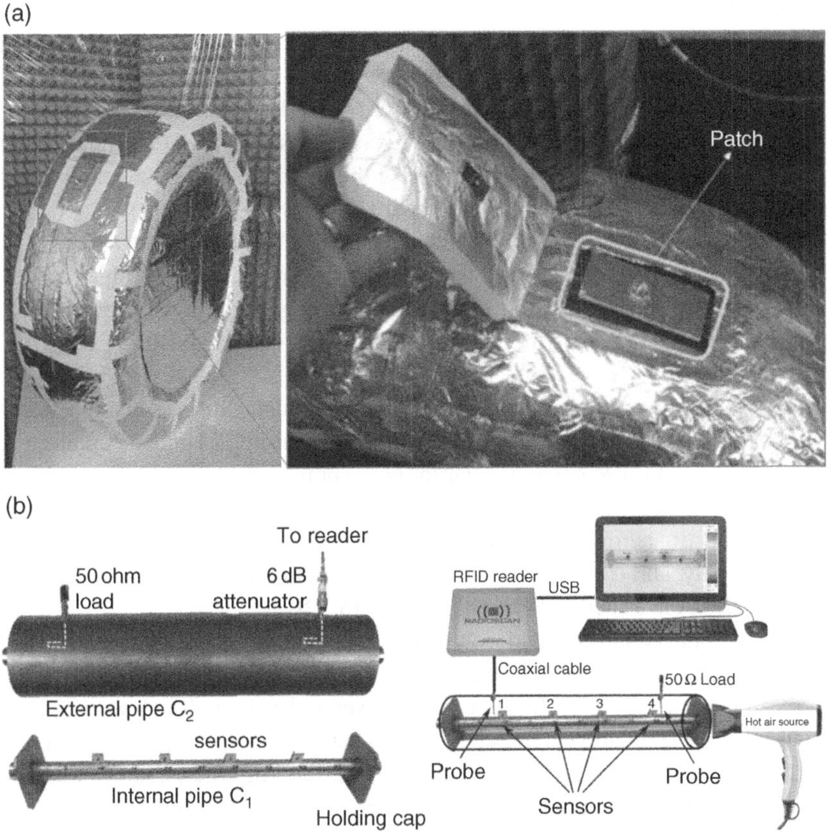

Figure 1.9 Examples of embedded reader's antennas: (a) patch antenna embedded into the metal mold to control tire vulcanization and (b) probe for thermal mapping of metal cavities.

lightweight elaboration that can be performed on-edge at the reader or even tag level, complex inferences and supporting actionable decisions demand advanced analytics. The latter is usually implemented at the Application layer, where RFID sensing data are merged with other relevant information of the target process.

The use of standardized protocols plays a key role in boosting the adoption of RFID tech in I-IoT, as it permits straightforwardly adding a new source of data to a preexisting monitoring layer. Accordingly, more and more industrial readers are available on the market and have a native industrial communication interface. Alternatively, an industrial converter, adapter, or intermediate gateway can be adopted at the system integration level to make the reader talk to the IoT ecosystem via industrial protocols.

The choice of communication architecture and a particular protocol is driven by the field of application and its specific requirements (transmission rates, packet size, latency, reliability, and power consumption).

In the automotive industry, Controller Area Network (CANBUS) is the preferred option for interconnected vehicles where data measured by RFID transponder placed on tires, rim, brake chassis, or inside the motor compartment can be sent to the central control unit to enable on-board diagnostics [31].

In industrial automation and control systems, through the MODBUS dialogue protocol, RFID readers are interfaced with programmable logic controllers (PLCs) within a production line [33], or power generation [34], or a remote terminal unit (RTU) in supervisory control and data acquisition (SCADA) systems [16].

With the growing introduction of service-oriented architectures in manufacturing and operational systems, the Open Platform Communication Unified Architecture OPC-UA protocol is becoming popular for both communication between machines (M2M) and vertical connectivity to massive cloud infrastructure. Being platform-independent, it permits manufacturer-independent exchange of information, interoperability, and ease of integration. The first example of an RFID system that adopts OPC-UA in a chemical plant is given in [15, 32].

In more application-oriented solutions, Hypertext Transfer Protocol (HTTP) through Representation State Transfer (REST) is the natural choice for direct communication between the RFID gateway and server nodes through web services [35].

The usage of standardized protocols plays a key role in boosting the adoption of RFID tech in I-IoT, as it permits straightforwardly adding a new source of data to a preexisting monitoring layer. Accordingly, more and more industrial readers are available on the market and have a native industrial communication interface. Alternatively, an industrial converter, adapter, or intermediate gateway can be adopted at the system integration level to make the reader talk to IoT ecosystem via industrial protocols.

1.2.3 Data Management and Analysis

The management of large amounts of raw data generated by RFID platforms comprises sequential steps from data transmission to support decision at the high level.

First, *processing* (step 1) involves data cleaning and filtering, outlier removal, combination with external sources of information, and metadata insertion. The resulting structured or unstructured data then fed into a database.

The *archiving* (step 2) can be distributed or centralized, depending on the network architecture: data can be stored as time-series for faster retrieval and comparison or exported to external access points. Different storage policies are applied depending on the access-permission strategies.

Data Analytics (step 3) is then performed through ad hoc queries on the database (SQL queries). Depending on whether known data are provided with descriptive labels or, instead, the discovery is made directly from the dataset, supervised or unsupervised learning is adopted. Several mining techniques such as Support Vector Machine (SVM), k-means clustering algorithm, and Deep Neural Network (DNN) have been proposed to process large RFID-based production data.

The final stage of the *decision-making* and influential process is to interpret the information, knowledge, patterns, and rules achieved in a useful and straightforward way (Step 4).

1.3 RFID for Product Monitoring at Item-Level

According to the first monitoring approach, denoted as *"from inside,"* the RFID sensor tag is embedded or contained in an object, turning the tagged item into a digital entity capable of wirelessly transmitting relevant information about its internal status throughout its entire lifecycle.

The item-level monitoring path starts with manufacturing, where the item usually moves along the production line and is interrogated at one or more checkpoints by an external reader, which are connected to the machinery PLC through industrial protocols (see the scheme in Figure 1.10).

Once activated from the impinging electromagnetic field, the tag samples and transmits back to the reader the instantaneous state of the object

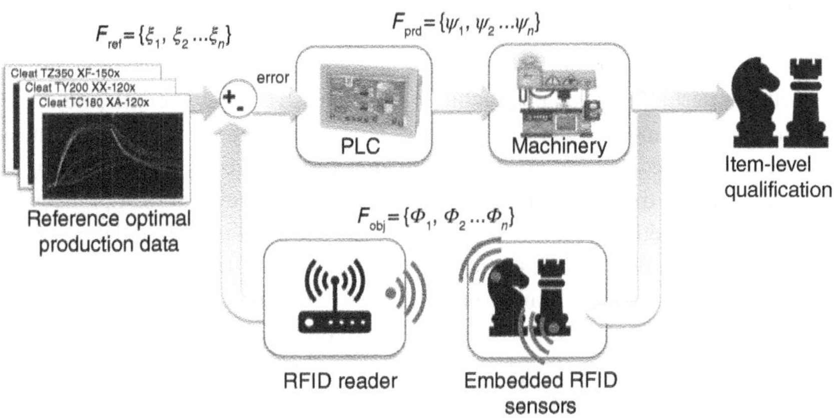

Figure 1.10 Block scheme of the monitoring and control platform *from inside.* RFID sensors embedded within the object enable the online control of the production process. Measured data F_{obj} is transmitted to the machinery PLC via the external reader and used for the real-time setting of the production parameters F_{prd}.

$F_{obj} = \{\xi_1, \xi_2 . . .\xi_n\}$, where ξ_n is any physical\chemical\mechanical parameter that is crucial for quality control of production, such as temperature or pressure. On the reader side, the retrieved information is sent to the PLC to feed real-time adaptive control systems. A typical feedback loop aims to minimize the discrepancy between the reference optimal production data $F_{ref} = \{\Phi_1, \Phi_2 . . .\Phi_n\}$ and the measured data, which are unpredictably affected by variable boundary conditions, by adjusting the machine parameters $F_{prd} = \{\Psi_1, \Psi_2 . . .\Psi_n\}$ at runtime.

Therefore, due to embedded sensing enabled by RFID sensors, the object itself tells the machinery how to optimize its manufacturing process. A practical example is the control system to increase/reduce the temperatures of a curing press or to increase/reduce the exposure time of a chemical treatment. Multiple reader antennas can extend the monitoring area or clusterize the different plant sectors, as discussed above.

After manufacturing, the same RFID sensor can be exploited for additional purposes, for example, the item-level monitoring of the product's KPIs during its specific activity. Therefore, this capillary source of information can be used for very accurate data-driven models of target entities (digital twins) to support design optimization loops, empower predictive maintenance strategies, and estimate the residual life.

Finally, at end of the item's life, sensors data can assist a sustainable recycling or dismantling.

Pioneering examples of RFID systems for item-level monitoring in civil, heavy industry, pharmaceutical, and food are described next.

1.3.1 Construction

In civil engineering, the supervision of the temperature *inside* the concrete during its pouring period allows one to achieve sufficient structural strength and avoid shrinkage stress and aging of the structure. This is even more crucial for the mass production of precast reinforced concrete elements where accelerated curing is applied to reduce the maturation period from 28 days to a few hours. A pioneering example of embedded RFID sensors for concrete is described in [36]. Here, a passive RFID tag, denoted as *T-Tag*, integrating an EM4325 [37] sensing IC is designed ad hoc for integration within precast segments for tunnels (Figure 1.11). The shape of the tag allows for the insertion in fresh concrete, the measurement of the temperature at approximately 15 cm depth, and the transmission of data to an external reader at a distance of up to 2 m.

In industrial carousel plants, hundreds of fresh ashlars are poured into caissons and then pass through one o more curing chambers or heated warehouses at controlled pace to induce accelerated maturation. By installing an RFID interrogating network inside the carousel oven, we can control in real time the thermal profile

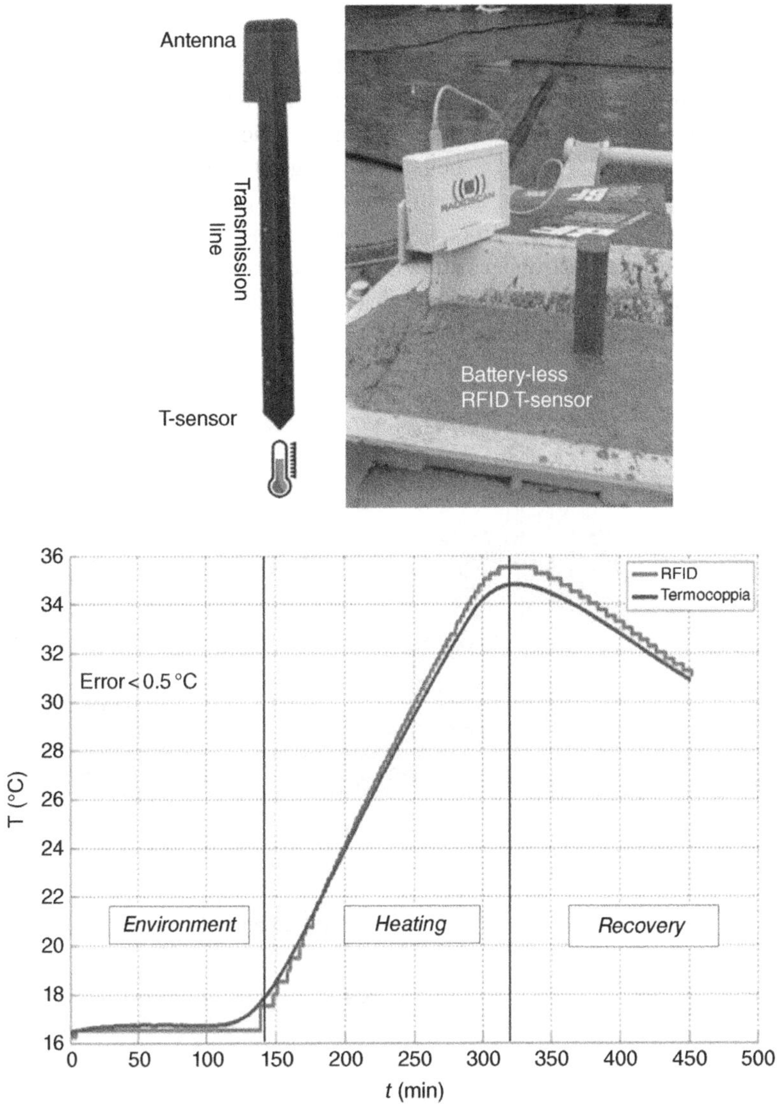

Figure 1.11 T-tag embedded within the concrete precast (left) and an example of temperature measured inside concrete measured by the T-tag and the thermocouple, during the heating/cooling process inside industrial oven.

of each precast and adjust the chamber settings to force precasts to follow their ideal hydration curve. As a result, the production process is qualified by a one-by-one (item-level) high-quality check of the produced precasts.

1.3.2 Conveyor Belt

A second example of RFID sensors embeddable within an object is shown in Figure 1.12a. The transponder, which integrates Magnus S3 IC [38] is designed to drown in the compound within the elastomers. The application concerns rubber-based industrial conveyor belts consisting of base belts, sidewalls, corrugated edges, and cleats that are hot vulcanized to guarantee adhesion and durability. During the curing process, monitoring the internal temperature between different layers of rubber is fundamental to assess if the "core" rubber undergoes the proper curing profile, which is correlated with the mechanical properties of the resulting product.

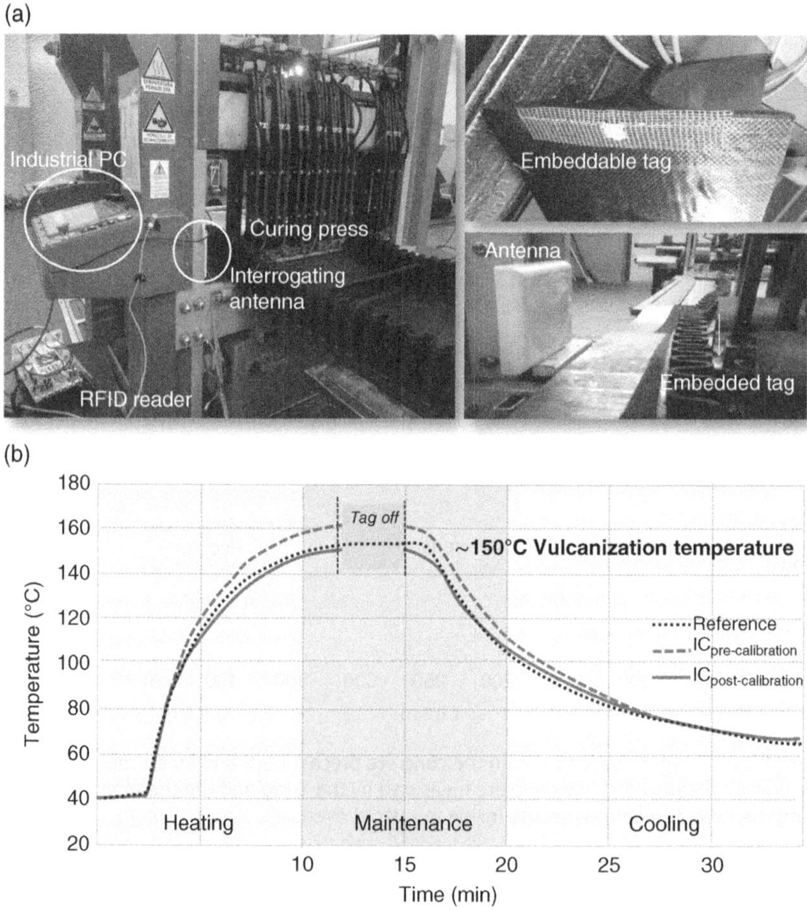

Figure 1.12 (a) RFID passive tag on polymeric mesh integrated within the rubber layer of a conveyor belt. (b) Temperature profile of a typical vulcanization cycle.

Communication and mechanical performance of the sensor have been jointly optimized to achieve the multi-fold purpose of (i) enabling the wireless reading in very complex EM environment (high loss of rubber, metal reinforcement inclusions within base belts, proximity metal press), (ii) withstanding the harsh conditions of manufacturing in terms of applied temperature and pressure, and (iii) seamless integration within the rubber compound, avoiding structural defects or air bubbles that might compromise mechanical performance. The resulting tag is made of conductive fibers sewed over a permeable and flexible polymeric net, which allows the fluid elastomer to flow through its mesh to transparently and permanently incorporate the sensor.

The conveyor production line is retrofitted by integrating the RFID acquisition system. The interrogating antenna is mounted in a lateral niche aligned with the curing press and the reader is connected to the PLC via the RTU MODBUS protocol. Figure 1.12b shows a temperature record of the typical vulcanization measured from the embedded sensors and used from the control system at runtime to ensure the ideal bonding curve.

1.3.3 Pharmaceuticals

Pharma is a highly regulated industry where capillary control of factors altering the effectiveness and safety of drugs is utmost importance at any time, from the late-stage development of new pharmaceuticals up to the large-scale manufacturing and distribution. For example, Accelerated Stability Studies (ASAP), which involve short-term aging under multiple storage conditions (50–80 °C and 10–75% RH), require recording the temperature and humidity in the packaging without opening the sealed bottle to accurately model the degradation of internal drugs and measure its water activity [39].

For this purpose, a miniaturized RFID probe for combined measurement of temperature and moisture has recently been proposed in [14]. The tag includes the latest generation of multi-parametric sensing [40] and consists of a normal-mode helical antenna (NMHA) wrapped around a threaded Nylon Poliammide support (PA). The sensor is inserted into a low-water absorption polymeric case whose size and shape resemble a pill (Figure 1.13).

Figure 1.14 shows the experimental setup arranged in the analytical pharma laboratory where the climate chamber is loaded with several vials filled with different drugs under an accelerated testing program and equipped with RFID probes. An example of temperature and humidity profiles measured through five different prototypes is also reported with a reference T-RH sensor, revealing a temperature accuracy of ±1 °C in the range of 20–60 °C and an RH accuracy of ±10% between 20 and 80 % of relative humidity. RH accuracy can be further

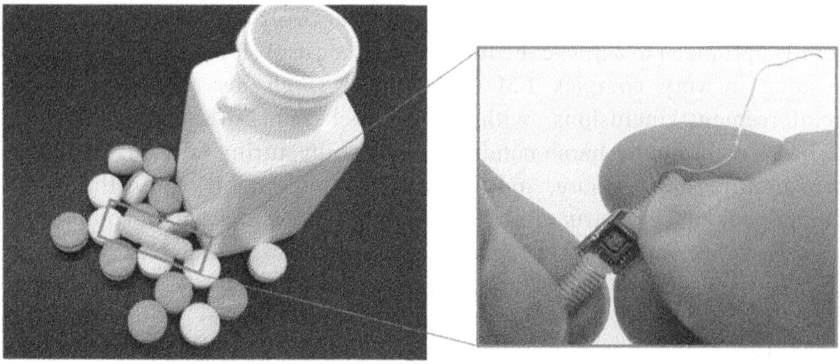

Figure 1.13 Prototype of miniaturized rH-T RFID sensor.

improved by proper calibration to ± 5% in the same range of interest. These values are in accordance with the requirements imposed by pharmaceutical standards.

1.4 RFID for Plant and Processes Monitoring

The second architecture is conceived to monitor an industrial environment *"from outside"* (scheme in Figure 1.15). The reader interrogates a multiplicity of sensor tags properly dispersed over the plant environment. The interaction between the reader and the tags is *one-to-many* and the information about the operations $F_{mac} = \{\xi_1, \xi_2 ... \xi_n\}$ is given by simultaneously processing data from several sampling points. The approach is particularly useful in the case of complex installations, where several thermal gradients, fluxes, and pressures must be controlled at the same time, or to test and qualify large systems such as engines, satellites, and cars. In the former case, a traditional feedback scheme can be adopted, similar to the previous scenario. In the latter case, instead, the RFID system is used for diagnostic operations and/or implementing predictive maintenance procedures.

Here, three examples of real implementation of RFID systems for plant and process monitoring are introduced. The described applications are showcased of a complete and integrated platform involving all layers of the Industry 4.0 solution, as described in Figure 1.1.

1.4.1 Filter Press in Chemical Industry

The filter press is a large industrial machinery that is used to separate solids from liquids by pressure filtration. It is widely used in chemical, pharmaceutical, coat plants, and wastewater treatment. Frequent damages of key components

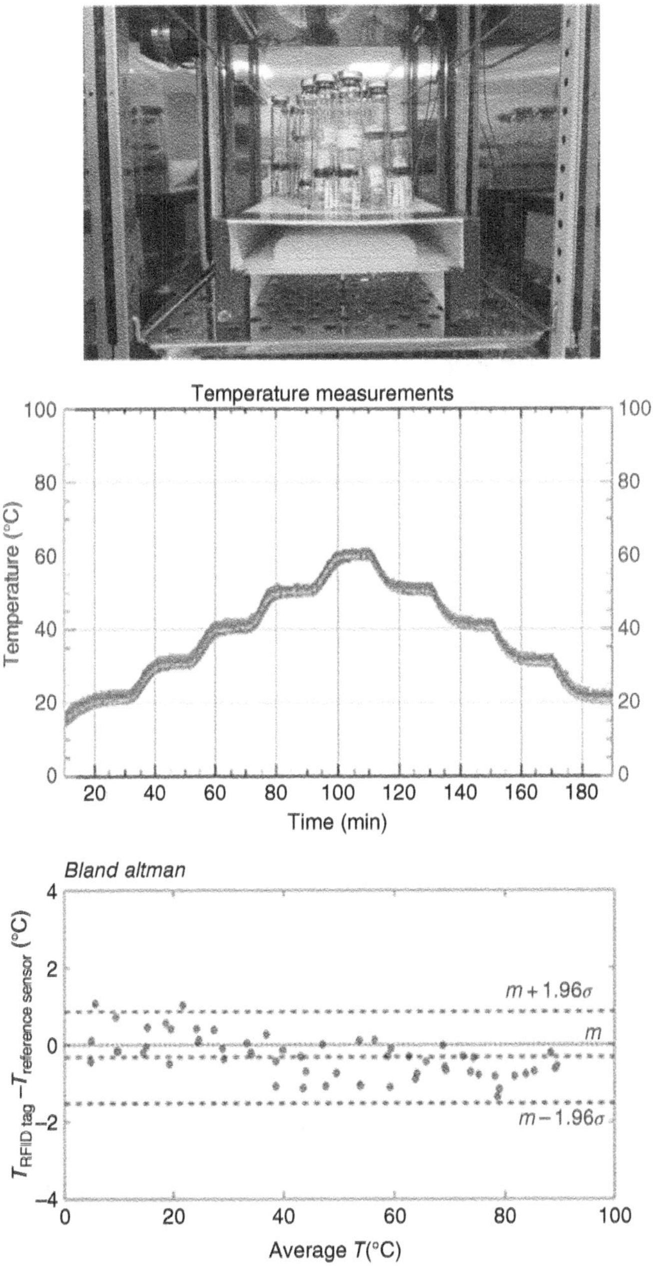

Figure 1.14 Temperature and humidity profiles measured inside vials through RFID inside a climate chamber.

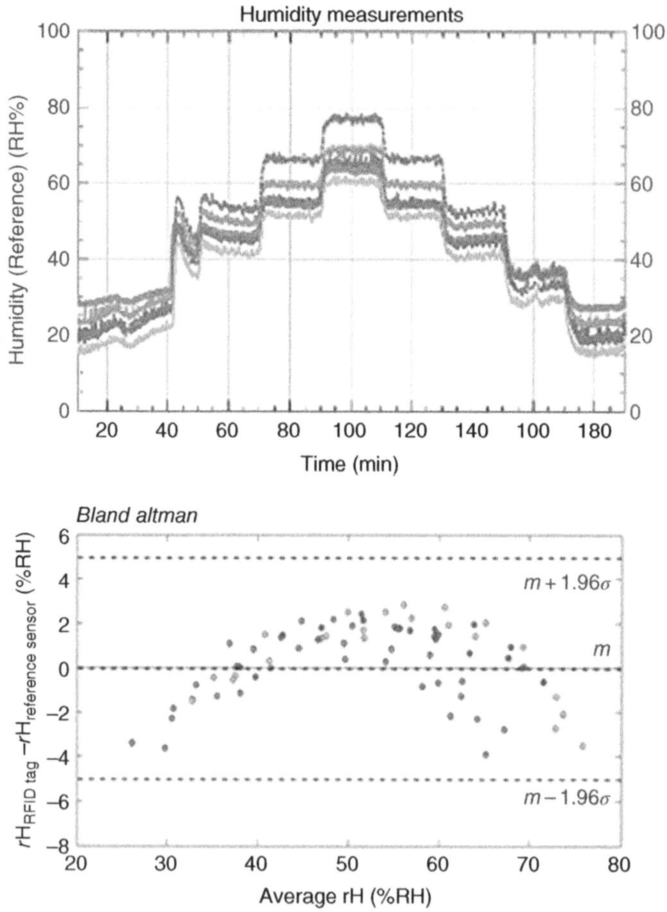

Figure 1.14 (Continued)

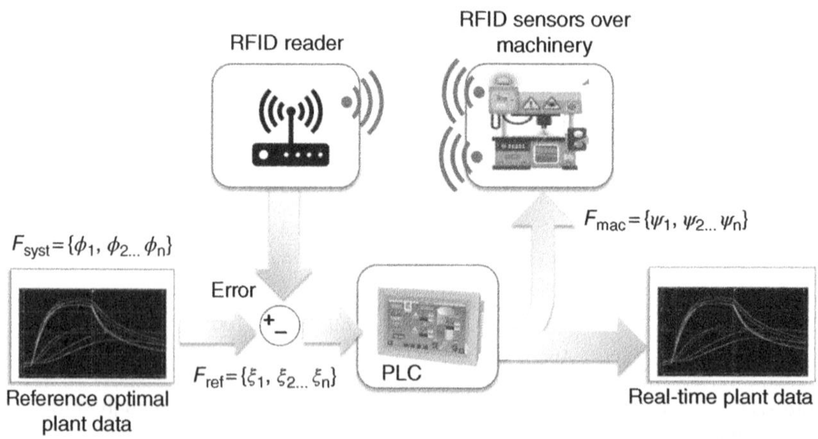

Figure 1.15 Functioning scheme of the "from outside" monitoring platform. RFID sensors dispersed over the machinery enable an online control/monitoring of the plant operations. Measured data F_{mac} can be used for correcting the plant parameters F_{sys} or simply monitoring purposes.

(e.g. plates and membranes breakage or leakages in hydraulic lines) may trigger cascade failures up to downtime of the production, especially in case of old equipment. The RFID-based system in [32] addresses the challenge of anticipating machine failure through continuous and capillary monitoring of the temperature of hundreds of inflating hoses (vertical rubber tubes). Early detection of anomalous thermal gradients (∇T 5 °C) over one out of any of the hoses is, in fact, an important predictor of future problems.

Figure 1.16 shows a pioneering RFID platform that has been operating in a silica production site since March 2021. The installation and commissioning of the pilot were performed during the short out-of-services scheduled for ordinary maintenance and did not require any modifications to existing machinery (retrofit approach).

The physical layer of RFID includes more than 150 temperature sensors attached to each pipe in a butterfly configuration using plastic ties and deployed in triplets vertically staggered by few centimeters to mitigate mutual coupling between adjacent sensors (Figure 1.16). The sensing tag is an adapted version of the COTS inlay (RFM3200 by Axzon) fully encapsulated for corrosion protection. A redundant multiantenna system consisting of six fan-beam arrays (Advantenna-p16 by Keonn) ensures the robust coverage of the 15 m-length target area and a reliable communication link with a multitude of sensors despite the EM-hostile environment (full of metals, vapors, and RF interferences).

Two PoE-powered readers (Izar by Jadak ThingMagic) installed in IP66 boxes close to the machine are controlled by a single-board industrial PC located in the rack area. The readers are configured to receive through the GPIO interface operational information about the filter press (i.e. the working phase) to be edge-processed in combination with temperature trends.

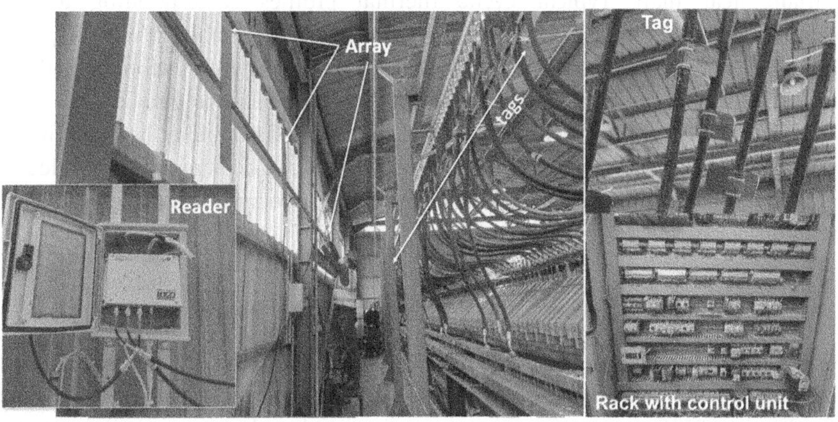

Figure 1.16 Physical layer of the RFID system installed in chemical plant for thermal monitoring of the filter press.

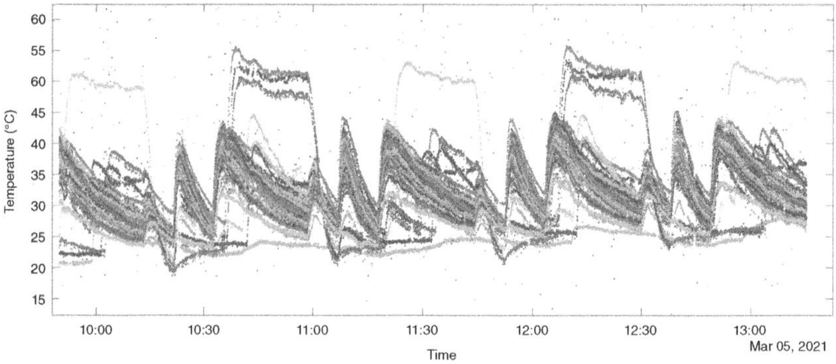

Figure 1.17 Trace of raw temperature data measured by the reader during 3 hours of filter-press operations.

A Java-based custom software SW handles the multiplexed interrogation according to a time-division scheme, the preprocessing (**outliers removal and applying power-based correction for improving temperature accuracy [41]**) and the data transmission (Figure 1.17).

The RFID physical layer is interfaced with the existing SCADA system of the plant through the Open Platform Communications Unified Architecture (OPC UA) standard. The OPC-UA server running on the control unit streams the data every 30 seconds to the InfluxDB database for the persistence of the accumulated time series of data. Finally, temperature trends are visualized on a Graphana dashboard specifically designed for providing the end-users with high-level meaningful data, analytics, and alerts.

Figure 1.18 reports an example of the views remotely accessible by operators in the control room, including the temperature trace of all hoses and a 2D-time-variant heat map. Typical cyclic thermal profiles corresponding to the different working phases (filling, washing) are clearly detectable, as well as spatial gradients due to the hole position within the press. Any deviation from these baseline thermal fingerprints can be interpreted as potential anomalies.

The described platform is modular and highly configurable at both the hardware and software levels. Accordingly, the solution can be adapted and scaled to any kind of filter-press layout as well as to similar industrial machinery provided with rubber pipe.

1.4.2 Electrical Equipment and Renewable Energy Plants

Contact faults, material fatigue, and aging of insulators, which are well-known common causes of electrical systems' failures, early provoke localized and subtle overheating. RFID temperature sensors can be distributed in large numbers at any

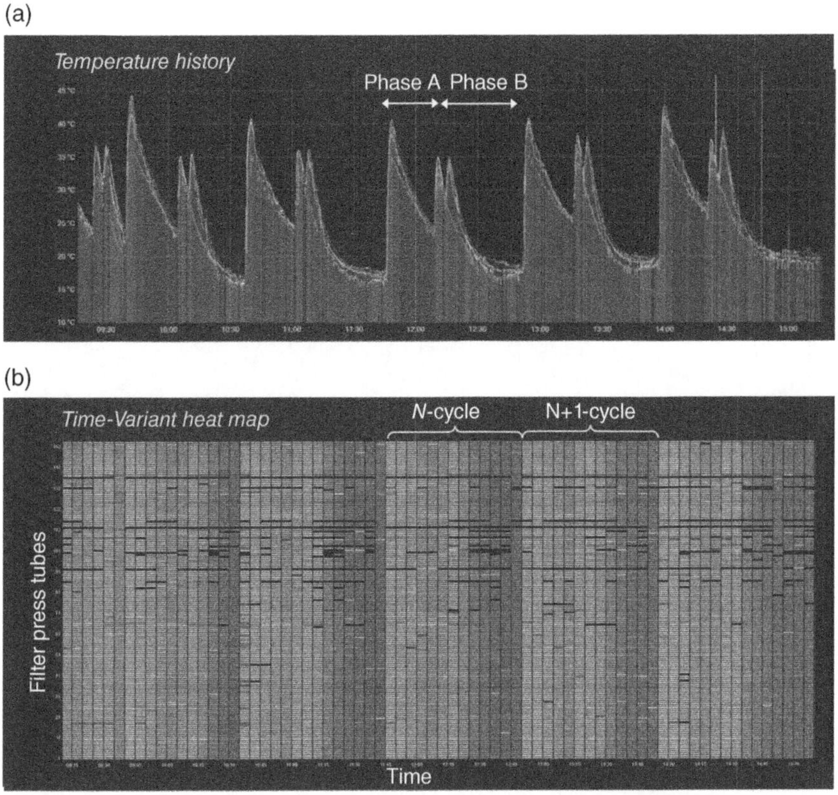

Figure 1.18 Example of some graphical views accessible though the web-dashboard of the filterpress monitoring platform: (a) Thermal Profile versus time of all the hoses with superimposition of the color corresponding to the working phase and (b) 2D-time-variant heat map of the filter press with the input channels (hoses) on the *y*-axis. Each row shows the thermal history of a single hose, while grouped columns depict the instantaneous thermal status of the entire machine.

susceptible points (cable connections, cable junctions, cable lugs, copper bar connections) to maximize the likelihood of capturing anomalous thermal patterns due to system overload, inefficiency, or upcoming failures, and timely troubleshooting.

Figure 1.19 shows an example of an RFID monitoring layer integrated within a medium-voltage (MV) metal-clad switchgear consisting of four compartments, partially segregated with metal enclosures (circuit breakers, low-voltage control cabinet, transformer, and busbar). Many cubic-like COT temperature sensors are mounted using plastic rings, clips, or adhesive tape on target pints such as plum contacts or bushing (6×) of circuit breakers, on bus bars (3×), and power cables connection bars (3×), in the transformer area. Two circularly polarized patch

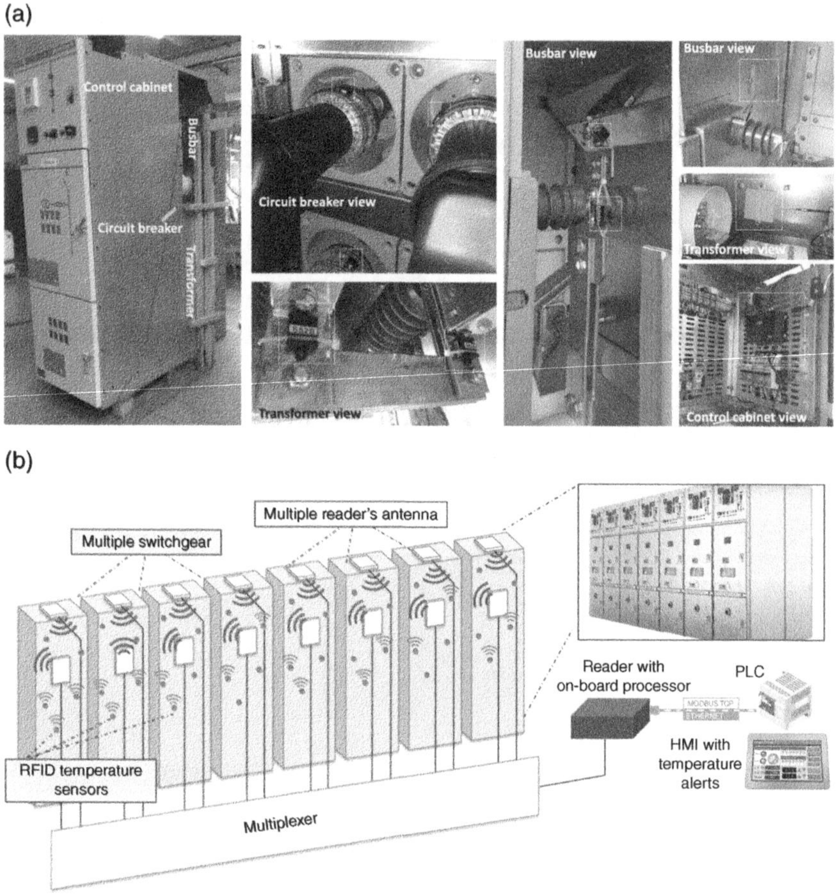

Figure 1.19 (a) RFID-based thermal monitoring system integrated within a medium-voltage (MV) metal-clad switchgear. (b) Multiplexed RFID architecture for modular industrial electrical panels.

antennas installed inside the cabin by means of a magnets - in particular over side wall of the busbar area and on ceiling of transformer compartment - are sufficient to read all the twelve sensors due to not perfect shielding among adjacent compartments and reflections occurring inside the metal cavity. The upper LV control cabinet host the industrial 8-port reader provided with two native MODBUS RTU interfaces for connection to PLC. It is worth mentioning that all the HW components have flat layouts with minimum protruding corners to minimize the risk of electrical discharges.

Industrial electrical panels usually consist of multiple switch gears modules (typically 6–8) connected to each other through transversal busbars. The described

RFID layer can be accordingly scaled by repeating in series the described rollout in a multiplexed architecture, as schematized in Figure 1.19b.

When the RFID system is installed directly during the manufacture of an electrical panel, it can be exploited before product commissioning, that is, during standardized testing and qualification procedures. Figure 1.20b reports the thermal profiles measured during a *high-voltage test* - that return as expected at a constant temperature - and during a *temperature rise test* showing the temperature increase of the target points is coherent with the applied current.

Switchgears are an integral part of more complex industrial electrical systems that involve power generation, transformation, and distribution networks. In these operative scenarios, the modularity of the RFID platform can be further exploited to achieve the remote protection of several electrical systems distributed over wide areas. For instance, renewable energy plants comprise a grid of electrical cabins located in the production site or even off-shore.

1.4.3 Fruits Monitoring in Ripening Rooms

Accelerated ripening enforced by exposure to environmental conditions and gases is nowadays one of the most assessed food processing methods, especially for climacteric and exotic fruits. However, industrial plant control still resorts to qualitative, sporadic, and destructive tests based on humans, producing waste of food and money. To address this challenge, an RFID-based system has recently been developed to achieve fine-grained and automatic monitoring of avocados within industrial ripening cells [35]. Following the sensor-less approach, the temporal variation of EM signals along with the maturation process is exploited to retrieve indirect information on the ripeness stages.

At the physical layer, the multi-shelves setup schematized in Figure 1.21 includes multiplexed RFID networks comprising 1× UHF reader, 4× multiplexers, 32× near-field UHF antennas, and industrial grade processor uni provided with Internet connection.

The resulting trolley can house up to 128 fruits, grouped by origin and size on the different shelves. The sensing components are made up of the 384× loop tags (three for each fruit) attached to the leaflets of the fruit lodges whose geometry has been properly shaped to ensure the adhesion of the tag to the peel of the fruits.

In general, the smart trolley is conceived as a web application using a three-tier architecture that organizes applications into the three logical and physical computing tiers. The data acquisition process is delegated to SW running on board the trolley, which implements ad hoc measurement protocols involving different EPC Gen 2 standard commands based on interrogation modalities (power and frequency sweeps). The trolley can be remotely controlled through a

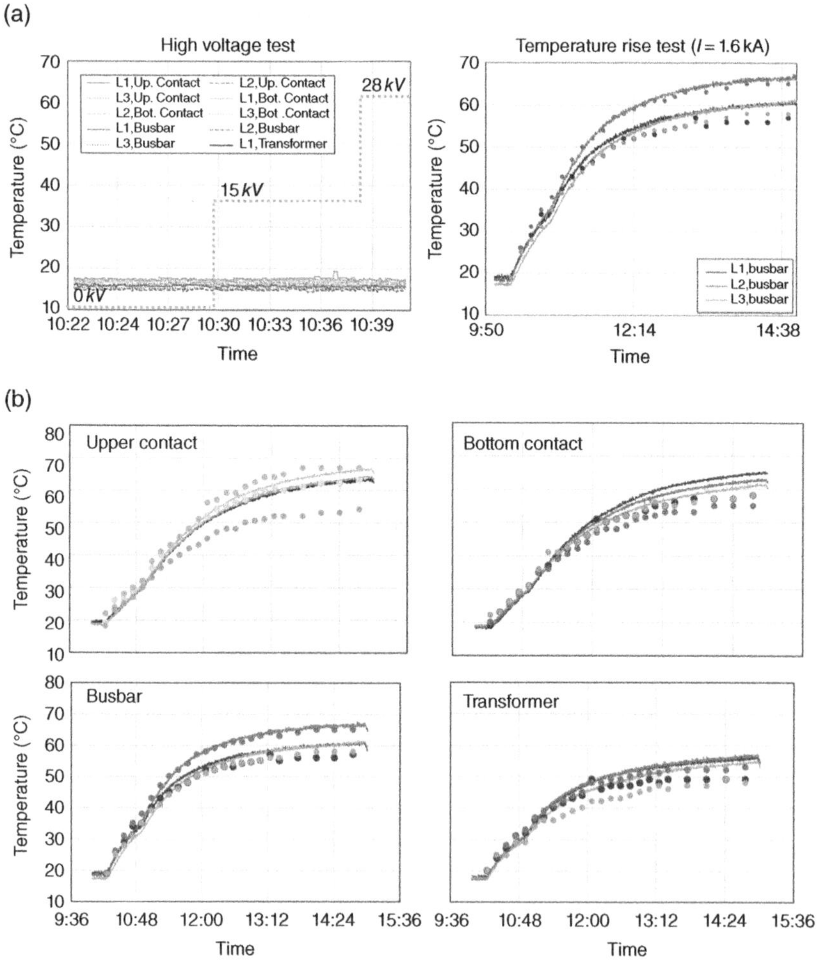

Figure 1.20 Thermal profile measured by RFID sensors inside the switchgear during (a) *Dielectric test* consisting of stepped voltage applied up to 28 kV at industrial frequency and during (b) *Temperature rise test* involving application of nominal (1.2 kA) and overload (up to 1.6 kA) currents. Dots refer to reference thermocouples installed in proximity of RFID sensors.

dedicated interface using any device connected to the Internet (mobile, tablet, or desktop) to start/stop the measurement session (by operators), as well as to tune/update the system configuration (by developers). The raw measured data are then sent via HTTP protocol to the AI module, which, after a cleaning step and feature extraction, classifies the ripeness level of each fruit (ripe versus unripe) by using an enforced SVM. Being supervised, the classification algorithm must

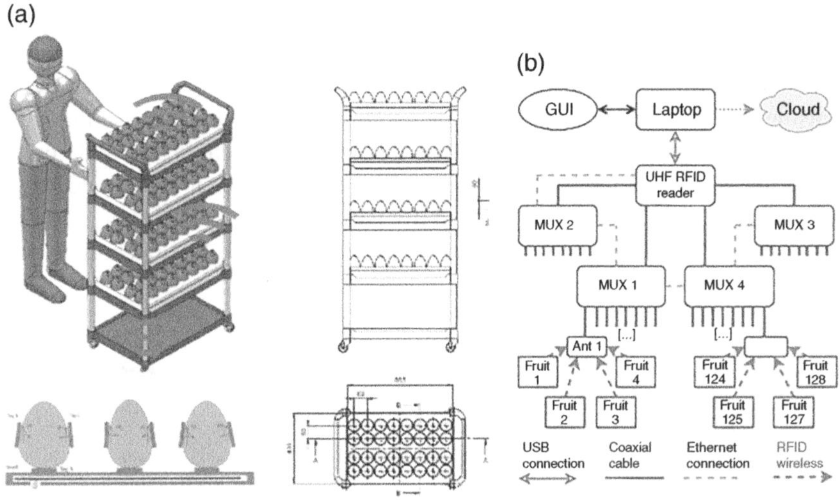

Figure 1.21 RFID-based monitoring of the maturation of tropical fruits. (a) The measurement trolley, and (b) the system architecture.

be previously trained versus the golden standard procedure - here destructive tests with penetrometers - according to predefined threshold levels for discriminating different classes. Subsequently, the AI module sends the classification output to the server (application tier) that writes the result to the database.

All communications with the database are managed by the application layer that handles the permissions according to an authentication and authorization system and exposes the allowed functions through API REST. The query and display of high-level information are delegated to the presentation layer, that is, through an interface that can be navigated via a web browser (Figure 1.22). The aggregate data are shown through advanced dashboards that use charts and search filters tailored for a specific user experience. The platform also provides a Telegram notification system to alert users when a specific percentage of loaded fruits have moved into the ripe state. Whenever new data comes to the server, this takes care of sending the notification to a dedicated Telegram group that will inform all the subscribers.

The entire platform is currently at an advanced validation stage through a pilot deployment within the ripening chambers located at the European Avocado Import Center. Experimental campaigns involving more than 4500 corroborate the accuracy of the target greater than 85% as well as the usability of the system in accordance with the normal operating procedures of the ripening site.

1.5 Challenges and Countermeasures

Establishing a robust communication link and retrieving precise and accurate sensing measurements are the most demanding challenges when dealing with RFID sensor networks both *from inside* and *from outside* scenarios. In addition to the issues related to wireless powering of the miniaturized tag surrounded by a lossy medium and metallic objects, additional detrimental effects may arise in industrial applications from the operating conditions at high temperatures and

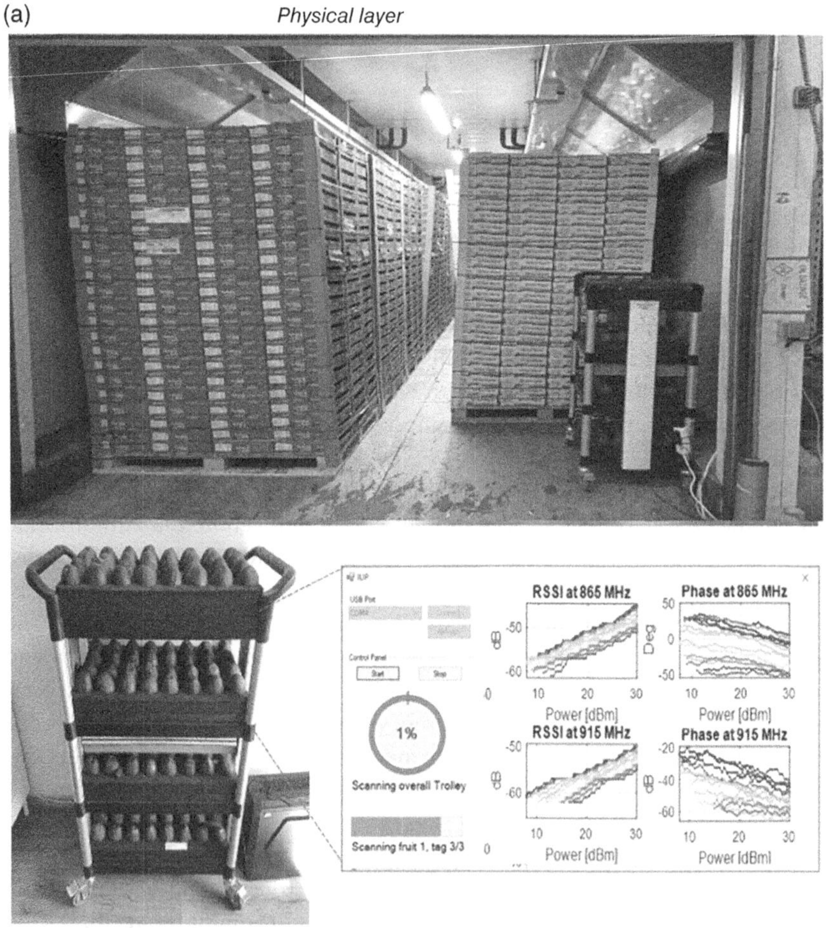

Figure 1.22 (a) The RFID-based ripeness monitoring system deployed into an industrial plant. (b) Screenshots from the web application.

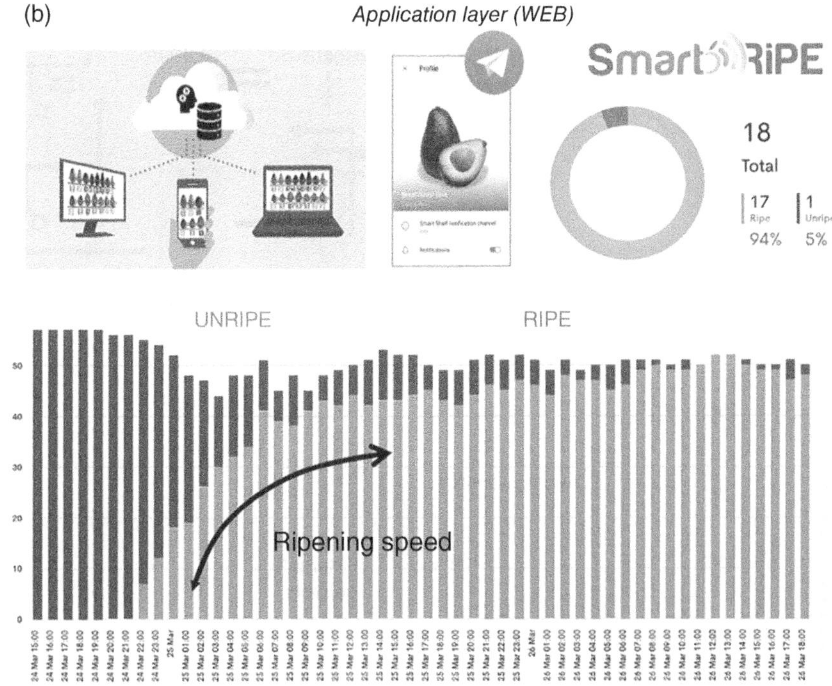

Figure 1.22 (Continued)

the placement of sensors on high-speed moving mechanical components (both linear and rotating movements).

The problem was first addressed in [42]. Here, the sensing and communication performance of RFID tags subjected to high-temperature stress is experimentally investigated with the aim of assessing the effective operability of conventional RFID technology in Industry 4.0 scenarios. Typically, sensing ICs suffer from a temperature increase with a negligible reduction of the maximum communication ranges. As visible in Figure 1.23a, the effect is monotone and progressive; at 130 °C, the achievable reading range for three classes of standard ICs (Monza6, Magnus S3, and SL900A) is approximately 42%, 56%, 50% of the reference room temperature value. Accordingly, the link budget for high-temperature applications must account for these detrimental effects.

Temperature also impacts the sensing performance of the IC. Generally, on-chip sensors exhibit a drift for temperature beyond 80–90 °C (Figure 22b). Nevertheless, a linear calibration is usually enough to compensate for the overestimation error. When external probes are connected to the IC through the

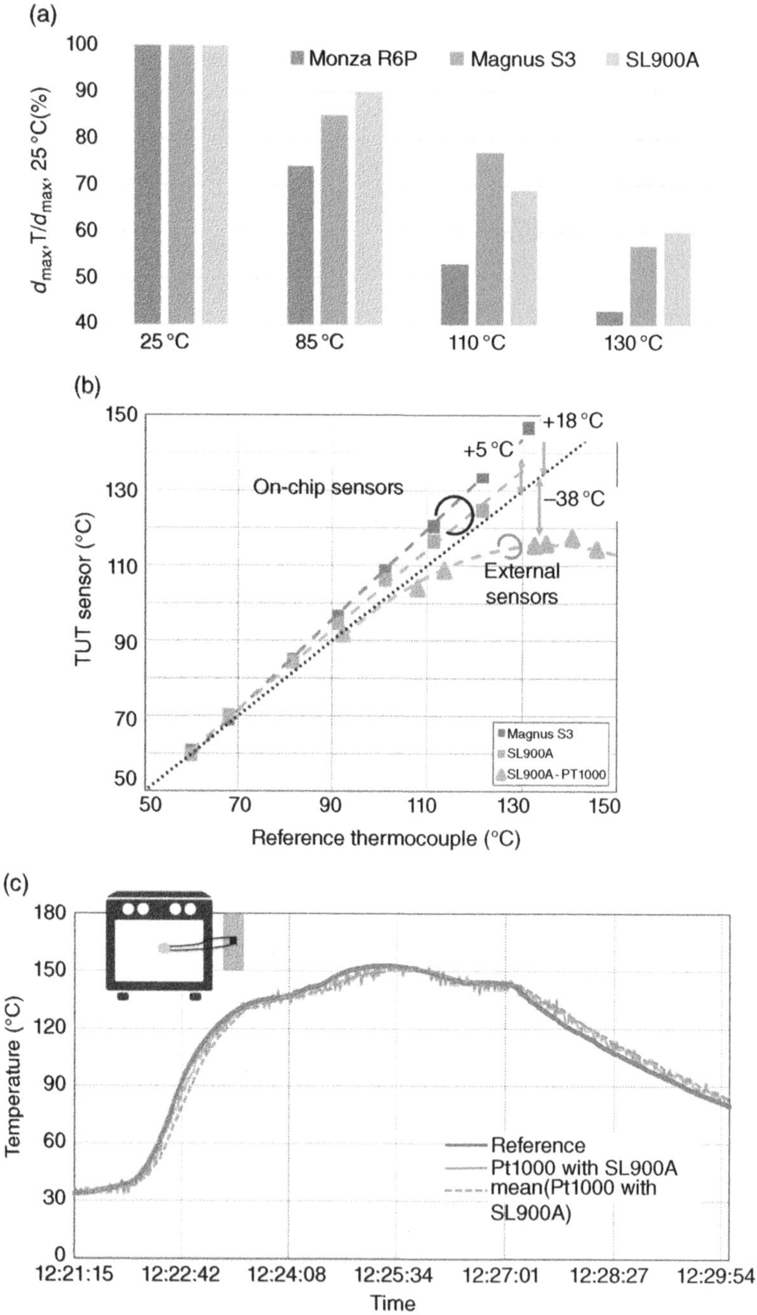

Figure 1.23 (a) Percentage reduction of the reading distance of RFID sensor tag exposed to high temperature condition. (b) RFID sensor data versus a reference thermocouple. (c) Temperature profile measured by a PT1000 connected to the SL900A IC with the electronics maintained at environmental conditions.

analog front end, the sensitivity to the temperature could be much stronger. Severe nonlinear and nonmonotonic drift (underestimation) arises as a result of the temperature dependence of the DC current that the IC injects into the external probe to derive the resistance. In the latter case, protecting/insulating the IC to maintain its internal circuitry below the critical temperature is a viable countermeasure to overcome the problem. In Figure 1.23c, the temperature profile measured by a PT1000 connected to the SL900A IC matches well with the reference thermocouple, provided that the IC is kept outside the thermal chamber, that is, at room temperature.

When a sensor tag is in charge of monitoring components moving at high speed for example, rotating parts of engines and turbines, communication, and sensing performances could also be severely affected. At least two phenomena should be considered: (i) the robustness of the backscattering link; (ii) the robustness of data transmission/decoding. In Figure 1.24, the RSSI and the temperature values detected by RFID sensor tags rotating in front of a circularly polarized reader antenna are visible. In the first case, the sensor is placed close to the center of the

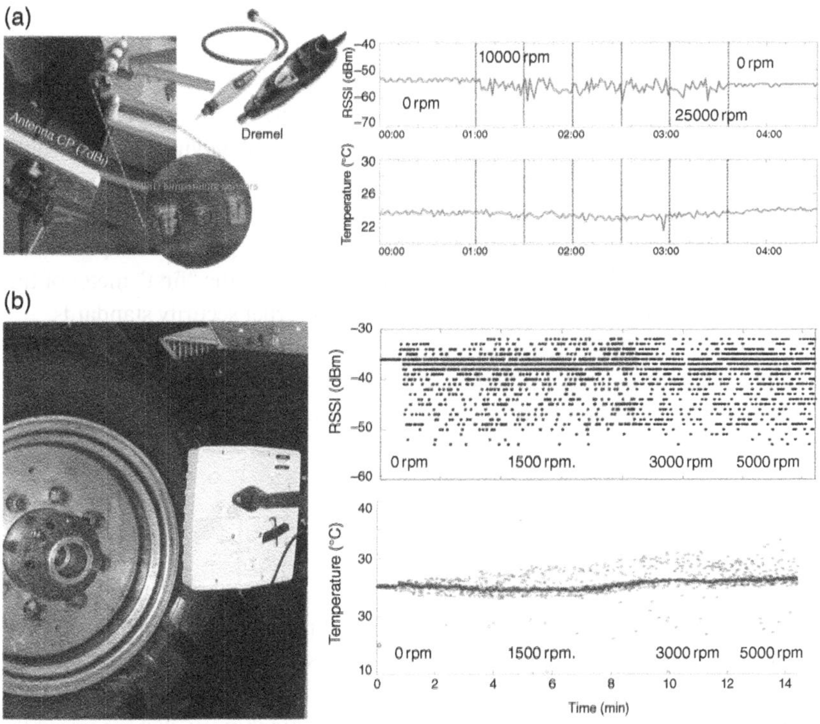

Figure 1.24 (a) RSSI and Temperature retrieved by a passive sensor rotating up to 25 000 rpm. (b) RSSI and Temperature retrieved by a passive sensor rotating up to 5000 rpm.

rotating element, and therefore the mutual position between the reader antenna and the tag can be considered stationary throughout the test. In the second example, instead, the tag is placed laterally with respect to the rotating center (about 17 cm) so that the mutual position between the reader and the tag sensibly changes while rotating, with the tag even outside the reader antenna beam for certain radial positions.

In both cases, rotation does not disable the communication, as tags are readable even at the highest speed. RSSI oscillates around the steady-state value registered for 0 rpm; the spread is particularly evident in the second test, due to the different reader/tag mutual position during rotation. While in the first case, the temperature acquisition looks robust and stable, in the second case, as the speed increases, several measurement outliers progressively appear. This behavior can be related to the fact that temperature measurement and transmission require longer communication between the reader and the tag than a simple inventory cycle, and hence a more stable backscattering link. If during the complex operations sequence the power (and hence the circuitry voltage) levels at the reader and tag side oscillate, some processes can be corrupted (incomplete data packet, single-bit, and burst errors).

Finally, an additional source of error could be the electromagnetic noise produced by the activation and operation of the high-power electric motor (see again Figure 1.24b). Despite the critical aspects, the correct thermal profile can be easily retrieved by standard data processing, such as filtering and moving averages (blue continuous line in Figure 1.24b).

Security and vulnerability assessments are needed at multiple levels to ensure the security and integrity of the data during all levels of the IoT platform. In particular, the physical tier from the sensor to the control unit - that is, the "first" meter of the wireless link - is still not in the crosshairs of the latest cyber security standards.

The feasibility of adopting new-generation microchips with sensing capabilities that support low-power encryption algorithms for data encryption and protection of their integrity must be evaluated, with particular attention to the corresponding impact on the power budget and data rate.

1.6 Conclusions

Real-time monitoring of industrial platforms and processes using RFID passive technology is a valid and applicable tool. Thanks to new-generation sensing-oriented ICs, deploying complex wireless sensor networks directly within objects in manufacturing, or over equipment and machinery, is now possible. The ubiquitous nature of RFID technology enables innovative monitoring paradigms in which the network is reconfigurable, the number of acquisition points is highly scalable, and single-use applications are feasible. Readers can be directly integrated within the factory standard monitoring architectures by exploiting several communication protocols. As in other IoT systems, the monitoring experience can be sensibly enriched by adopting AI frameworks to process data in real time, classify operating conditions, and detect anomalous events early. Finally, web interfaces make the industrial user experience highly fast, straightforward, and "modern."

Overall, RFID sensor platforms, similar to SCADA systems, have earned the role of one of the pillars of Industrial 4.0 monitoring.

References

1 Ng, I.C.L. and Wakenshaw, S.Y.L. (2017). The Internet-of-Things: review and research directions. *International Journal of Research in Marketing* 34 (1): 3–21.

2 Andrews, J.G., Buzzi, S., Choi, W. et al. (2014). What will 5G be? *IEEE Journal on Selected Areas in Communications* 32 (6): 1065–1082.

3 Agenda, I. (2015). *Industrial Internet of Things: Unleashing the Potential of Connected Products and Services*. World Economic Forum.

4 Brynjolfsson, E. and McAfee, A. (2014). *The Second Machine Age: Work, Progress, and Prosperity in a Time of Brilliant Technologies*. W. W. Norton and Company.

5 Amendola, S., Lodato, R., Manzari, S. et al. (2014). RFID technology for IoT-based personal healthcare in smart spaces. *IEEE Internet of Things Journal* 1 (2): 144–152.

6 Occhiuzzi, C., Vallese, C., Amendola, S. et al. (2014). NIGHT-care: a passive RFID system for remote monitoring and control of overnight living environment. *Procedia Computer Science* 32: 190–197.

7 Occhiuzzi, C., Caizzone, S., and Marrocco, G. (2013). Passive UHF RFID antennas for sensing applications: principles, methods, and classifications. *Antennas and Propagation Magazine* 55 (6): 14–34.

8 Marrocco, G., Occhiuzzi, C., and Amato, F. (2010). Sensor-oriented passive RFID. In: *The Internet of Things*, 273–282. New York: Springer.

9 Occhiuzzi, C., Amendola, S., Nappi, S. et al. (2019). RFID technology for industry 4.0: architectures and challenges. *2019 IEEE International Conference on RFID Technology and Applications (RFID-TA)*, 181 186.

10 Dargie, W. and Poellabauer, C. (2010). *Fundamentals of Wireless Sensor Networks: Theory and Practice*. Wiley.

11 Occhiuzzi, C. and Marrocco, G. (2016). Precision and accuracy in UHF-RFID power measurements for passive sensing. *IEEE Sensors Journal* 16 (9): 3091–3098.

12 Miozzi, C., Nappi, S., Amendola, S. et al. (2019). A general-purpose configurable RFID epidermal board with a two-way discrete impedance tuning. *IEEE Antennas and Wireless Propagation Letters* 18 (4): 684–687.

13 Occhiuzzi, C., D'Uva, N., Nappi, S. et al. (2020). Radio-frequency-identification-based intelligent packaging: electromagnetic classification of tropical fruit ripening. *IEEE Antennas and Propagation Magazine* 62 (5): 64–75.

14 Masi, D., Amendola, S., D'Uva, N. et al. (2022). Radio-pill: RFID miniaturized and battery-free humidity probe for pharmaceutical moisture analysis. *2022 IEEE 12th International Conference on RFID Technology and Applications (RFID-TA),* 153 154.

15 Nappi, S., Amendola, S., Ramacciotti, M. et al. (2021). UHF RFID system for the predictive maintenance of a filter press: a real use case. *2021 6th International Conference on Smart and Sustainable Technologies (SpliTech),* 1 4.

16 Amendola, S., Occhiuzzi, C., and Marrocco, G. (2017). RFID sensing networks for critical infrastructure security: a real testbed in an energy smart grid. *2017 IEEE International Conference on RFID Technology & Application (RFID-TA),* 106 110.

17 Qi, Q. and Tao, F. (2018). Digital twin and big data towards smart manufacturing and industry 4.0: 360 degree comparison. *IEEE Access* 6: 3585–3593.

18 Li, N., Tan, J., and Zhu, Z. (2010). Monitor and control system with RFID technology in discrete manufacturing line. *2010 IEEE International Conference on RFID-Technology and Applications*, 71 76.

19 Lee, S.H. and Jin, I.S. (2011). Interoperation of an UHF RFID reader and a TCP/IP device via wired and wireless links. *Sensors (Basel)* 11 (11): 10664–10674.

20 de Melo, P.F.S. and Godoy, E.P. (2019). Controller interface for industry 4.0 based on RAMI 4.0 and OPC UA. *2019 II Workshop on Metrology for Industry 4.0 and IoT (MetroInd4.0&IoT)*, 229 234.

21 Abdulkadhim, F.G., Zhang, Y., and Khalid, M. (2020). Smart pharmacy monitoring system based on MQTT protocol using RFID and raspberry Pi. *EUREKA: Physics and Engineering* 1: 98–104.

22 Gutierrez, A., Nicolalde, F.D., and Ingle, A. (2013). High-frequency RFID tag survivability in harsh environments. *2013 IEEE International Conference on RFID (RFID),* 58 65.

23 Lahokallio, S., Kiilunen, J., and Frisk, L. (2014). Performance of passive RFID tags in a high temperature cycling test. *Proceedings of the 5th Electronics System-integration Technology Conference (ESTC),* 1 5.

24 Le Breton, M., Baillet, L., Larose, E. et al. (2017). Outdoor UHF RFID: phase stabilization for real-world applications. *IEEE Journal of Radio Frequency Identification* 1 (4): 279–290.

25 Fu, Y.Y., Chan, Y.L., Yang, M.H. et al. (2015). Experimental study on the washing durability of electro-textile UHF RFID tags. *IEEE Antennas and Wireless Propagation Letters* 14: 466–469.

26 Yu, X., Yu, Y., Wang, D. et al. (2016). A novel temperature control system of measuring the dynamic UHF RFID Reading Performance. *2016 6th International Conference on Instrumentation & Measurement, Computer, Communication and Control (IMCCC)*, 322 326.

27 Occhiuzzi, C., D'Uva, N., Nappi, S. et al. (2020). Radio-frequency-identification-based intelligent packaging: electromagnetic classification of tropical fruit ripening. *IEEE Antennas and Propagation Magazine* 62 (5): 64–75.

28 Manzari, S., Occhiuzzi, C., Nawale, S. et al. (2012). Humidity sensing by polymer-loaded UHF RFID antennas. *IEEE Sensors Journal* 12 (9): 2851–2858.

29 Nappi, S., D'Uva, N., Amendola, S. et al. (2017). A near-field RFID sensor network for the realtime monitoring of tire vulcanization. 2017 *IEEE International Conference on RFID Technology & Application (RFID-TA)*, 69 73.

30 Lopez-Soriano, S., Spassovsky, I.P., Parron, J., and Marrocco, G. (2017). Electromagnetic feasibility of a passive wireless sensor network for temperature mapping inside a shielded enclosure. *2017 11th European Conference on Antennas and Propagation (EUCAP)*, 1995 1998.

31 Caccami, M.C., Amendola, S., and Occhiuzzi, C. (2019). Method and system for reading RFID tags embedded into tires on conveyors. *2019 IEEE International Conference on RFID Technology and Applications (RFID-TA)*, 141 144.

32 Nappi, S., Amendola, S., Ramacciotti, M. et al. (2021). RFID based predictive maintenance system for chemical industry. *2021 IEEE International Workshop on Metrology for Industry 4.0 & IoT (MetroInd4.0 & IoT)*, 654 658.

33 Mazzarini, M., Amendola, S., Ricci, E. et al. (2018). Metodo di realizzazione di giunzioni di nastri trasportatori in gomma ed applicazione di profili mediante vulcanizzazione a caldo. 102018000004542.

34 CR Technology Systems. PR-E-SE: RFID Technology to monitor electrical systems. https://www.crtsgroup.com/en/engineering_talks/rfid-technology-to-monitor-electrical-systems/ (Accessed January 2022).

35 Occhiuzzi, C., Camera, F., D'Orazio, M. et al. (2022). Automatic monitoring of fruit ripening rooms by UHF RFID sensor network and machine learning. *IEEE Journal of Radio Frequency Identification* 6: 649–659.

36 Manzari, S., Musa, T., Randazzo, M. et al. (2014). A passive temperature radio-sensor for concrete maturation monitoring. *IEEE RFID Technology and Applications Conference (RFID-TA)* 121–126.

37 EM Microelectronic, EM4325. https://www.emmicroelectronic.com/sites/default/files/products/datasheets/4325-ds_0.pdf (Accessed January 2022).

38 Axzon (2019). RFM3300-E Magnus-S3 M3E. https://axzon.com/rfm3300-e-magnus-s3-m3e-passive-sensor-ic/ (Accessed July 2019).

39 Baertschi, S.W., Dill, A.L., Kramer, T.T. et al. (2019). Degradation rate observations as a function of drug load in solid-state drug products. *Journal of Pharmaceutical Sciences* 108 (5): 1746–1755.

40 Asygn (2022). AS321X. https://asygn.com/as321x/ (accessed January 2022).

41 Camera, F. and Marrocco, G. (2021). Electromagnetic-based correction of bio-integrated RFID sensors for reliable skin temperature monitoring. *IEEE Sensors Journal* 21 (1, 1): 421–429.

42 Occhiuzzi, C., Amendola, S., Nappi, S. et al. (2018). Sensing-oriented RFID tag response in high temperature conditions. *2018 3rd International Conference on Smart and Sustainable Technologies (SpliTech)*, 1 4.

2

RFID Sensing in Power-Plant Generators and Power Transformers

Konstantinos Zannas[1], Yvan Duroc[2], and Smail Tedjini[1]

[1] *LCIS, University of Grenoble-Alpes, 50 Rue Barthelemy de Laffemas, Valence, France*
[2] *Univ Lyon, Université Claude Bernard Lyon 1, INSA Lyon, Ecole Centrale de Lyon, CNRS, Villeurbanne, France*

2.1 Introduction

The radio-frequency identification (RFID) technology can be considered one of the most prominent wireless technologies as far as identification is concerned. With multiple applications, in traceability, security, and payments, to name a few, it has been around for several decades. Despite being a mature technology, research activity on RFID applications has been quite strong with applications in new domains such as sensing and IoT [1–3].

RFID sensing, specifically in ultra high frequencies (UHF), has been one of the most prominent research fields, with several papers published over the past few years [4]. Temperature sensing, light sensing, and gas sensing are some of the example applications which were the research subjects in [5–14]. Different approaches in RFID sensing have been used but in general can be summarized in two main groups: (i) sensing using an RFID tag as a transducer; (ii) using actual external sensors connected to an RFID chip or to an RFID module in a more "traditional" approach. More information on the RFID sensing techniques can be found in [15]. At the time of writing, besides the ongoing research efforts around the globe, several commercially available RFID sensing products can be found in the growing RFID market, such as in [16–18]. The RFID technology, as a sensing solution, can leverage its passive (battery-less) operation, the well-defined regulations from the corresponding institutions (e.g. European Telecommunications Standards Institute—ETSI, Federal Communication Commission—FCC) as well

Wireless Identification and Sensing Systems for Harsh and Severe Environments,
First Edition. Edited by Smail Tedjini and Valentina Palazzi.

Published 2025 by John Wiley & Sons, Inc.

as the low cost of the RFID tags versus Wi-Fi or Bluetooth solutions. All these attributes can potentially offer to the RFID technology a major slice of the wireless sensor market in the future.

The advantages which the RFID technology is offering as a wireless sensing solution can be a considerable asset specifically for industrial applications and the inherited harsh environment that they are usually subject to. High dense metallic environment, high-temperature variations, vibrations, non-line-of-sight (NLOS) communication, and moving parts to mention a few can degrade any wireless communication link and the operation of wireless sensors in this environment can become rather a challenging task.

In this chapter, the main goal is to present the impact of the harsh environment on RFID sensors and the RFID communication, explore how these as well as to present results from a project where an RFID sensing system was tested to monitor temperature in a real use case example, the power transformers of a hydroelectric power plant. In Section 2.2, the impact of the harsh environment on the RFID sensors and on the RFID system overall. In Section 2.3, the design process of RFID sensors capable to operate in the industrial environment is discussed and their initial measurement results while in Section 2.4 the results from the practical implementation of such an RFID sensing system are presented.

2.2 Harsh Environment

The RFID tags, and consequently RFID sensor tags, are purposed to be attached to specific items which are meant to be monitored. In an industrial environment, it is very important to study and understand the impact of the very unhospitable environment of the interior of a big-scale generator or the high-voltage power transformers, and generally of the industrial environment. For example, the interior of a generator has some, very specific properties, which can pose great limitations on the propagation of electromagnetic waves, while dramatically impacting the operation of the RFID tags, specifically in the UHF band.

Large industrial machinery is usually entirely made from different kinds of metallic parts (iron, copper, steel, silicon steel, etc.). The presence of metal has a detrimental effect on the antenna structures (detuning, degradation of radiation efficiency) as well as on the RFID communication due to multipath phenomena. The multipath phenomena can have destructive or constructive interference on the propagated RF signals. It is very difficult to predict or to model how the RF signals are affected by the multiple reflections on the conductive surfaces and how strongly received by the RFID sensor tags signal. Moreover, the impact of the high electric and magnetic fields, which are present around high current conducting parts, must be explored and assessed regarding their effects on the performance of either the RFID sensor tags or of the RFID communication. Also, one important thing to keep in mind is that the RFID communication should be compliant with RF and RFID regulations that impose the frequency band, the maximum power and communication protocol, since it must be safe for employees working in the premises.

In addition, in cases where there are moving parts, such as the case of an electrical generator, additional contributors can impact the RFID communication. The overall system (RFID sensor tags-RFID reader) is in a dynamic state in such a rotating scenario where their relative positions are quickly and periodically varying. This behavior can introduce small time windows where the RFID communication is possible, with limitations depending on the rotational speed and the delays due to the processing capabilities of the individual subsystems and the communication protocols.

In the Section 2.2.1, the focus will be mainly on the impact of the conductive environment on the antenna part of an RFID sensor tag. For context, the reader should be familiar with antenna theory, electromagnetic theory, and simulation techniques. For general information about these topics, someone could refer to [19–21].

2.2.1 Conductive Environment

The effects of a conductive surface on the antenna of an RFID sensor tag will be discussed by demonstrating its impact on a proposed UHF RFID antenna (Figure 2.1). This UHF RFID antenna is designed specifically as a solution to be

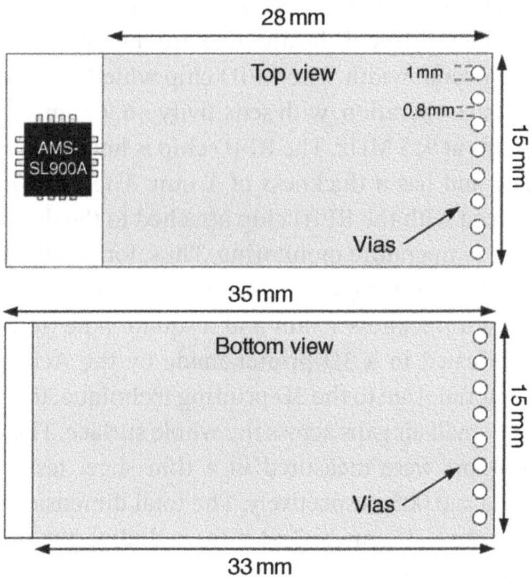

Figure 2.1 Proposed UHF RFID sensor tag (UHF RFID antenna with the RFID chip visible).

implemented in a temperature sensing application in power plants, something that will be discussed in more extent later through this chapter.

In general, the design of the RFID sensor tags is a non-trivial task since it is depending on the nature of the intended application. For instance, when the application would require the RFID sensor tag to operate on non-conductive materials, the constraints imposed on the design of the RFID antenna are very few and a variety of different antenna configurations (i.e. meandered dipole antennas) can be selected without impacting greatly the performance of the RFID antenna. On the contrary, when the RFID sensor tag is designed for operation in a metallic environment, then the design specifications become much stricter and every small touch can impact, positively or negatively, the performance. One common practice to overcome the impact of the effects of conductive items on the UHF RFID tags is to take advantage of metallic surfaces and adopt different antenna tags designs such as Inverted-F or patch-based antennas such as in [22–24].

For the proposed UHF RFID sensor tag, the antenna part is designed to exhibit the best performance when it is placed on a metallic surface. To minimize the effects due to the impact of the nearby metal, the electric field in this antenna structure is confined between two parallel, with uneven length, copper plates. A set of metalized via are connecting the top and the bottom layer of the RFID antenna to short-circuit these two layers and extend the length of the current path. The overall length of the tag antenna corresponds to $\lambda/4$, where λ is the wavelength of the frequency of operation. A substrate with a thickness 2.54 mm, permittivity $\varepsilon_{(r)} = 10.2$ and losses $\tan\delta = 0.0023$ (Rogers RT6010.2LM) was used.

The main geometrical parameters which can be exploited to tune the resonant characteristics of the antenna are the overall length, the position of the vias and the overall width. The RFID chip which was used is the AMS SL900A [16] under passive operation, with sensitivity −6.9 dBm and input impedance of (123 − j303) ohms at 915 MHz. The RFID chip is housed in a Quad Flat No-leads (QFN) package and has a thickness of 1 mm. The UHF RFID sensor tag is designed to be placed with the RFID chip attached to the desired metallic surface, for more precise temperature monitoring. Thus, for practical reasons and in order for the UHF RFID sensor tag to be positioned on a flat metallic surface seamless, a support with a thickness 1 mm and a square hole (to accommodate the RFID chip) was fabricated in a 3D printer made by the Acrylonitrile Butadiene Styrene (ABS) material. Due to the 3D printing technique, the support is not homogeneous, having small air gaps across the whole surface. The permittivity and losses of the ABS support were measured in a thin sheet tester cavity [25] to be $\varepsilon_r = 2.29$ and $\tan\delta = 0.005$, respectively. The total dimensions of the UHF RFID sensor tag are 35 mm × 15 mm × 3.54 mm, including the thickness of the RFID chip and the ABS support.

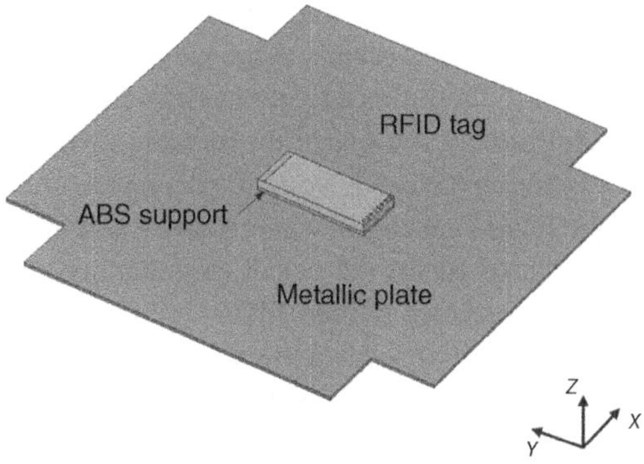

Figure 2.2 The UHF RFID antenna on a metallic surface.

A numerical model of the UHF RFID antenna was designed and simulated in the CST Microwave Studio 2016. In the simulation, a metallic plate surface with dimensions of 140 mm × 140 mm was used as demonstrated in Figure 2.2.

The exploration of the impact of the conductive surface on the UHF RFID antenna will begin by examining how the impedance at the input port of the antenna varies in relation to the distance between antenna and the conductive surface. The cases examined here are for distance of 0 to 1 mm with 0.25 mm increments. Figure 2.3 demonstrates how the real and imaginary part of the impedance is varying based on the distance between metallic surface and the UHF RFID antenna.

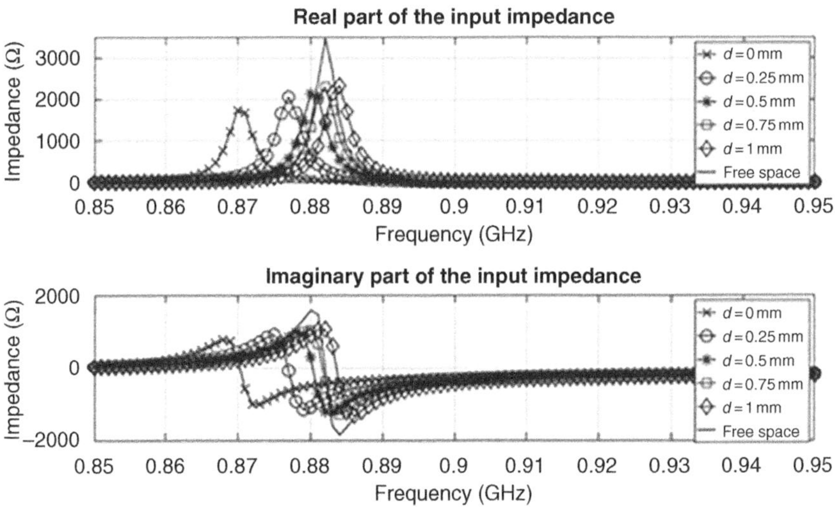

Figure 2.3 Simulated real and imaginary parts of the input impedance of the UHF RFID antenna.

For this UHF RFID antenna, there is a clear dependency of the resonance frequency with the distance to the metallic surface. As the distance increases from 0 to 1 mm the resonance frequency also increases from 870 to 884 MHz.

Additionally, the size of the metallic surface where the UHF RFID antenna lies is a factor which can have an impact on the antenna performance. This is explored in the Figure 2.4 below, where the real and imaginary part of the input impedance is simulated versus the size of the metallic surface.

The impedance of the RFID antenna and the resonance frequency for metallic rectangular ground plane sized between 60 mm × 60 mm and 200 mm × 200 mm presents a very limited shift, 0.75 MHz. For smaller size of the ground plane (20 mm × 20 mm and 30 mm × 30 mm) the impedance variation is considerable higher. In this case the size of the ground plane is comparable to the size of the RFID antenna, which is 35 mm × 15 mm.

A greater impact of the variation of the size of the ground plane can be detected on the radiation efficiency. As it is presented in Figure 2.5, the radiation efficiency is varying between 11% and 28% at 868 MHz.

Based on the results of Figure 2.5 when the RFID antenna is attached on a comparable sized ground plane, it presents a much lower radiation efficiency. The best performance is presented for ground plane size of 140 mm × 140 mm which corresponds to an even multiple of the biggest dimension (35 mm) of the RFID antenna and the resonant phenomena are stronger. As the size of the ground plane will increase further, the radiation efficiency will acquire a constant value.

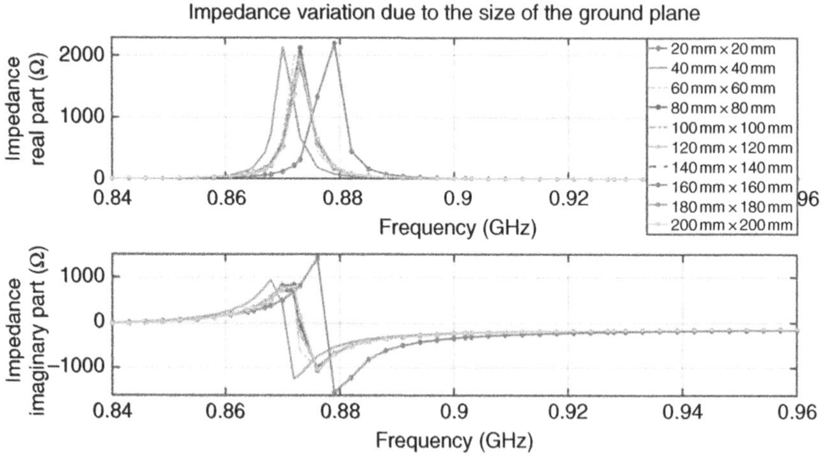

Figure 2.4 Input impedance variation versus the metallic surface area (i.e. ground plane).

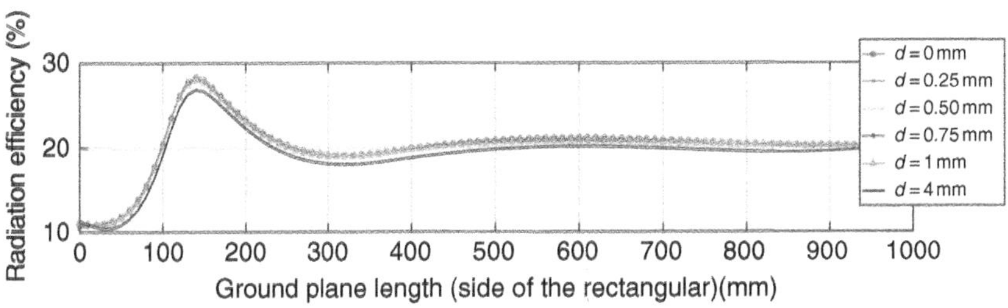

Figure 2.5 Radiation efficiency variation according to metallic surface size and distance between UHF RFID antenna and metallic surface.

In addition, the impact on the radiation efficiency and the distance between the RFID antenna and the size of the metallic surface on the radiation efficiency is presented in Figure 2.5. The values of the considered distances in each scenario were selected to be consistent with the previous simulations concerning the effect of the metallic plane on the input impedance (Figures 2.3 and 2.4). The results show that the maximum variation of radiation efficiency is 1.15% (28% for $d = 0$ mm and 26.85% for distance 4 mm for a ground plane of size 140 mm). In general, the radiation efficiency is acquiring a stable value as the metallic surface becomes larger and larger.

From the results presented so far for the specific UHF RFID antenna, someone could argue that for specific circumstances, the input impedance of the antenna could be a linear function of the distance between antenna and metallic surface. Indeed, the UHF RFID antenna and the impact on it can be represented by an equivalent circuit as the one presented in Figure 2.6.

Figure 2.6 Model of the UHF RFID antenna and the impact of the metallic surface.

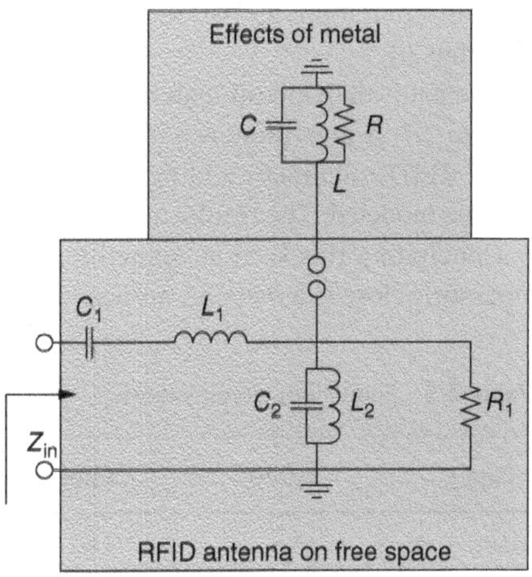

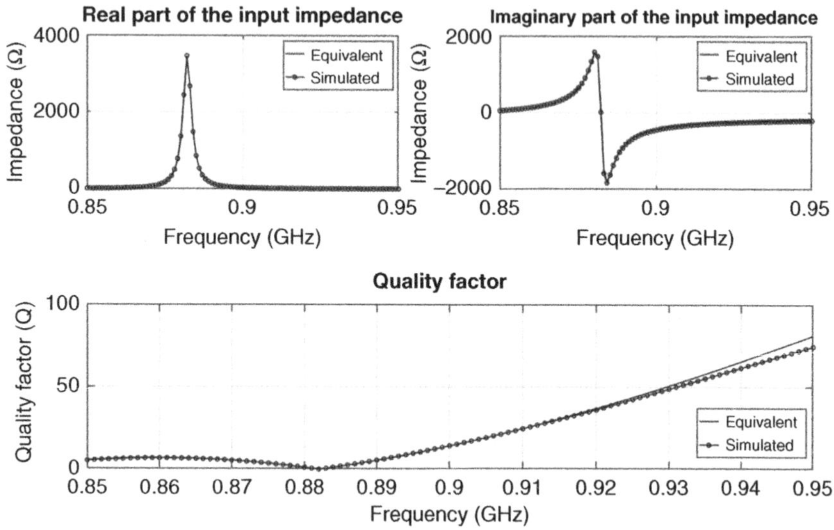

Figure 2.7 Equivalent and simulated UHF RFID antenna behavior.

By utilizing the equivalent model of the RFID antenna in free space presented in Figure 2.6, the effect of the metallic environment can be calculated for the specific antenna with quite good approximation near the resonance frequency. For example, in Figure 2.7 it is presented as the real part, imaginary part of the input impedance, and the unloaded quality factor Q (where $Q=\frac{X}{R}$, with X the imaginary part of the impedance and R the real part of the impedance) for the proposed UHF RFID antenna in free space. The results are from two sources: (i) the simulated numerical model of the UHF RFID antenna (a 50 Ω reference input port was used in simulation) and (ii) the equivalent circuit for the values that are presented in Table 2.1.

The equivalent circuit matches very well with the simulated as it can be seen in Figure 2.7. If now there is a metallic surface added with varying distance to the UHF RFID antenna by adding a parallel RLC circuit, the effects of the metal can also be modeled. The results of such an exploration are presented in Figure 2.8.

Concluding this section, something to remember is that when an antenna is operating close to a metallic surface (or to generalize near to a conductive surface)

Table 2.1 Equivalent circuit values.

C_1(pF)	L_1(nH)	C_2(pF)	L_2(nH)	R(kΩ)
0.9	15.6	14	2.325	3.4

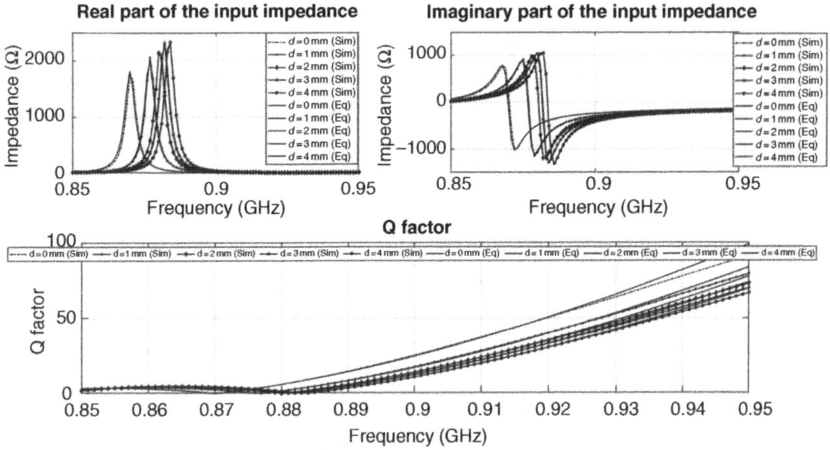

Figure 2.8 Comparison of the simulated numerical model versus the equivalent model (for appropriate component values in each case) of the UHF RFID antenna.

it should be expected that the input impedance will have a shift of its resonance frequency that can potentially be modeled by a parallel RLC circuit. More information and results of this approach are presented and discussed in [26].

2.2.2 Multipath Effects and Time-Variant Environment

When the communication channel between two antennas is considered, usually the analysis is based on a Line-of-Sight (LOS) scenario in free space propagation. In such a system, the communication is assumed to be carried out by a single path, connecting directly the two antennas. Besides the considered main single path, in reality, there are a number of paths, mainly due to reflections and diffractions, which form the communication channel [27, 28]. Each of the additional paths will result in a signal at the destination antenna with different amplitude and phase characteristics from the direct path. The phase difference between the direct signal and all the additional signals will result in a constructive or a destructive interference causing either higher or lower signal strength in the receiving position of the antenna. Such multipath scenario is very dependent on the geometry and the materials forming the environment of propagation between the two antennas. For instance, in an environment with absorbing materials the reflections and diffractions will be minimized (case anechoic chamber), while in an environment with metallic objects, like an industrial environment, the effect of multipath will be stronger.

The multipath phenomena should be taken into consideration for the RFID communication, especially in an environment with multiple metallic surfaces and objects. For a given RFID system, where RFID reader's antenna and RFID tags have fixed positions, a proper placement of each component can be achieved by finding the positions where there will be a constructive interference. This proper selection will enhance the read range of the RFID tags. In a moving system, where the antenna of the RFID reader is fixed but the RFID tags are moving or rotating, the multipath phenomena will be quite difficult to predict since the system is time variant.

In the latter case, where the RFID sensing is operating in a time-variant system such as inside an electrical generator, besides the multipath phenomena the capabilities of the RFID system should also be a point of interest. Let us consider a generator where an RFID sensor is attached on a rotating part which periodically enters the radiating field of an RFID reader. The time in which the RFID sensor can exchange information with the RFID reader might be very short, depending on the rotation speed of the generator. In this case, the time needed to have a successful communication exchange should be higher than the sum of the sensing time which includes the RFID chip delay, the propagation delay, and the RFID reader processing delay. More information about the timing constraints of an RFID sensing system in a rotating setup can be found in [29].

2.2.3 High Electric and Magnetic Fields

The impact of high electric and magnetic fields should also be a subject of interest. A detailed analysis of their impact on the RFID sensor tag would require extensive experimentation, something outside of the scope of the work presented here. To recreate in the lab, conditions that are met in power plants, it would require very high electric and magnetic fields. Especially for the case of the high electric field, it is difficult and dangerous to recreate an environment with tens of thousands of volts. Thus this condition shall be tested in the field and it will be commented on in the Section 2.4, when the application of RFID sensing in high-voltage transformers will be presented. For the case of the magnetic field, it was explored experimentally in the CREATIS/Platform PILoT of Lyon INSA University, where equipment capable of generating high magnetic field in the order of 4.7 Tesla was used to assess the impact on the RFID sensor tags. However, after testing several scenarios (e.g. different RFID tags, different distances between RFID tag—RFID reader) in the presence of the high magnetic field, it was concluded that there was no disturbing effect observed.

Concluding this part, it should be mentioned that even though there was not observed any impact on the operation of the RFID sensor tag from the high magnetic field it was taken into consideration as to what structure would be used for the proposed UHF RFID antenna. Specifically, it was decided to avoid any loop antenna design to prevent any unwanted currents from being generated by the induced magnetic field.

2.3 Design and Measurement of RFID Sensor Tag

2.3.1 Design Considerations

In this section, we will go one step back to describe how the proposed UHF RFID antenna and the UHF RFID sensor tag overall was designed and how the specifications were derived.

Initially, when aiming to use such an UHF RFID sensor for measurements in an industrial environment, someone needs to first identify what is needed to be monitored. In power-plant generators or in high-voltage transformers, there are multiple parts with different sizes and geometries. Due to this fact, and to monitor small areas, the UHF RFID sensor tag should maintain a small size and a low profile. Moreover, for the test case of a generator, multiple interpolar connections and the weight of each RFID sensor tag have to be low especially since the generator is rotating with high speed, presenting acceleration up to 1 000 G. Due to this acceleration, the actual weight of each RFID tag sensor can be considerably higher and if multiple RFID sensor tags are installed the overall weight needs to be accounted for.

Furthermore, this development should include the sensor characteristics which are required for the intended application, such as temperature range, temperature hysteresis, temperature accuracy etc. For each application, different temperature requirements apply. For instance, for temperature monitoring in the food industry, temperatures up to 40 °C are sufficient. On the other hand, monitoring a generator is required to measure temperatures more than 100 °C. Simultaneously, the expected variation of temperature is important if small variations can result in warning incidents. Based on this, the accuracy of the measured temperature is important. Finally, one more important consideration for the design of the RFID sensor tags is the need to avoid closed loop structures as stated in the Section 2.2.3. The very high magnetic field in the interior of the generators can result to the creation of very high currents and thus designs which would not result in closed loops should be selected.

Based on the discussion so far, the UHF RFID sensor tag, should respect specifications derived by its placement close to a metallic surface and to respect specifications about its sensing function. The first set of specifications mainly concerns the UHF RFID antenna and has been discussed already, while the second set of specifications are concerning mainly the UHF RFID chip which will compete with the UHF RFID antenna the UHF RFID sensor tag.

The UHF RFID chip, which was used to tackle between others, the specifications, was the AMS SL-900A [16] which has an integrated temperature sensor capable of sensing up to 125 °C. The sensitivity of the specific chip is rather low for passive, battery-less operation (it can operate either with a battery-assisted mode or completely passive) around −7 dBm. However, it was the only chip in the market at the time of the development (2017–2020) of the specific RFID sensor tags which was able to operate up to 125 °C and thus it was the best solution for the intended application.

2.3.2 Measurement of the Proposed UHF RFID Sensor Tag

Several prototypes of the designed UHF RFID antenna were fabricated, with the RFID chip SL900A soldered, as it is presented in Figure 2.9. Initially, one of the prototype RFID antennas was connected to an SMA connector to be tested with a VNA, aiming to compare the results of the measurement with the simulated results. To avoid discrepancies due to the existence of a metallic surface next to the SMA connector, the RFID antenna was measured and compared with the simulated behavior in a free-space scenario.

In Figure 2.10, the comparison between the S11 parameter and the real and imaginary part of the impedance of the measured and the simulated results are presented. There is a good agreement as far as the resonance frequency is concerned. However, there is a discrepancy regarding the maximum values of the real and imaginary parts around the resonance. This discrepancy is because of the SMA connector as well as due to the addition of solder on the RFID antenna. Additional information about the design and performance of the RFID antenna can be found in [26].

Figure 2.9 Fabricated UHF RFID sensor tag.

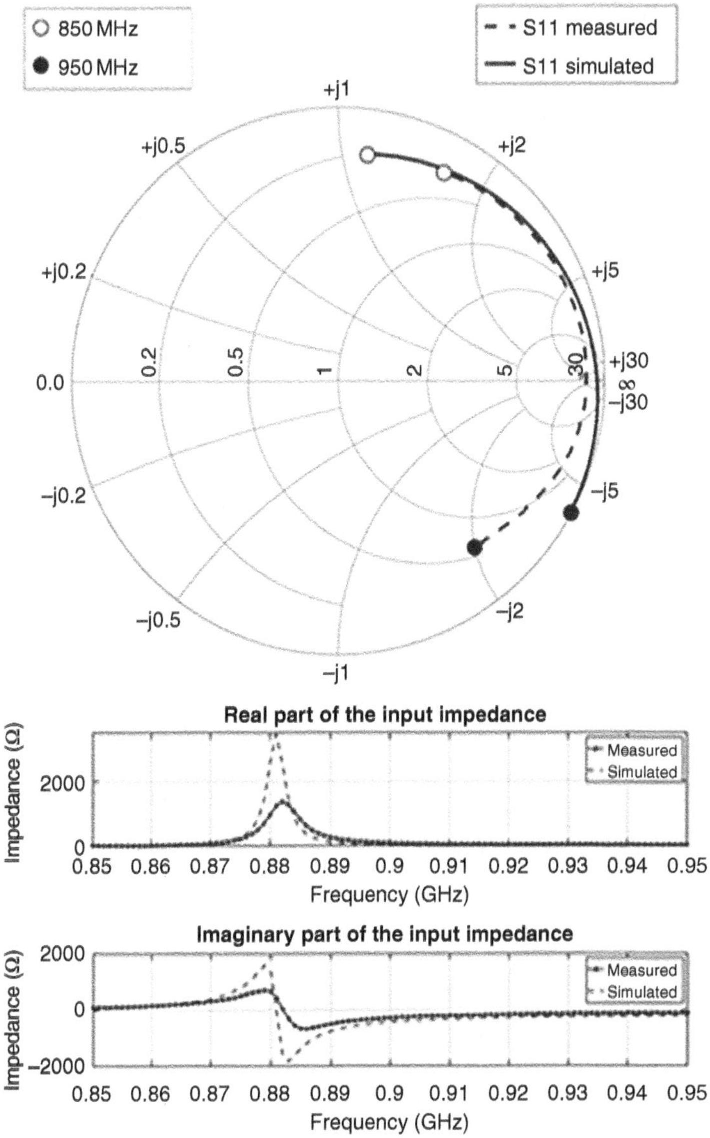

Figure 2.10 Measured and simulated results of the RFID antenna in a free space scenario.

Having concluded the final UHF RFID sensor tag by measuring its performance the next step would be to test it in a real industrial environment.

In the Section 2.4, the application of a system based on the proposed UHF RFID sensor tag is presented.

2.4 RFID Sensors: Application in Power Transformers

The UHF RFID sensor tags are designed focusing mainly on the interior of power-plant generators and the high-voltage transformers structures. However, when it comes to the practical testing in a real operating power plant it is not as straightforward. It needs to be an appropriate time window (e.g. during a scheduled maintenance) where the power grid is down and thus safe to work in the power plant's premises. Also, there need to be two time windows to complete a measurement round, one window for the installation and one window for the measurement. Due to these limitations, the specific UHF RFID sensing system was tested only in high-voltage transformers. The power plant of Grand' Maison situated in the south-east region of France was used for testing.

2.4.1 Installation Procedure

The presented UHF RFID sensor tag was used for an initial production of 50 RFID sensor tags. The transformer structure is presented in Figure 2.11. The transformer has three-phase connections and thus RFID sensor tags would be installed in each one.

The part where the RFID sensor tags needed to be installed was below a black rubber cover. This position was advantageous since after the installation it would be possible to use an antenna at this point to read the RFID sensor tags.

Figure 2.11 Transformers in hydroelectric power plants.

In each phase connection, 7 RFID sensor tags were installed, each one in different position. A strong glue (Loctite 9492) for metallic items was used to keep the RFID sensor tags in place. One of the fabricated RFID sensor tags and the placement with the glue is presented in Figure 2.12. In Figure 2.13 the placement of some RFID sensor tags in the transformer can be viewed.

Each RFID sensor tag had the EPC value renamed according to its placement. The name would include on which transformer it was installed and at which position. All seven positions which were utilized in this installation are presented in Figures 2.13 and 2.14.

One very important issue during the installation of the RFID sensor tags was the fact that each position had different geometry and the metallic surfaces in the proximity would pose a different effect on the RFID sensor tag antenna. This had an impact on the resonance frequency of the RFID sensor tag, resulting not to operate with the best performance in the ETSI frequency band. This issue had to be considered even more since the designed RFID sensor tags exhibit a narrow band operation and the fact that the ETSI frequency band has only 2 MHz of bandwidth. To address this problem, the following procedure was adopted. Initially, the RFID sensor tag was placed in the specific position it would be installed. The

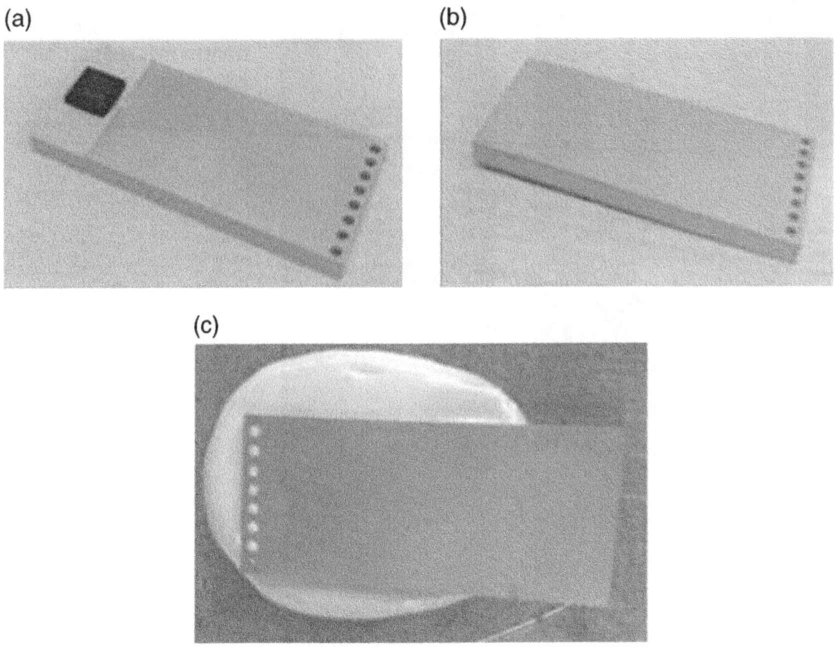

Figure 2.12 Fabricated RFID sensor tag (a) bottom view, (b) top view, and (c) placement with the glue.

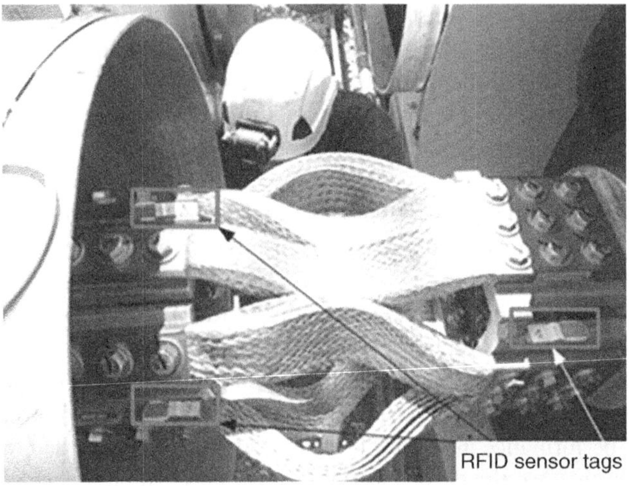

Figure 2.13 Power transformer with RFID sensor tags in various installation positions.

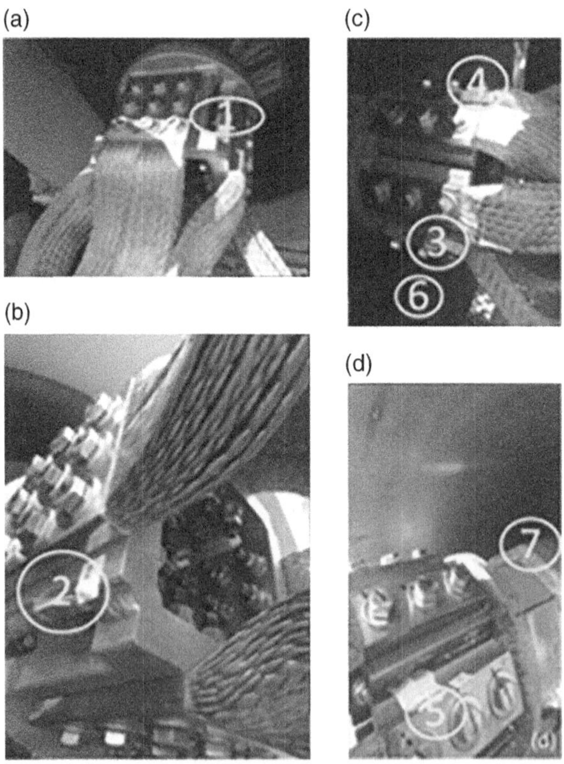

Figure 2.14 Installation positions of the RFID sensor tags in each transformer connection (seven positions overall).

performance of the RFID sensor tag would be evaluated by using the Tagformance Pro RFID reader. The RFID sensor tag would be trimmed (by removing small parts from the copper layer) gradually until the resonance would correspond to the ETSI frequency band. Then the RFID sensor tag would be permanently installed.

By trimming the RFID sensor tag antenna, the resonance frequency can only vary toward higher frequency values. The potential tuning of the RFID sensor tags was anticipated and thus the RFID sensor tags were fabricated to resonate a little lower of the ETSI frequency band (~860 MHz) considering the possibility of tuning them to the desired frequency.

In addition to the impact of the nearby metallic structures on the resonance frequency of the RFID sensor tag, the radiation properties of the RFID sensor tag would be affected, downgrading the performance and the read range of the RFID sensor. Specifically, the RFID sensor tags were positioned on different-sized metallic surfaces. Larger metallic surfaces are resulting in a higher directivity and thus higher read range.

After the installation of the RFID sensor tags and a time interval for the maturation of the glue, some measurements of the temperature were conducted. For this measurement the ThingMagic M6e reader [30] was used. The SL900A RFID chip to transmit the value of the output of the integrated temperature sensor, needs to receive a custom command. The M6e reader operates with the Mercury application programming interface (API) which already includes libraries with the custom commands that the RFID chip requires to respond to the value of the integrated temperature sensor. Hence, an interface including the standard EPC commands as well as the custom commands of the SL900A RFID chip was developed using the C# programming language for this specific application [31].

In Figure 2.15, a screenshot of the developed interface is presented when the M6e RFID reader was reading one of the installed RFID sensor tags. The interface contains two main windows. In the left graph the output of the ADC value of the RFID chip is appearing. In the right window, the value of the temperature can be seen after the appropriate data processing completes. For instance, the average value transmitted was *675* which according to the reference voltages of the RFID chip corresponds to 32.2 °C.

2.4.2 Temperature Measurement with the RFID Sensor Tags

The RFID sensor tags which were installed in the hydroelectric power plant of Grand' Maison stayed idle until the opportunity to conduct measurements while the transformers were operating arisen. This was a first-class opportunity to test the operation of the RFID sensor tags under high electric and magnetic field conditions.

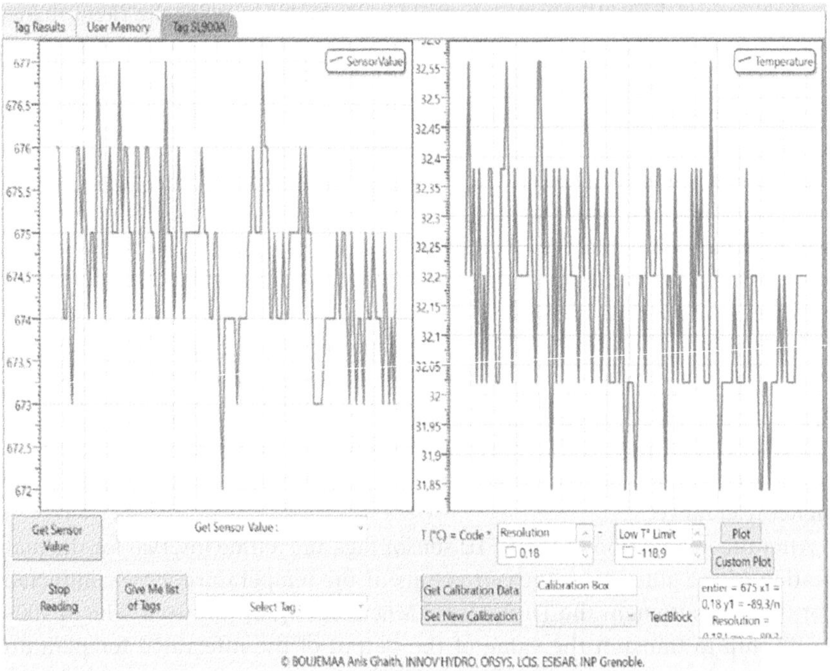

Figure 2.15 Developed interface for acquiring the output value of the integrated temperature sensor of the SL900A RFID chip.

To perform the measurements, some very important precautions should be considered. When the transformers are operating a very high electric and magnetic field surrounds them, thus the antenna and the RFID reader which is needed to be in proximity, should be shielded to avoid the generation of high current passing through the equipment. For this reason, a measuring setup was developed in cooperation with the team of the EDF company. The measuring setup is presented in Figure 2.16.

The measurement setup contains the RFID reader ThingMagic M6 with right-handed circular polarized antenna with gain 6 dBi, an Ethernet to optical fiber converter and a power supply which was powering up the RFID reader and the converter. The whole measurement setup was positioned on a wooden base attached to an insulated stick.

Moreover, the position of the antenna to read the RFID sensor tags was an issue which was given the appropriate attention. The antenna was positioned around the rubber cover since the rest of the transformer cover was metallic. However, as presented previously, the RFID sensor tags were installed on different sides of the transformer. With such placement, the RFID sensor tags positioned at the other side of the antenna could potentially be unreadable due to the intervention of the transformer body. For this issue, during the measurements, the antenna of the RFID

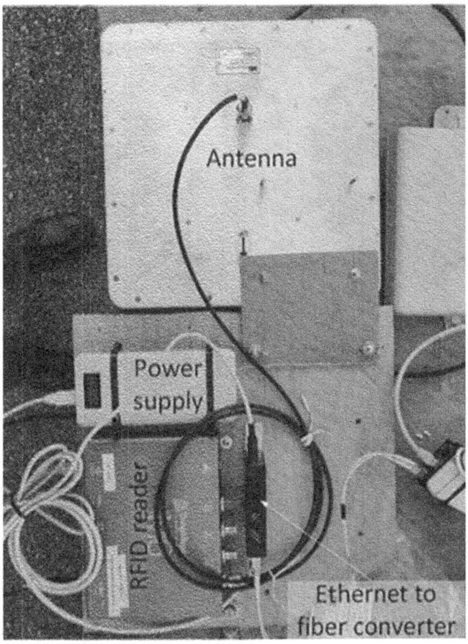

Figure 2.16 Shielded measurement setup.

reader was tested in different positions to identify the case where the highest amount of RFID sensor tags was read. One example position is presented in Figure 2.17.

The several different positions where the RFID reader's antenna was held in order to read the RFID sensor tags are summarized in the Figure 2.18 below.

Figure 2.17 Measurement of the RFID sensor tags from position A.

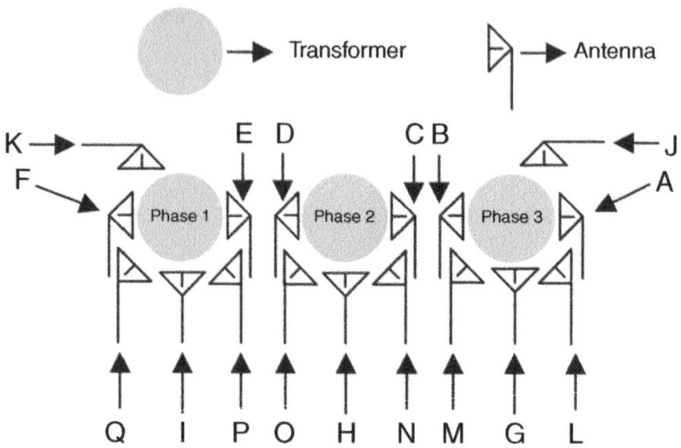

Figure 2.18 Tested positions of the RFID reader's antenna.

The shielded measurement setup was positioned in all the presented positions and every time a reading round of the temperature sensor value was initiated. In the following tables, the results of the measurements are presented (Tables 2.2–2.4).

Concluding this part, the results of the measurements in the high-voltage transformers of the hydroelectric power plant of Grand' Maison were encouraging. The UHF RFID sensor tags could operate under the effects of high magnetic and electric fields. Furthermore, the positioning of the antenna of the RFID

Table 2.2 Phase 1.

RFID sensor tags (3 last digits of the EPC code)	Average temperature measured for each antenna position (°C)[a]					
	E	F	I	K	P	Q
8A1	—	—	—	—	—	—
8A2	—	—	—	—	—	—
8A3	45.5	−2.1	43.4	—	2.4	40.2
8A4	—	—	—	—	—	—
8A5	—	—	—	—	—	—
8A6	26.9	—	—	—	—	—
8A7	—	45.2	—	47.4	47	48.1

[a] For the phase 1 the dedicated positions of the antenna were the: E, F, I, K, P, Q.

Table 2.3 Phase 2.

RFID sensor tags (3 last digits of the EPC code)	Average temperature measured for each antenna position (°C)[a]					
	C	**D**	**H**	**N**	**O**	**–**
0B1	40.5	—	41.1	41.9	—	—
0B2	40.3	40.3	40.3	41.3	41.5	—
0B3	—	—	41.7	45	—	—
0B4	44.1	—	46.5	48.9	—	—
0B5	—	—	—	—	—	—
0B6	—	—	—	—	—	—
0B7	46	45.7	—	46.9	—	—

[a] For the phase 2, the dedicated positions of the antenna were the: C, D, H, N, O.

Table 2.4 Phase 3.

RFID sensor tags (3 last digits of the EPC code)	Average temperature measured for each antenna position (°C)[a]							
	A	**A_H**[b]	**A_V**[b]	**B**	**G**	**J**	**L**	**M**
0A1	41	42.5	—	—	—	41.8	41.9	18.3
0A2	—	44.4	—	—	—	—	—	44.3
0A3	—	—	—	—	—	—	—	—
0A4	—	—	—	—	—	—	—	—
0A5	—	—	—	48.9	49.8	—	—	50.8
0A6	—	28.9	—	—	25.4	—	—	38
0A7	—	50.6	—	47.9	48.8	—	—	50

[a] For the phase 3, the dedicated positions of the antenna were the: A, B, G, J, L, M.
[b] In the position A, a linear antenna with gain of 8 dBi was also tested, where A_H when the antenna was held horizontally and A_V when the antenna was held vertically.

reader, as expected, played a major role in the readability of the UHF RFID sensor tags. Indicatively, for the phase 3, the position M was the most prominent for the antenna to be installed in a permanent measurement setup. Also, some RFID sensor tags yield quite different results for different positions of the RFID antenna. This is due to the operation of the RFID chip, where if the received power is not quite enough, the reference voltages are not acquiring the proper value and wrong data are transmitted. In addition, a few RFID sensor tags were not read, and this is due to the tuning process of trimming. The trimming procedure was performed manually, allowing a big margin of error at the resonance frequency of each RFID sensor tag. Also, another reason for not being able to read all the tags was the positioning of some of them where all the tested antenna positions might not be beneficial. Finally, the need to test the UHF RFID sensing system in real environment is a real cornerstone. Such an environment has a lot of unknown or time-dependent variables and regardless of how accurately the environment is modeled; it is very difficult to be simulated with high confidence. And so the best way to evaluate any system is an old traditional way of testing in the real intended environment.

2.5 Conclusion

The tests in the Grand' Maison power plant signified an important milestone: one of the earliest, if not the first, attempt to utilize a UHF RFID sensing system in an industrial application with quite encouraging results.

As it was demonstrated throughout the chapter, the challenges to implement the UHF RFID technology in a sensing application in an industrial environment includes several different aspects. From the main challenge, which is the UHF RFID antenna to be used, to the selected UHF RFID chip, to the other considerations which are adding constraints to the overall RFID sensing system and to the absolute necessity to test this sensing system in a real environment, this chapter might give the necessary insights for someone to use as a source of useful information.

This work to further explore the topic of RFID sensing in industrial environment should also dive into measurements in the generators of power plants. Even though, there was an installation at one generator, there were made no conclusions since the measurement was not totally performed within the duration of the overlining project. This is something which should be attempted to correctly evaluate the potential of the UHF RFID sensing technology for industrial application in power plants.

Finally, closing this chapter, the authors would like to underline the potential of the UHF RFID technology for sensing applications. Battery-less devices, which are mainly found in RFID technology, offer exceptionally high advantages to sensing applications and additional research both on application level and also in the development of the next generation of RFID integrated circuits IC chips can offer a real revolution to industrial monitoring.

References

1 Costa, F., Genovesi, S., Borgese, M. et al. (2021). A review of RFID sensors, the new frontier of internet of things. *Sensors* 21 (9): 3138.

2 Gupta, A. and Jha, R.K. (2015). A survey of 5G Network: architecture and emerging technologies. *IEEE Access* 3: 1206–1232. https://doi.org/10.1109/ACCESS.2015.2461602.

3 Baumbauer, C.L., Anderson, M.G., Ting, J. et al. (2020). Printed, flexible, compact UHF-RFID sensor tags enabled by hybrid electronics. *Scientific Reports* 10: 16543. https://doi.org/10.1038/s41598-020-73471-9.

4 Haibi, A., Oufaska, K., Yassini, K.E. et al. (2022). Systematic mapping study on RFID technology. *IEEE Access* 10: 6363–6380. https://doi.org/10.1109/ACCESS.2022.3140475.

5 Manzari, S., Caizzone, S., Rubini, C., and Marrocco, G. (2014). Feasibility of wireless temperature sensing by passive UHF-RFID tags in ground satellite test beds. *2014 IEEE International Conference on Wireless for Space and Extreme Environments (WiSEE)*, Noordwijk, Netherlands (30–31 October 2014), 1–6. https://doi.org/10.1109/WiSEE.2014.6973074

6 Rima, S., Georgiadis, A., Collado, A. et al. (2014). Passive UHF RFID enabled temperature sensor tag on cork substrate. *2014 IEEE RFID Technology and Applications Conference, RFID-TA*, Tampere, Finland (8–9 September 2014), 82–85. https://doi.org/10.1109/RFID-TA.2014.6934205.

7 Zannas, K., El Matbouly, H., Duroc, Y., and Tedjini, S. (2018). Self-tuning RFID tag: a new approach for temperature sensing. *IEEE Transactions on Microwave Theory and Techniques* 66 (12): 5885–5893. https://doi.org/10.1109/TMTT.2018.2878568.

8 Salmeron, J.F., Molina-Lopez, F., Rivadeneyra, A. et al. (2014). Design and development of sensing RFID tags on flexible foil compatible with EPC Gen 2. *IEEE Sensors Journal* 14 (12): 4361–4371. https://doi.org/10.1109/JSEN.2014.2335417.

9 Manzari, S., Occhiuzzi, C., Nawale, S. et al. (2012). UHF RFID humidity sensor tag based on hygroscopic polymeric load. *2012 6th European Conference on Antennas and Propagation (EUCAP)*, Prague, Czech Republic (26–30 March 2012), 3030–3033. https://doi.org/10.1109/EuCAP.2012.6206150.

10 Gonçalves, R., Pinho, P., Carvalho, N.B., and Tentzeris, M.M. (2015). Humidity passive sensors based on UHF RFID using cork dielectric slabs. *2015 9th European Conference on Antennas and Propagation (EuCAP)*, Lisbon, Portugal (13–17 April 2015), 1–4.

11 Colella, R., Tumolo, M.R., Sabina, S. et al. (2021). Design of UHF RFID sensor-tags for the biomechanical analysis of human body movements. *IEEE Sensors Journal* 21 (13): 14090–14098. https://doi.org/10.1109/JSEN.2021.3069113.

12 Soodmand, S., Zhao, A., and Tian, G.Y. (2018). UHF RFID system for wirelessly detection of corrosion based on resonance frequency shift in forward interrogation power. *IET Microwaves, Antennas & Propagation* 12: 1877–1884. https://doi.org/10.1049/iet-map.2018.0143.

13 Luo, C., Gil, I., and Fernández-García, R. (2020). Wearable textile UHF-RFID sensors: a systematic review. *Materials* 13 (15): 3292. https://doi.org/10.3390/ma13153292.

14 Tentzeris, M.M. and Nikolaou, S. (2009). RFID-enabled ultrasensitive wireless sensors utilizing inkjet-printed antennas and carbon nanotubes for gas detection applications. *2009 IEEE Conference on Microwaves, Communications, Antennas and Electronic Systems COMCAS*, Tel Aviv, Israel (9–11 November 2009), 1–5. https://doi.org/10.1109/COMCAS.2009.5385940.

15 Zannas, K., Matbouly, H.E., Duroc, Y., and Tedjini, S. (2020). From identification to sensing. In: *Wireless Power Transmission for Sustainable Electronics* (ed. N.B. Carvalho and A. Georgiadis). https://doi.org/10.1002/9781119578598.ch8.

16 AMS. SL900A EPC Sensor Tag. https://ams-osram.com/products/interfaces/sensor-interfaces/ams-sl900a-epc-gen2-sensor-tag (accessed 8 July 2024).

17 ASYGN. AS321X. https://asygn.com/as3213t_temperature/ (accessed 8 July 2024)

18 Isabel Freitas, Load Carrier Tracking with RFID in Wine Production, 27 May 2021. https://wiot-group.com/think/en/articles/load-carrier-tracking-with-rfid-in-wine-production/ (accessed 8 July 2024)

19 Balanis, C. (2005). *Antenna Theory, Analysis, and Design*, 3e, 184–204. New York, NY: Wiley.

20 Stutzman, W.L. and Thiele, G.A. (2012). *Antenna Theory and Design*, 3e, 304–308. New York, NY: Wiley.

21 Gibson, W.C. (2008). *The Method of Moments in Electromagnetics*, 1e. Boca Raton, FL: Chapman & Hall/CRC.

22 Chen, S. (2009). A miniature RFID tag antenna design for metallic objects application. *IEEE Antennas and Wireless Propagation Letters* 8: 1043–1045. https://doi.org/10.1109/LAWP.2009.2032252.

23 Zhang, J. and Long, Y. (2014). A novel metal-mountable electrically small antenna for RFID tag applications with practical guidelines for the antenna design. *IEEE Transactions on Antennas and Propagation* 62 (11): 5820–5829. https://doi.org/10.1109/TAP.2014.2354412.

24 Barman, B., Bhaskar, S., and Singh, A.K. (2018). Spiral resonator loaded S-shaped folded dipole dual band UHF RFID tag antenna. *Microwave and Optical Technology Letters* **61** (3): 720–726.

25 Model 08 Thin Dielectric Sheet Tester system, MuEpsln, LLC. https://muepsln.com/cavities/ (accessed 8 July 2024).

26 Zannas, K., Matbouly, H.E., Duroc, Y., and Tedjini, S. (2021). A flipping UHF RFID sensor-tag for metallic environment compliant with ETSI/FCC bands. *IEEE*

Transactions on Antennas and Propagation 69 (3): 1283–1292. https://doi.org/10.1109/TAP.2020.3026869.

27 Rummler, W.D. (1980). Time- and frequency-domain representation of multipath fading on line-of-sight microwave paths. *Bell System Technical Journal* 59 (5): 763–796.

28 Wiesbeck, W., Sturm, C., Soergel, W. et al. (2007). Influence of antenna performance and propagation channel on pulsed UWB signals. *Proceedings of International Conference on Electromagnetics in Advanced Applications (ICEAA)*, Turin, Italy (17–21 September 2007), 915–922.

29 El Matbouly, H., Zannas, K., Duroc, Y., and Tedjini, S. (2017). Analysis and assessments of time delay constrains for passive RFID tag-sensor communication link: application for rotation speed sensing. *IEEE Sensors Journal* 17 (7): 2174–2181. https://doi.org/10.1109/JSEN.2017.2662058.

30 Jadak. ThingMagic M6e UHF RAIN RFID. https://www.jadaktech.com/product/thingmagic-m6e-uhf-rain-rfid/ (accessed 8 July 2024).

31 Boujemaa, A.G. (2018) Project report, "Développement d'une Application Windows pour la Commande et a Communication d'un tag RFID dans un Environnement Electromagnétique Sévère", 29 September 2018.

3

Design of Passive UHF RFID Sensors Meeting Food Industry Regulations

Benjamin Saggin[1], Arnaud Vena[1], Brice Sorli[1], Fanny Coffigniez[2], and Camille Ramade[3]

[1] *IES, Univ Montpellier, CNRS, Montpellier, France*
[2] *Institut d'Électronique et des Systèmes, Univ Montpellier, Montpellier, France*
[3] *Tageos, Montpellier, France*

3.1 Introduction

In 2015, the United Nations adopted sustainable development goals (SDG) that target to reduce food loss and waste, and indicate to reduce wastes that occur "from retail to the final consumption/demand stages" [1]. To this end, food market must use strong logistic management and extend shelf life of products. But, to do so needs strong precautions to preserve human health. World Health Organization estimated that more than 600 million foodborne diseases have been declared in 2010 and some of them led to numerous fatalities including death and disabilities [2]. In the United States of America, only, 37% of foodborne diseases were due to improper storage temperature between 1993 and 1997 [3]. To this point, the responsibility could be shared between producer, carrier, seller, and customer. This led to the success of time-temperature indicators (TTI) that monitor time when products were exposed to high temperature. Among the remaining cases, only 11% foodborne diseases are due to improper cooking. So, lots of contaminated food should have been discarded before cooking because of other factors (contaminated equipment, unsafe source, poor personal hygiene, ...). Because it poses a risk to health, sensors must focus on food products microbiology.

In this chapter, we will discuss each step that drives the development process of our food spoilage sensor in the context of "granting society with low

Wireless Identification and Sensing Systems for Harsh and Severe Environments, First Edition. Edited by Smail Tedjini and Valentina Palazzi.

Published 2025 by John Wiley & Sons, Inc.

environmental impact innovative packaging" (GLOPACK) project, backed by the European Horizon 2020 framework. This project was held by numerous partners that work for three main actions: to promote biopolymer food packages, to trick food product storage conditions using natural solutions to extend their shelf life, and to precisely monitor food spoilage using the most environmentally aware solution.

To cross food packages with air-tight barriers and address sensors that are stacked in shelves, wireless solutions are needed. This needs a more complex sensor design and tends to conflict with SDG target 12 fulfillment which states to use less materials. Furthermore, to reduce materials needed and to avoid food contamination by hazardous chemicals, batteries-powered must be banned. Only composed of a millimeter-cube size electronic component and a micrometer-thin metallic antenna, passive RFID (radio-frequency identification) transponder solution ticks the boxes until now.

To present our work, we first pose, in the next section, arguments toward replacement of simple barcodes by advanced RFID transponders to embed sensors. In the next section, after a quick portrayal of food spoilage process happening in packages, we display several works that are about to RFID food spoilage sensors. We discuss their advantages and drawbacks to shape our solution. Then, we first highlight our RFID sensors' main features before detailing their structure. Because of their "analog" sensing technique, we need to present how the sensed data will be given to the end-user and the underlying software that drives RFID readers. Finally, we show results in a mocked industrial setup to confirm our solution is up to work as planned.

3.2 RFID Sensors

3.2.1 Interest in RFID

To assess the amount to be invested in a system that makes identification more automatic, the examination of acquisition costs is essential. Manufacturers first turned to very inexpensive identification solutions using the printing of sequences of numbers and/or letters. These indications are adapted for reading by a human operator. In this case, to allow automation of the process, the machine will need an optical character recognition system. The disadvantages of this technique are its higher error rates and its prohibitive cost. Thus, automation will prefer "barcode" systems, invented in the 1950s [4]. These systems encode the identifier in their 2D structure. The reading is done by comparing the reflectance of the surface. Several barcode systems exist and are divided into two categories: linear codes (one dimension) and matrix codes (two dimensions). For the former, the universal product code (UPC) largely dominates the market, especially for marketed products. For the latter, the quick response (QR) code was imposed. Thanks to its two dimensions, the encoded information can be more important. The QR code often encodes a uniform resource identifier (URI) pattern that allows it to be used in a variety of ways. Global Standard 1 (GS1) is a global nongovernmental, nonprofit organization that tries to harmonize the identification of objects. It created the Electronic Product Code (EPC) which is a standard for identifying objects more precisely. Indeed, it contains a series of information on the nature of the object as well as the unique serial number of the latter. With EPC, identification not only becomes more accurate but also takes up more space. Storing information digitally should then be considered. The main argument in favor of using RFID is that this technology does not require line-of-sight to be read and allows scanning of many stacked objects.

GS1 first chose ultra-high frequency (UHF) RFID. Today, this technology is called RAIN and is based on EPC UHF Generation 2 Air Interface Protocol. Even though alternatives exist [5, 6], RAIN standard matches logistic needs according to its first authors [7]. Communications are now up to approximately 10 m and allow passive (i.e. batteryless) operations while reader provides continuous wave. When requested by reader, transponders first broadcast their EPC which contains information about their attached product (producer, reference, serial number, expiry date, ...). Furthermore, RAIN-embedded anti-collision algorithms let transponders (or tags) answer in sequence. A conventional off-the-shelf (COTS) reader is able to read up to 750 tags per second [8]. In addition, it offers dynamic electronic storage, i.e. data can be modified during the transit of the identified object. Finally, with security options, technology makes tampering impossible.

3.2.2 Types of RAIN Sensors

In the development of RAIN solutions, it did not take long to add sensor circuits to RAIN tags. In January 2005, Intel Research and the University of Washington published a paper presenting the open-source wireless identification and sensing platform [9] which has given rise to many sensor works.

Among the RAIN sensor tags, two basic architectures stand out. The sensor tags can transmit the measured values to the reader either by:

- The digital way, the value will be encoded in a response of the RAIN protocol,
- The analog way, the quality of communication between tag and reader will be modulated according to the measurement.

To power advanced features, RFID chip manufacturers choose to use a more energy-efficient integrated circuit (IC), a greater directivity of the antenna of the reader and tag, or to harvest energy from surrounding sources (solar, thermal, mechanical, etc.). To require less energy and less material, tags should not require additional circuits for sensing. Besides, RAIN tags based on RFID chips connected to digital sensors (i.e. SPI, I^2C, ...) are expensive. Consequently, such design is limited to proof of concept prior to the design of IC, or they can be used for reusable concepts. For these reasons, the analog way looks more promising.

Several analog ways exist to transmit measurements: switched tag to detect a threshold overtaking, frequency or amplitude modulation due to changes in antenna tuning. Generally speaking, integrating sensor in the antenna reduces IC power consumption, and thus, the activation power of the full tag so that the reading distance is improved. Furthermore, the cost of the final RFID-enabled sensor is reduced to compare with a digital solution. However, adding a sensor function in the antenna circuit can have drawbacks to its effectiveness in transmitting digital information and complying with standards. Furthermore, analog RFID sensors are sensitive to their surrounding environment (proximity of walls reflecting the signal, nature of medium in which the signal is transmitting). Some solutions are emerging to increase tag's reliability: distance estimation, use of a reference tag near the sensing tag, radar techniques, ...

Finally, adoption by industry will be facilitated by reducing the resources required at the interrogator levels. To make possible the use of COTS readers with the chosen measurement principle, it is necessary to consider the regulatory framework that limits final resolution. Indeed, the readers will be constrained in frequency and transmission power. This will have consequences on reader's ability to correctly estimate the original measurement. To overcome this, we will determine an appropriate combination of techniques.

3.3 Monitoring Food Spoilage

3.3.1 Food Spoilage Process

Because of the nature of life to grow, it is difficult to keep things sterile. Especially, if some food is present and human handling is needed. And, even if these kinds of contamination are not considered, viruses and bacteria can be contained because foodstuff is or was alive and healthy, and many fungi reproduce by spraying spores in the environment [10]. To avoid human health problems, some regulations were developed to establish a tolerance level for each type of microorganism in food. Beyond this level, food cannot be consumed anymore.

Microorganism growth can be slowed down or simply stopped by several methods: temperature control, irradiation, antimicrobial treatments, or modified atmosphere package (MAP), which are often used during storage of food products. However, these strategies are sometimes undermined during storage, as a break in the cold chain, or a leaking in packaging. It means that the limit of consumption will be crossed depending on product history from process to distribution.

Therefore, foodstuff stock management is crucial in food distribution. Expiry dates mentioned on food packages help to guarantee consumer safety. However, these dates are not related to the history and/or the state of real packaged products. They are estimated beforehand during aging tests and according to an arbitrary scenario, legislated or not (microbial load of the raw materials, predictive microbiology, safety margin considering, for example, a temperature overrun over a certain period, ...) [11]. Even today, no indication of expiry date is easily digitized to allow a simple management. Moreover, expiry dates do not consider conditions of transport and storage of food products.

As a workaround, sensors were developed to monitor different aspects of food quality, such as milk freshness, bacterial growth, or fish spoilage, to reduce waste, and inform food logistics actors about product spoilage [11]. For microorganisms to stay alive and grow, the host environment needs to fulfill some conditions (temperature, humidity, ...). Therefore, some sensors monitor these to just evaluate microorganisms' presence. To have good information on the freshness of the product, some sensors prefer to monitor microorganisms' supplies and/or exhausts (dioxygen, carbon dioxide, ethanol, ...) that were due to their growing process [12].

Finally, to preserve taste and food safety, agri-food industrials refuse to add novel compounds in packaging. Consequently, a contactless solution is preferred. Therefore, if food products are sealed in an air-tight package avoiding gas loss through packaging, monitoring gases avoid a direct contact with food and can inform on contamination status.

3.3.2 State of the Art on RFID-based Food Sensors

Table 3.1 shows current state of the art on RAIN RFID-based sensor to monitor food degradation/maturation.

Digital-based RFID sensors allow greater measurement reliability but are often more expensive. Such technology is more suitable for solutions where the sensor can be reused as in Ref. [13]. In this case, the idea is to inform about the degree of maturation of a cheese by monitoring the pressure of the air-tight packaging [13]. Pressure variations depend on the production of carbon dioxide (CO_2) generated by bacteria. The sensor is long-lasting and is reusable on another cheese. However, digital solutions are unsuitable for irreversible measurement, especially in the agri-food sector, where simple and low-cost solutions are required. The irreversibility is particularly interesting because it avoids continuous measurement using a constant power source (external source or constant RFID reading) which can be unfeasible in some situations. Logistics is the main

Table 3.1 Food spoilage sensing RAIN RFID solutions.

Sensed value	Sensed value transmission	Sensor	Application démontrée	Food contact	Reference
Pression	Digital	TE MS5837	Cheese ripening	Yes	Abdelnour et al. [13]
Temperature	Switching between two tags	Shape memory polymer	Temperature excess	No	Bhattacharyya et al. [14]
Temperature	Switching between two modems	Shape memory metal	Temperature excess	No	Caizzone et al. [15]
Temperature	Turn-on power modulation	Hiding the antenna triggered by solid/liquid phase change	Temperature excess	No	Ali Ziai et Batchelor [16]
Temperature	Digital	Bandgap temperature sensor in silicon	—	No	Kay Law et al. [17]
Humidity	Turn-on power modulation	Silver-nanoparticle antenna on cock stopper	—	Yes	Gonçalves et al. [18]
Meat permittivity	Turn-on power modulation	Antenna on meat	Meat spoilage	Yes	Nguyen et al. [19]
Humidity, volatile organic compounds	Digital	—	—	No[a]	European project FlexSMELL [20] Kapucu et al. [21]

[a] The sensitive polymer must be inside the food packaging.

sector of RAIN technology; however, during its journey, the food can hardly be followed without costly solutions financially and environmentally. To solve this, the works of Bhattacharyya et al. [14], Caizzone et al. [15], and Ziai and Batchelor [16] present irreversible temperature threshold exceedance detectors. However, exposing the food to an increase in humidity/temperature during a short time does not mean that one can observe a significant degradation. To focus only on the effective degradation of the food, we need to measure more representative indicators. This is the idea developed in Ref. [19], which links the permittivity of meat to its degradation, as well as in the European Project FlexSMELL [20], which targets volatile organic compounds released by the chemical activity that accompanies decomposition [22].

It is to be noted that solutions requiring direct contact with food or using a test body based on materials are not compatible with the food sector and should be avoided. Indeed, they risk coming into conflict with food safety standards. For example, Gonçalves et al. use inkjet printing of silver nanoparticles on an epoxy substrate [18]. The solvents used can induce irritation if not properly removed and the use of nanoparticles is controversial [23].

3.4 Food Spoilage Sensitive RFID Tag Design

3.4.1 Hardware Design

3.4.1.1 Position

To meet the multiple objectives of the GLOPACK H2020 project, we designed a hybrid tag, supporting RAIN and near-field communication (NFC) protocols, which is presented in Figure 3.1.

The first objective was to satisfy logistical needs and is addressed by the RAIN protocol. EPC is intended to be more descriptive of the unique serial number for each product sold. Thus, in the context of food products, we will be able to store standardized information in the EPC memory of the IC such as [24]: the supplier, the producer, the date of production, the expiry date, ... Furthermore, we will use RAIN tag "user memory" to store "proprietary" data. These calibrations help us determine the food freshness according to the type of product.

The second objective is to make information about the purchased product accessible to the consumer via a smartphone, using the NFC technology, because no smartphone currently marketed to public has a RAIN reader. To simplify the encoding of information, the tag will only require one chip, the EM4423 from EM

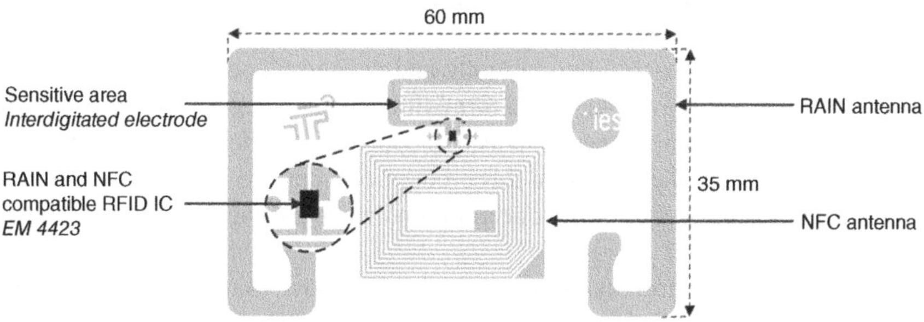

Figure 3.1 Sensing RFID tag layout.

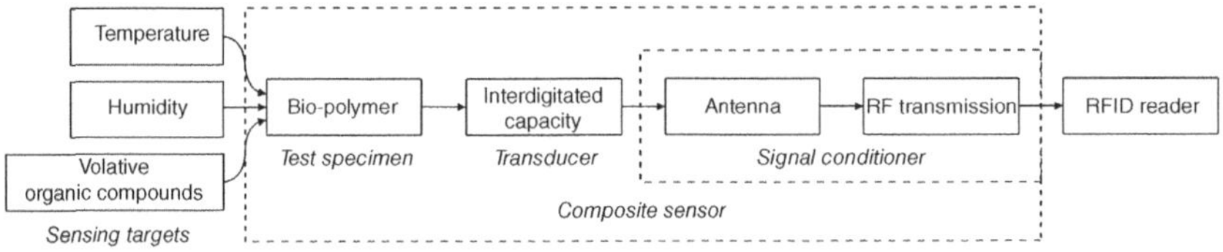

Figure 3.2 Sensing concept.

Microelectronics, as shown in Figure 3.1. Its memory can be made accessible from both RAIN and NFC protocols.

Finally, the third and main objective is to add food spoilage monitoring when the tag is in a sealed package. To promote the use of such a tag, food spoilage monitoring must induce a minimum of additional cost compared to a conventional RFID tag. We follow the measurement chain shown in Figure 3.2. The sensitive layer is based on a biopolymer film which is bio-based, biodegradable, and compatible with food contact. Indeed, this biopolymer modulates its dielectric properties with environmental conditions, and therefore the sensor must be of the capacitive type. To be compatible with classical manufacturing process, the RFID sensor will be realized according to a planar geometry. As shown in Figure 3.1, an interdigitated capacitive sensor coated with the biopolymer layer is directly coupled with the antenna matching circuit. Impedance mismatch between the RFID chip and the antenna will be correlated to the physical parameter to be detected. This design turns the whole tag into an "analog" RFID-enabled sensor.

Partner of the GLOPACK project, Tageos [25], a French RFID tags manufacturer, was involved onto the conception and manufacturing to this food spoilage sensing RFID tag. They help us to choose the correct materials and methods to industrially manufacture the tag. Finally, they provide thousands of individual tag prototypes.

3.4.1.2 Biopolymer Test Specimen

Some biopolymers (gluten, casein, gelatin, soy protein, ...) are sensitive to environmental conditions through their dielectric permittivity [26]. These organic molecules are proteins that are long chains of amino acid residues. These residues can form links with each other naturally and depending on external factors. In this process, proteins are folded, and form different shapes, even with same original sequence. The complex permittivity of biopolymers is affected by the composition of proteins and their links. In this proof of concept, we choose wheat gluten because of its sensitivity to CO_2 concentration, inside packaging, and in the presence of food [11].

The dielectric permittivity of biopolymers is sensitive to temperature. Indeed, by increasing the latter, it leads to increased mobility of molecules. In addition, biopolymers approach their glass transition temperature and impact the ionic behavior of permittivity. The sensitivity to temperature is also explained by the presence of residual water due to the preparation of the biopolymer gel. Indeed, the permittivity of water is highly dependent on temperature [27]. The presence of water or plasticizers leads to a lowering of the glass transition temperature and leads to an increase in permittivity.

The sensitivity of the permittivity of biopolymers to moisture is mainly due to the variation in the glass transition temperature, and the variation of carriers' mobility. Here again, the moisture sensitivity of biopolymers is due to the presence of water. In small quantities, water could interact more strongly with available binding sites of the material [36], and thus be less available for external dielectric stresses. From 90% RH, the water brought by the humidity will no longer bind with the material and will be able to restore its usual behavior according to the time-variable electric field. Concerning wheat gluten biopolymer, moisture sensitivity does not have a linear behavior as shown in Ref. [28].

We chose CO_2 concentration as the main target to be measured with the proposed RFID sensor. Because of their aerobic respiration, some microorganisms and food spoilage can be detected by CO_2 sensor. In a hermetically sealed package such as the one targeted by the application, CO_2 will accumulate, and the concentration will rise.

The swelling of the biopolymer in the presence of moisture also makes it possible to observe other environmental variables including the presence of CO_2. When CO_2 is dissolved in water, it is in balance between its dissolved form and the bicarbonate ion (Eq. (3.1)). The release of the proton (denoted H^+) leads to acidification of the medium. By using a material with a higher isoelectric point (pH_i), an increase in the pH of the biopolymer gel is expected. According to Le Chatelier's principle, the dissolution of CO_2 and then its transformation into carbonate will be favored by the basic character of the gel that will capture the protons formed. The increase in carbonate concentration within the gel and the uptake of the protons formed should induce greater differences in the observed dielectric and ionic mechanisms and allow an increase in sensitivity to CO_2.

$$CO_2 + H_2O \Leftrightarrow HCO_3^- + H^+ \tag{3.1}$$

In the case of a significant breach in the packaging, the humidity will fall back to the ambient value. Here, assessing the sensitivity of the sensor tag to moisture can be crucial to detect a breach.

To coat the sensitive layer on the sensor tags, it is necessary to prepare a liquid solution of wheat gluten according to the following protocol [29]. A sodium sulfite solution (0.06 g/50 mL water) is mixed with a gluten powder solution (30 g/50 mL sodium sulfite) and waited for 30 minutes. After adjusting the pH to 4 with the addition of acetic acid (3.4 mL at 50/50 v/v), distilled water is added to obtain a final volume of 130 mL. The solution must be sufficiently well mixed to be homogeneous. For the sensitive layer coating process, the industrial partner will use a technique already well known, that is, flexography. Over eight measurements made on the limit of the gluten layer and over a surface of $0.561\,mm^2$, the pitch height (coated layer) is, on average, $14.6\,\mu m$.

3.4.1.3 Interdigitated Capacitor

To capture the variations in permittivity of biopolymers, we then chose to integrate an interdigitated capacity, illustrated in Figure 3.3. With its planar design, it is perfectly suited to the manufacture of RAIN tag. In addition, by multiplying the electrodes next to each other, the sensitivity of this component can be modulated without adding too much space.

For a planar design like ours, a guard electrode, or ground plane (connected to ground), would reflect electric fields created by interdigitated electrodes. It strengthens electric fields toward the sensitive layer. However, for the current design depicted in Figure 3.4, we cannot add a ground plane because the biopolymer is deposited on the other side of the substrate than the aluminum circuit layer. In the future, we plan to directly coat the biopolymer layer inside the lid film of the package. In this case, the sensing RFID tag could be glued with circuits facing the lid, releasing the other face of the tag to add ground plane.

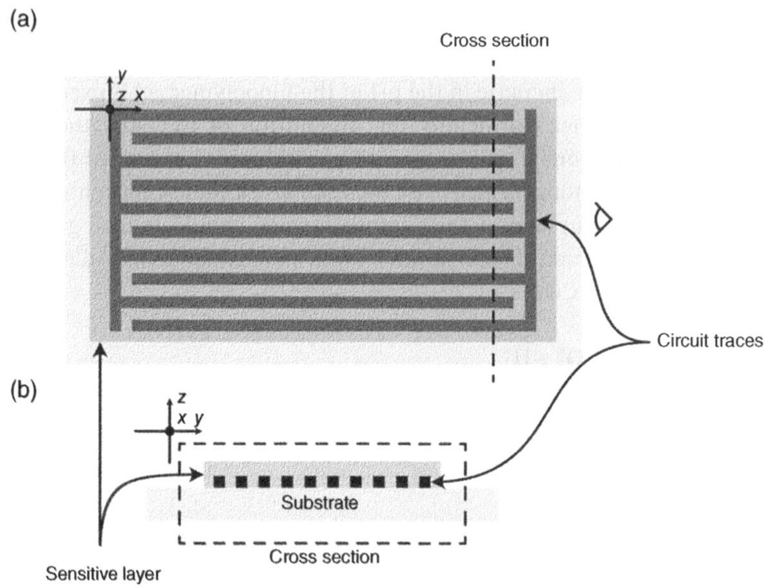

Figure 3.3 Top view (a) and cross section (b) of an interdigitated electrode.

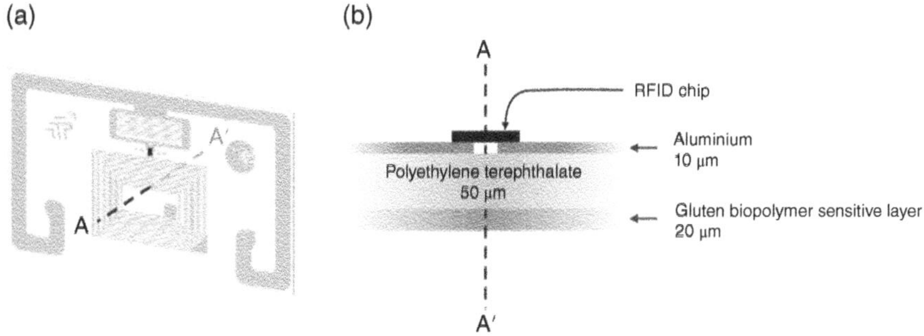

Figure 3.4 The sensing RFID tag (a) and its cross-section view (b).

For the first design step, we simulate the value of the capacity according to its dimensions and surrounding materials using a mathematical model [30]. In Figure 3.5, we swept several parameters (length of fingers, number of fingers, width between them, and thickness of the sensitive layer). Our goal is to characterize a sensitive layer according to the parameters of the structure. Thus, at each step of the parameter scan, we will calculate 2 capacity values corresponding to the minimum and maximum permittivity measured on gluten.

As observed in Figure 3.5, sensitivity to permittivity evolves linearly with the length of the fingers and with their number. Nevertheless, multiplying the number of fingers seems to have a greater impact on sensitivity. The widening of the space between the fingers reduces sensitivity. And, to grow the thickness of the sensitive layer increases sensitivity, but with a slowdown. We can explain this by a better confinement of the field lines created between the electrodes.

As mentioned earlier, the confinement of electric field lines depends on the spacing between the electrodes and the thickness of the sensitive layer. To maximize sensitivity to the coated material, we propose to reduce the inter-electrode spacing as much as possible. Most of the electric field would be contained in the coated material. In addition, this new design brings greater independence to the external magnitudes of influences. Indeed, if the field lines are not confined to the intended materials, they can interact with neighbor materials that constitute external noises.

3.4.1.4 Sensing RFID Transponder

The design, as depicted in Figure 3.4, has been realized while considering high-scale manufacturing constraints. A conventional RFID tag is made of an aluminum antenna on thin and flexible substrates, as illustrated by the cross-sectional view of the proposed RFID sensor in Figure 3.4. It is to be noted that the thinnest the substrate material is, the fewer resources are used. This leads to RFID tag cost reduction.

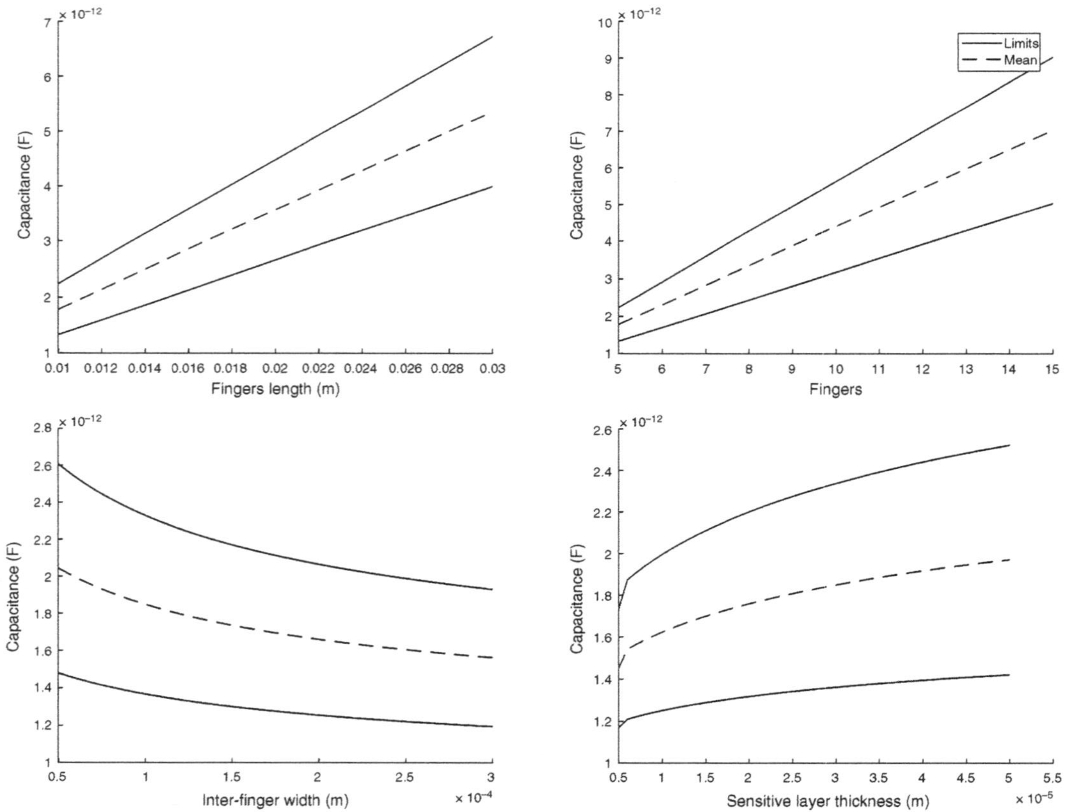

Figure 3.5 Interdigitated capacity model parameter sweeps.

In order not to short-circuit the IC or break it during the coating process, the sensitive layer is coated on the side opposite to the circuit, as shown in Figure 3.4.

The sensor is integrated in the T-match structure typically used for impedance matching, in order to obtain a compact design. Indeed, the antenna (Z_D) must be complex conjugated and matched to the parallel circuit made of the IC RAIN interface (Z_{RAIN}) and the sensor (Z_C) which are both capacitive (Figure 3.6). An optimization of the antenna dimensions was performed to obtain a low reflection coefficient in the targeted frequency band (low ETSI, 866–868 MHz).

To meet GLOPACK project requirements, we must provide dynamic information on food products to customers. As most of them own a NFC-compatible smartphone, we chose an IC with an NFC interface, the EM Microelectronic em|echo (EM4423). To communicate with users' devices, a loop antenna is connected to this interface and its impedance (Z_L) is the complex conjugate of NFC interface impedance (Z_{NFC}).

We noted that the antenna de-tuning due to the presence of meat and plastic separating the sensor-equipped tag from the ambient air is important. Indeed, the model of the RAIN interface alone is adapted to a frequency of more than 1.1 GHz. In the final model with meat and plastic packaging, shown in Figure 3.7, the resonance frequency is around 850 MHz.

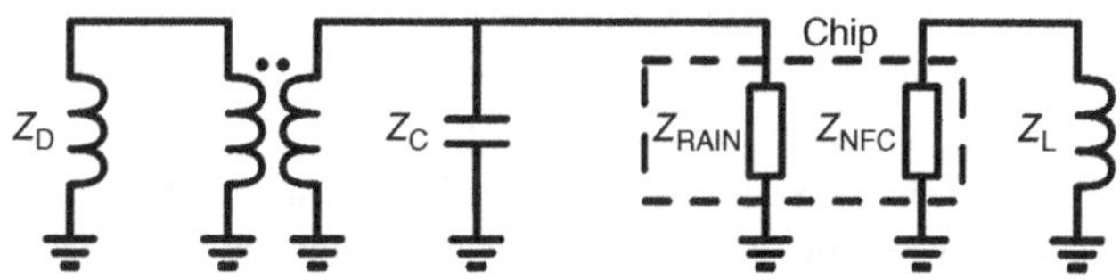

Figure 3.6 Sensing RFID tag electrical equivalent diagram.

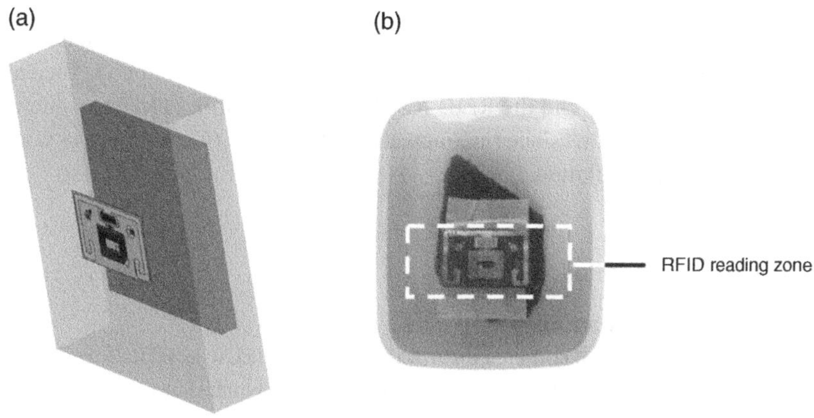

Figure 3.7 Simulation model (a) and photograph (b) of the tag sensor in its environment.

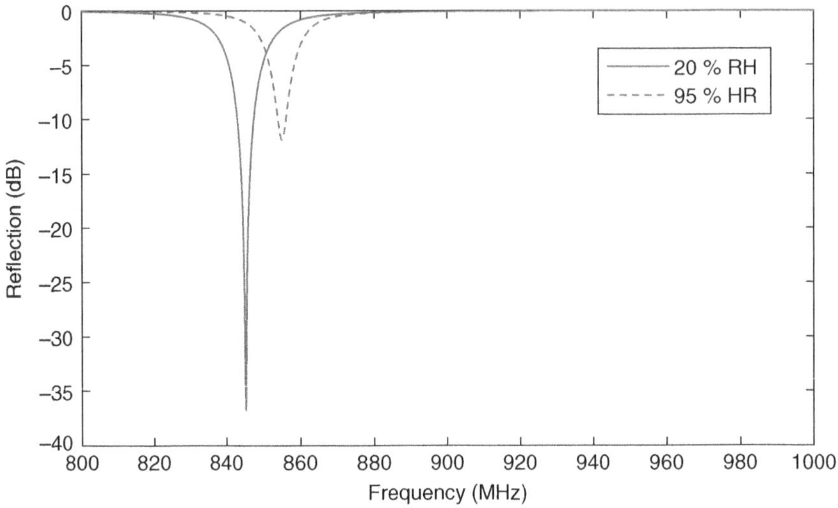

Figure 3.8 Reflection coefficient of the sensor tag.

In all the experiments, carried out by us in the laboratory or by our industrial partners, we used 100-gram beef (muscle) samples cut into 1 cm slices. In Figure 3.7, the meat is represented in red and is simulated by a material having a permittivity of $\varepsilon_r = 56 + j21$ for a frequency of 916 MHz [31].

For laboratory experiments, the plastic packaging used was made of PP (polypropylene) with a thickness of 550 μm. The capacity of this packaging is 781 mL (156 × 143 × 35 mm). A 70 μm thick transparent PP film is heat-sealed on the tray to close the package. In the model, the PP is simulated having a permittivity of $\varepsilon_r = 2.26 + j7 \cdot 10^{-5}$ [32].

As shown in Figure 3.8, variations of the relative permittivity of the gluten layer (here, with moisture) induce a significant variation in the reflection coefficient between the tag antenna and the RFID IC. This result combines electromagnetic and electronic simulations since the impedance of the chip ($Z = 18.1 + \text{j}248\ \Omega$) is considered for the calculation.

3.4.2 Measurement Reading

3.4.2.1 Common Software Framework

End users need to get food spoilage information easily. And, to get data from RFID tags, you need a reading software built on top of a framework that manages RFID communication. The framework's goal is to abstract many tedious operations and leave the programmer free to design more advanced features.

The framework used integrates the management of tags and readers as virtual "objects" in C# programming language context. It is designed to handle all the operations necessary to make an RFID reader compatible with RAIN standards. As virtual objects can be extended with methods or properties, it is easy to implement manufacturer-specific features for tags or readers.

To enable cheap analog RFID sensors, extensions of the library are developed to help in measuring. Our choice to delay the digitization of measurements leads to greater complexity in making the signal reliable. Due to the EM environment, measurements will be affected by the distance between the reader and a tag, medium characteristics, and fading... . Fortunately, some of them can be reduced using techniques featured in following paragraph and implemented in our library.

3.4.2.2 Turn-on Power Algorithm

Analog sensing tags that use ordinary RFID modem mainly rely on tag gain, $G_{T_{rx}}$ in Eq. (3.2).

$$P_{R_{rx}}\left[W\right] = P_{R_{tx}} \cdot G_R^2 \cdot \left(\frac{\lambda}{4\pi \cdot d}\right)^4 \cdot PLF^2 \cdot G_T^2 \cdot K_{bs} \tag{3.2}$$

$$= P_{T_{rx}} \cdot K_{bs} \cdot G_T \cdot \left(\frac{\lambda}{4\pi \cdot d}\right)^2 \cdot PLF \cdot G_R$$

$P_{R_{rx}}$ = received power on reader $\left[W\right]$, $P_{R_{tx}}$ = reader sent power $\left[W\right]$,
G_R = reader's antenna gain, G_T = tag's antenna gain,
λ = wavelength [m], d = distance [m], PLF = polarization loss factor,
K_{bs} = tag's chip backscattering coefficient, $P_{R_{tx}}$ = received power on tag $\left[W\right]$

With fixed setup conditions (reader antenna gain, wavelength used, distance and relative orientation between reader and tag), power exchanged by the reader is linear with tag gain variation. To restore targeted signal out of power exchanged, measurement needs to be calibrated with tag gain.

To estimate tag gain and, hence, targeted signal, most authors measure $P_{R_{rx}}$, known as Return Signal Strength Indicator (RSSI). To this end, power sent by readers must stay constant, and hence, be stored with calibration conditions. In contrast, some authors use $P_{R_{tx}}$ at tag power threshold to activate the chip $\left(P_{T_{rx}}\left(\text{threshold}\right)\right)$, known as "turn-on power." Using turn-on power variable helps to save memory gas power sent does not need to be saved. Furthermore, the chip threshold is known as its "read sensitivity" and the gain of the reader and the tag are known as well. Readers and tags could use circularly polarized that cancels polarization loss factor. Distance could be estimated using phase values over frequency sweeps (not detailed in this chapter).

$$P_{R_{rx}}(\text{threshold}) = \frac{P_{T_{tx}}(\text{threshold})}{G_R \cdot \left(\frac{\lambda}{4\pi \cdot d}\right)^2 \cdot \text{PLF} \cdot G_T} \tag{3.3}$$

Extensions of the RFID framework implement the algorithm to get a tag's turn-on power. As illustrated in Figure 3.9, this algorithm searches for the minimum power that the reader must send to get a response from the tag.

To make the resulting sensing solution affordable, chosen readers must be COTS. Typically, their resolutions, for both RSSI and set power, are 1 dBm. Sensitive tag resolution is limited by reader accordingly as they are part of the measurement chain in case of designing an analog sensing RFID tag.

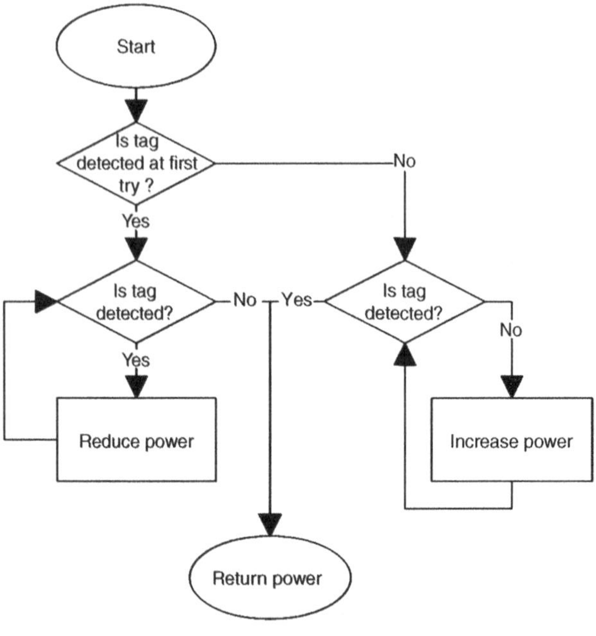

Figure 3.9 Turn-on power algorithm.

3.4.2.3 Frequency Sweeps

If some variable exerts an influence on the resonance frequency of the antenna, this implies that a measurement of an amplitude parameter at a fixed frequency will also undergo a variation. As described in Figure 3.10, depending on the position of the resonance, the variation may change its direction. The measured parameter does not have a monotonous variation with the target variable.

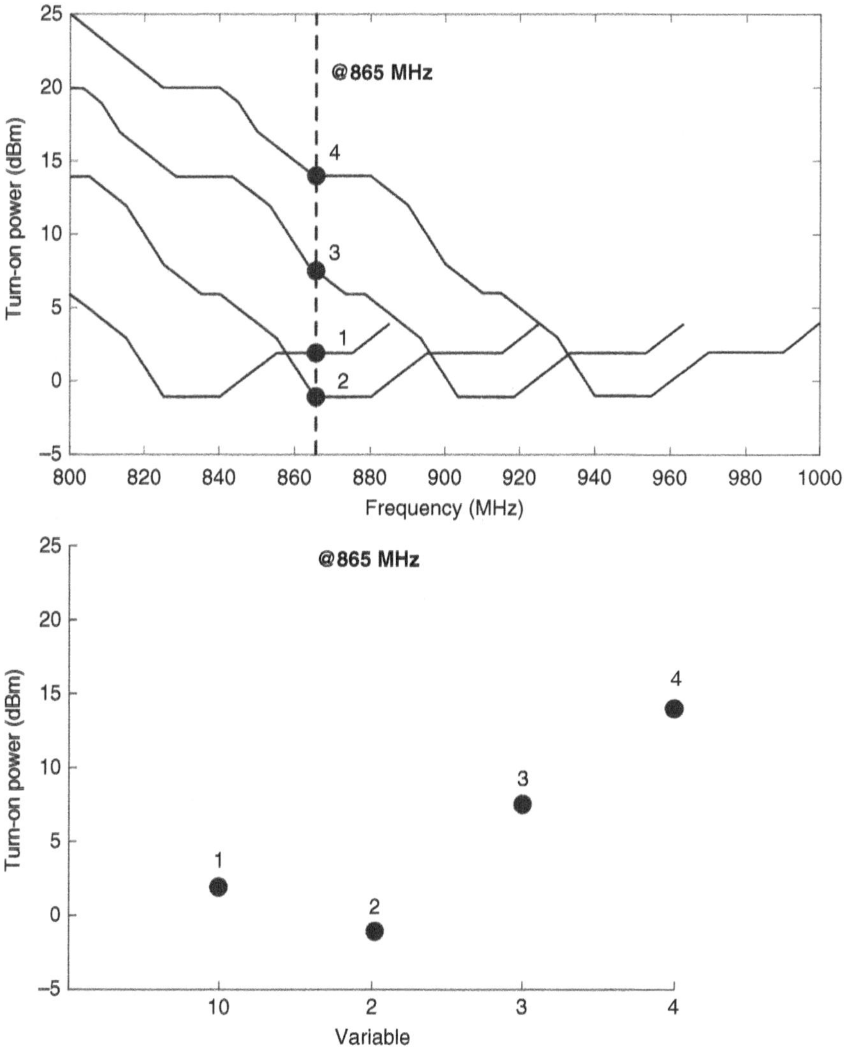

Figure 3.10 Frequency shift due to a variable.

A frequency scan can help to correct this problem. Indeed, a derivative of the amplitude parameter with respect to the frequency can help to locate the variable. The width of the frequency band on which the scan is done is important. The smaller the band, the more difficult it will be to estimate the variation. This problem is more present on tags targeting several frequency bands (with a lower quality factor). In this case, amplitude variation over frequency is small. Currently, few European countries have implemented the high ETSI frequency band [33] but many manufacturers provide tags targeting the two RFID UHF bands. Therefore, a frequency scanning algorithm has been implemented in extensions of the library.

3.4.2.4 Transponder-Embedded Calibration

Besides EPC memory zone, some RAIN tags offer a dynamic memory structure called "User memory." This structure could be used for measurement-related data such as calibration points. In this configuration, tag calibration is embedded and prevents readers to fetch data from remote databases. When calibration conditions are fixed, only coefficients must stay associated. It is to be noted that calibration coefficients are associated to a unique tag or a batch of tags. In the last configuration, tag user memory must be large enough to store data. Nowadays, RAIN tag manufacturers propose user memories that fill what is needed to store calibration coefficients, as stated by Table 3.2. Even with the smallest memory size presented in this table, that is 128-bit, 8 half-precision floating-point numbers can be stored.

Extensions of the library offer to calibrate a specific setup over a variable range and to store the resulting calibration point in tags memory. Readers run a calibration procedure that gathers tag turn-on power depending on sensitive states. Its initial transmission is saved and compared to new states to obtain measurements. Finally, we develop an Android application, based on our algorithms. As shown in Figure 3.11, the application is used to estimate food spoilage states in an industrial context. It compares new turn-on power value to a previous value stored in user memory. For now, it simply indicates variations that state an event

Table 3.2 RAIN tag manufacturers' user memory for best-in-class tags.

Model	User memory (bits)
EM Microelectronic em|echo (EM4423)	128
Impinj Monza 4QT	512
Alien Technology Higgs-9	688
EM Microelectronic aura-sense	2048
NXP UCODE 7xm+	2048
AMS SL900A	8416

happened such as rising contamination or package opening.

3.5 Validation

3.5.1 In Simulated Conditions

Figure 3.11 GLOPACK application screenshot. *Source:* GLOPACK.

Sensing tags were fixed on the cover plate inside a cylindrical (∅ 10×10 cm) hermetic container made of polyoxymethylene. Facing the tags, outside the container, a Voyantic Snoop Pro antenna was attached to communicate through the lid. The antenna was connected to a Voyantic TagFormance® Pro laboratory reader. Gas concentration monitoring is performed by the wireless and compact PreSens oximetry solution which gives both CO_2 and O_2 concentrations. As the sensing tag is also sensitive to temperature [6], we also controlled the temperature during the experimentation. For this purpose, a climatic chamber was used (Memmert HPP110). To maintain a high relative humidity as for a classic food packaging, we placed a container of water in the air-tight container to saturate the environment with moisture (Figure 3.12).

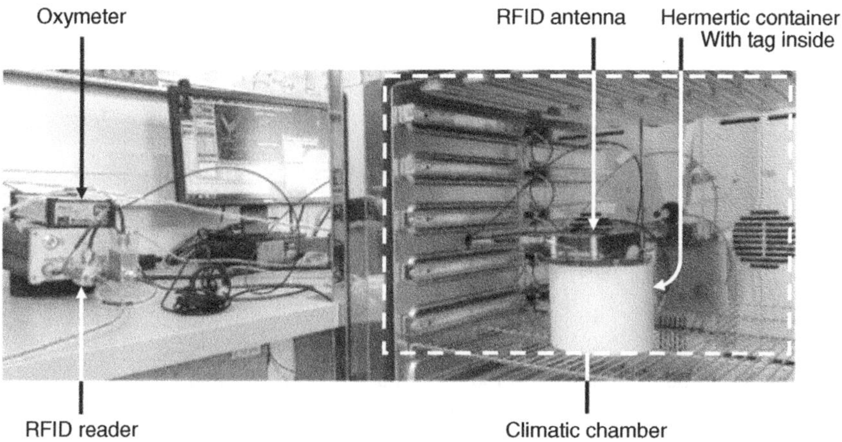

Figure 3.12 Simulated spoilage experiment setup.

First, a "protective" environment consisting of 30–70% CO_2–O_2 mix was applied for approximately 150 hours. Then, this mix was changed to 50–50% CO_2–O_2 for the same period. The first mix is suited for red meat packaging as it slows down the bacterial growth while preserving the red color of the meat. The second mix simulates an important microbial activity that is induced by the spoiled meat, packed in the first mix.

A Voyantic TagFormance® Pro is used to measure the turn-on power over frequencies and through the time. For this experiment, three frequencies (860, 900, and 920 MHz) correlated with limits of the main RFID bands are displayed in Figure 3.13.

After the initial flush of the protective environment, turn-on power goes up. This evolution depends on the frequency: −3 dBm at 860 MHz, +3.5 dBm at 900 MHz, and +6.5 dBm at 920 MHz. Without real foodstuffs in the headspace of the package, high frequencies show most variations. When the "spoiled" environment is then flushed, the preceding conclusion is again right with +1.5 dBm at 860 MHz, +3.0 dBm at 900 MHz, and +3.0 dBm at 920 MHz. The signal is still distinguishable for the 860 MHz frequency as conventional UHF RFID readers can detect these steps because their resolution is typically of 1 dBm. Nevertheless, we plan to measure signals at multiple frequencies to enhance differentiation.

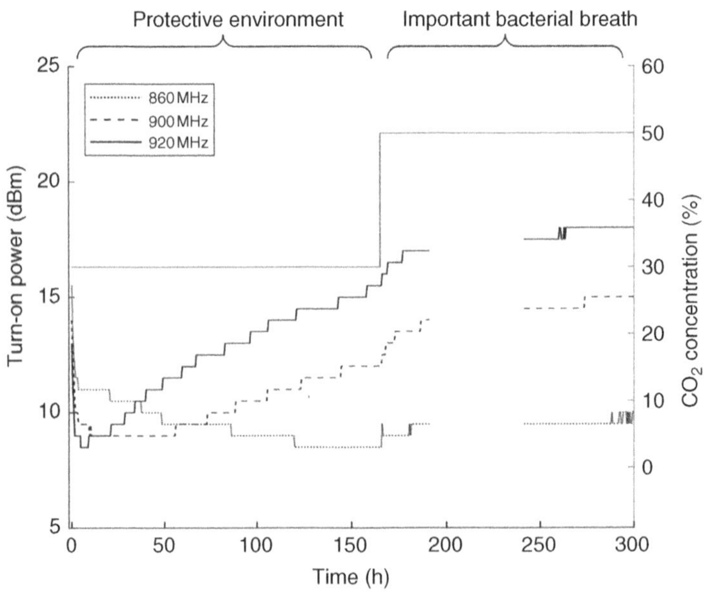

Figure 3.13 Simulated spoilage experiment RFID signals

3.5.2 In Real Conditions

As depicted by Figure 3.7a, the sensing RFID tags are now glued inside the lid of food packages containing 100 g of beef steak. Packages are made of polypropylene with a thickness of 550 μm (70 μm for the lid) and are filled with a protective environment (70% O_2 and 30% CO_2).

This experiment is conducted at 5 °C on 17 identical samples in the same climatic chamber as previously mentioned experiment. Three were set aside at the end of their preparation to measure contamination generated by this step. And, each day, two more samples were removed from the set to measure the evolution of the contamination. As microbiological analysis is destructive, a decreasing number of samples are present in the set as time passes. Microbial contamination is measured by counting viable microorganisms per gram of the food product using the "Miles and Misra" method. It consists of mixing a certain amount of analyte in an inert solute for it to be spread onto a "growing media" [34]. Then, viable microorganisms will reproduce and start to become visible. Finally, we counted colony-forming units (CFU) and applied a factor given by the dilution.

In Figure 3.14, the three samples that were analyzed on the first day do not reveal any major dispersion. In this figure, the expiration date of the meat samples is represented by a vertical line.

In France, regulation authorities defined a value of 10^7 CFU per gram as the contamination limit [35]. This limit is reported in Figure 3.14 as the horizontal

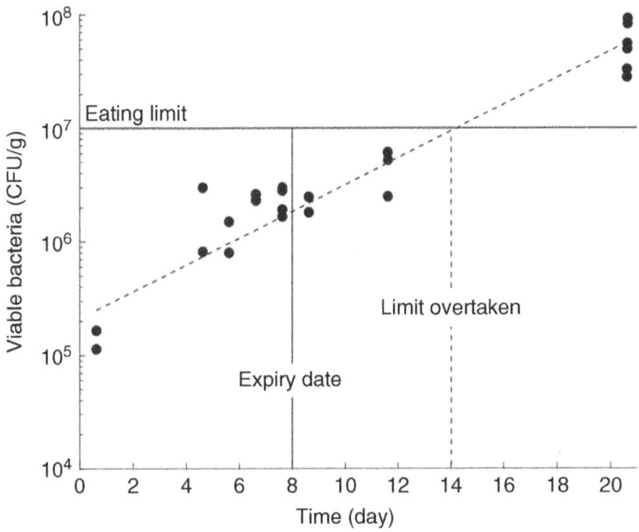

Figure 3.14 Contamination of meat samples.

red line, labeled as "eating limit." As accidents in foodstuffs storage (bad protective gas environment, too high temperature, ...) accelerate bacterial growth, a margin is added to expiry date to prevent consumers from eating spoiled food. As depicted in Figure 3.14, the exponential trend line (increasing dotted line) crosses the eating limit 6 days after the expiration date. Then, our sample preparation step did not induce a bias in the experiment.

As before, sensing RFID tags are read by a Voyantic TagFormance® Pro RFID reader connected to a Voyantic Snoop Pro antenna. The distance between read tag and the reading antenna stays the same throughout the experiment and across the measured sample thanks to few guides. In Figure 3.15, turn-on power signals from individual samples are averaged because sensing RFID tags are sensitive to their immediate environment and meat samples can differ. In this figure, the expiry date, and the date when consumption limit is reached are replicated.

Sensing RFID tags turn-on power fluctuates by more than one decibel between the first 2 days. This relates to a stabilization of the gas environment in the food container. During this period, the meat absorbs O_2 and desorbs nitrogen. Successively, the turn-on power increases with time of exposure to meat samples, which shows a good correlation with the growth and respiration of bacterial colonies. By applying a linear regression, we obtain a growth rate of 0.24 dBm per day. In this computation, the first-day measurement is not considered because it corresponds to the stabilization stage. In this specific case, by precisely calibrating turn-on power growth limits, we could have extended these samples' shelf life to 6 days more.

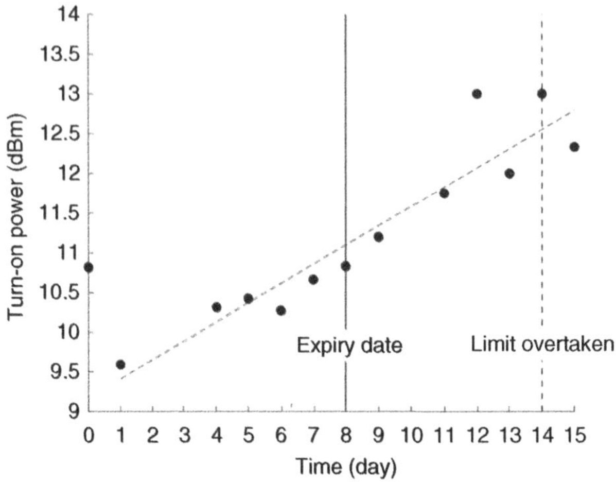

Figure 3.15 Sensing RFID tags turn-on power signal along meat spoilage.

3.6 Conclusion

In this chapter, we choose to develop "analog" RFID tags to sense food freshness. This is motivated by the search for a budget solution in food packaging sector that sells large volumes and low-priced objects. For the sensor to stay low-cost and biocompatible, a bio-based polymer provides gas composition sensing. The RFID classic planar conception simplifies the polymer coating operation. To maximize sensing interaction with antenna gain, we then introduce an interdigitated capacitor in the antenna design. As measured values are distorted by other factors too before they get to readers, this design involves a heavier signal processing that is solved in this chapter. As this solution intends logistic actors, we provide them with a user-friendly application and process.

Currently, RFID tags need to be inside the packages for the sake of sensing, so the food industry players will have to attach them inside the package. To correctly label the product, they define the EPC and encode tags to facilitate inventory throughout the supply chain. A delay of one day is required for the sensor tag to be fully operational and to achieve stability. Then, professionals in charge of the packages at this time must initiate the calibration. Finally, the following actors in the supply chain will have access to the estimation of the food maturity.

As demonstrated, a margin exists between the expiry date written on the product and the date when the product is spoiled. For now, our sensing RFID tags can inform the user of a poor storage condition at a low price and can bring more valuable information than optical barcode. In future developments, these RFID tags can help to reduce the gap between the real maturation state of the food and its legal expiry date.

References

1 Food and Agriculture Organization of the United Nations (2019). State of Food and Agriculture 2019. Moving forward on food loss and waste reduction. Food and Agriculture Organization. http://www.fao.org/policy-support/tools-and-publications/resources-details/fr/c/1242090/ (accessed 21 January 2021).

2 World Health Organization and Foodborne Disease Burden Epidemiology Reference Group (2015). *WHO* estimates of the global burden of foodborne diseases. http://apps.who.int/iris/bitstream/10665/199350/1/9789241565165%5F eng.pdf (accessed 20 January 2021).

3 Yiannas, F. (2009). Looking back to shape the future. In: *Food Safety Culture: Creating a Behavior-Based Food Safety Management System*, Food Microbiology and Food Safety (ed. F. Yiannas), 1–10. New York, NY: Springer. https://doi.org/10.1007/978-0-387-72867-4_1.

4 Bernard, S. and Woodland, N.J. (1952). Classifying apparatus and method. US2612994A, October 07, 1952.

5 Shein, E. (2021). A battery-free internet of things. *Communications of the ACM* 64 (7): 16–18. https://doi.org/10.1145/3464937.

6 Kellogg, B., Talla, V., Gollakota, S., and Smith, J.R. (2016). Passive wi-fi: bringing low power to wi-fi transmissions. *Presented at the 13th USENIX Symposium on Networked Systems Design and Implementation (NSDI 16)*, pp. 151–164. Santa Clara, California (16–18 March 2016). USENIX Association. https://www.usenix.org/conference/nsdi16/technical-sessions/presentation/kellogg.

7 Sarma, S. (2012). How inexpensive RFID is revolutionizing the supply chain: (Innovations Case Narrative: The Electronic Product Code). *Innovations Technology Governance Globalization* 7 (3): 35–52. https://doi.org/10.1162/INOV_a_00138.

8 OEM embedded ThingMagic UHF/RAIN RFID Module/M6e|JADAK, Jadak – A Novanta Company. https://www.jadaktech.com/products/thingmagic-rfid/thingmagic-m6e-uhf-rain-rfid/.

9 Philipose, M., Smith, J.R., Jiang, B. et al. (2005). Battery-free wireless identification and sensing. *IEEE Pervasive Computing* 4 (1): 37–45. https://doi.org/10.1109/MPRV.2005.7.

10 O. Erkmen, and T. F. Bozoglu, Eds., 'History and development of food microbiology', in *Food Microbiology: Principles into Practice*, vol. 1, Wiley, 2016, pp. 3–12. https://doi.org/10.1002/9781119237860.ch1.

11 Guillaume, C., Sorli, B., Guillard, V., and Gontard, N. (2017). Emballage intelligent et RFID. *Innovations Agronomiques* 58: 21–30. https://doi.org/10.15454/1.513783292735162E12.

12 Erkmen, O. and Bozoglu, T.F. (2016). Microbial metabolism of food components. In: *Food Microbiology: Principles into Practice*, vol. 3, 217–227. Wiley https://doi.org/10.1002/9781119237860.ch38.

13 Abdelnour, A., Fonseca, N., Rennane, A. et al. (2019). Design of RFID sensor tag for cheese quality monitoring. In: *2019 IEEE MTT-S International Microwave Symposium (IMS)*, Boston, MA, USA (2–7 June 2019), 290–292. IEEE https://doi.org/10.1109/MWSYM.2019.8700769.

14 Bhattacharyya, R., Di Leo, C.V., Floerkemeier, C., et al. (2010) RFID tag antenna based temperature sensing using shape memory polymer actuation. *Presented at the SENSORS, 2010 IEEE*, Kona, HI, USA (1–4 November 2010). IEEE. https://doi.org/10.1109/ICSENS.2010.5690951.

15 Caizzone, S., Occhiuzzi, C., and Marrocco, G. (2011). Multi-chip RFID antenna integrating shape-memory alloys for temperature sensing. *Proceedings of the 5th European Conference on Antennas and Propagation (EUCAP)*, Rome, Italy (11–15 April 2011), 2810–2813. IEEE. http://ieeexplore.ieee.org/document/5782163/.

16 Ziai, M.A. and Batchelor, J.C. (2014). Multi-state logging freeze detection passive RFID tags. *IEEE Transactions on Antennas and Propagation* 62 (12): 6406–6411. https://doi.org/10.1109/TAP.2014.2359474.

17 Law, M.-K., Bermak, A., and Cam Luong, H. (2010). A sub-μW embedded CMOS temperature sensor for RFID food monitoring application. *IEEE Journal of Solid-State Circuits* 45 (6): 1246–1255. https://doi.org/10.1109/JSSC.2010.2047456.

18 Gonçalves, R., Rima, S., Magueta, R. et al. (2015). RFID-based wireless passive sensors utilizing cork materials. *IEEE Sensors Journal* 15 (12): 7242–7251. https://doi.org/10.1109/JSEN.2015.2472980.

19 Nguyen, S.D., Le Nguyen, N., Lam, P.T. et al. (2015). A wireless sensor for food quality detection by UHF RFID passive tags. *2015 International Conference on Advanced Technologies for Communications (ATC),* Ho Chi Minh City, Vietnam (14–16 October 2015), 258–263. IEEE. https://doi.org/10.1109/ATC.2015.7388330.

20 FlexSmell (2014). FlexSMELL – FP7 Marie Curie ITN Project. FlexSmell. http://www.flexsmell.eu/ (accessed 17 May 2014).

21 Kapucu, K., Merino Panadés, J.L., and Dehollain, C. (2013). Design of a passive UHF RFID tag for capacitive sensor applications. *Proceedings of the 2013 9th Conference on Ph.D. Research in Microelectronics and Electronics (PRIME)* (24–27 June 2013), 213–216. IEEE. https://doi.org/10.1109/PRIME.2013.6603157.

22 Erkmen, O. and Bozoglu, T.F. (ed.) (2016). Principles of food spoilage. In: *Food Microbiology: Principles into Practice*, 269–278. Wiley https://doi.org/10.1002/9781119237860.ch15.

23 National Institute for Occupational Safety and Health (2009). Approaches to safe nanotechnology managing the health and safety concerns associated with engineered nanomaterials. https://doi.org/10.26616/NIOSHPUB2009125.

24 GS1 (2021). GS1 general specifications. https://www.gs1.org/standards/barcodes-epcrfid-id-keys/gs1-general-specifications (accessed 19 September 2021).

25 Tageos (2021). Tageos. Tageos. https://www.tageos.com/ (accessed 21 January 2021).

26 Bibi, F., Villain, M., Guillaume, C. et al. (2016). A review: origins of the dielectric properties of proteins and potential development as bio-sensors. *Sensors* 16: 1232. https://doi.org/10.3390/s16081232.

27 Kaatze, U. (1989). Complex permittivity of water as a function of frequency and temperature. *Journal of Chemical & Engineering Data* 34 (4): 371–374. https://doi.org/10.1021/je00058a001.

28 Bibi, F., Guillaume, C., Vena, A. et al. (2016). Wheat gluten, a bio-polymer layer to monitor relative humidity in food packaging: electric and dielectric characterization. *Sensors and Actuators A: Physical* 247: 355–367. https://doi.org/10.1016/j.sna.2016.06.017.

29 Bibi, F. (2015). Study of dielectric properties at high frequency of plant polymer: the case study of wheat gluten proteins to be used as sensing materials of environmental markers of food quality in passive RFID systems. Thesis. University of Montpellier, Montpellier, France. http://www.theses.fr/2015MONTS069.

30 Igreja, R. and Dias, C.J. (2011). Extension to the analytical model of the interdigital electrodes capacitance for a multi-layered structure. *Sensors and Actuators A: Physical* 172 (2): 392–399. https://doi.org/10.1016/j.sna.2011.09.033.

31 Farag, K.W., Lyng, J.G., Morgan, D.J., and Cronin, D.A. (2008). Dielectric and thermophysical properties of different beef meat blends over a temperature range of −18 to +10°C. *Meat Science* 79 (4): 740–747. https://doi.org/10.1016/j.meatsci.2007.11.005.

32 Riddle, B., Baker-Jarvis, J., and Krupka, J. (2003). Complex permittivity measurements of common plastics over variable temperatures. *IEEE Transactions on Microwave Theory and Techniques* 51 (3): 727–733. https://doi.org/10.1109/TMTT.2003.808730.

33 European Comission (2018). Commission Implementing Decision (EU) 2018/1538 of 11 October 2018 on the harmonisation of radio spectrum for use by short-range devices within the 874–876 and 915–921 MHz we display several works frequency bands (notified under document C(2018) 6535) (Text with EEA relevance.) http://data.
europa.eu/eli/dec_impl/2018/1538/oj/eng (accessed 6 May 2021)

34 Institut national de recherche et de sécurité (2015). Microorganismes aérobies M-147. https://www.inrs.fr/publications/bdd/metropol/fiche.html?refINRS=METROPOL_147(accessed 25 February 2021)

35 Fédération du Commerce et de la Distribution (2020). Critères microbiologiques applicables à partir de 2021 aux activités de fabrication, préparation, découpe ou simple manipulation de denrées nues en rayon « à la coupe » et en atelier magasin. Fédération du Commerce et de la Distribution. http://www.fcd.fr/media/filer_public/2d/53/2d53c9bb-5ff8-4ec3-bcf5-e202d15dcd08/fcd_criteres_microbiologiques_2021_ateliers_rayons_coupe_091020208_vdef.pdf (accessed 18 August 2021).

4

Challenges of Using RFID for Outdoor Environmental Monitoring

Mathieu Le Breton[1,2], *Rahul Bhattacharyya*[3], *and Mathieu Cassel*[4]

[1] *Géolithe Company, Crolles, France*
[2] *ISTerre, Université Grenoble Alpes, Grenoble, France*
[3] *Auto-ID Labs, Massachusetts Institute of Technology, Boston, MA, USA*
[4] *ENS, Université de Lyon, Lyon, France*

4.1 Versatile Data Acquisition Approaches

This chapter focuses primarily on the challenges of RFID tag deployment in different outdoor applications. While the work is focused on the tag side, it is also important to understand, and acknowledge, the role that the RFID reader monitoring infrastructure plays in outdoor asset management. We, therefore, begin the paper with a brief synopsis of how RFID readers (and reader antennas) are deployed in the field.

RFID allows versatile configurations with a choice between mobile or fixed readers and mobile or fixed tags. A fixed reader configuration (Figures 4.1 and 4.2) can monitor an area continuously, but it requires a costly hardware infrastructure setup and the area is restricted by the read range <100 m [3–5]. On the contrary, a mobile reader configuration (Figures 4.3 and 4.4) needs almost no infrastructure and can monitor much larger areas but requires periodic surveys that are time consuming and provide only temporally discontinuous data. Therefore, RFID devices offer the ability to pick the most appropriate configuration for each kind of use case, as described below.

When continuous and timely data is critical, a permanent infrastructure with fixed tags and fixed readers is often most appealing. For example, monitoring landslides and rockfalls often requires a high time resolution, so that potential

Wireless Identification and Sensing Systems for Harsh and Severe Environments,
First Edition. Edited by Smail Tedjini and Valentina Palazzi.

Published 2025 by John Wiley & Sons, Inc.

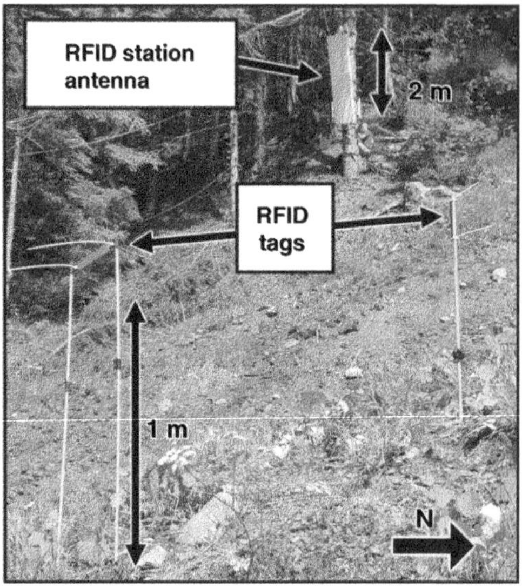

Figure 4.1 Fixed reader and quasi-fixed tag configuration, used here to monitor the slow displacement of a landslide. *Source:* Mathieu Le Breton et al. [1]/ Reproduced with permission from Elsevier.

Figure 4.2 Fixed reader and mobile tag configuration, used here for tracking fishes or pebbles passing in a small river. *Source:* Mao et al. [2]/Reproduced with permission from Elsevier.

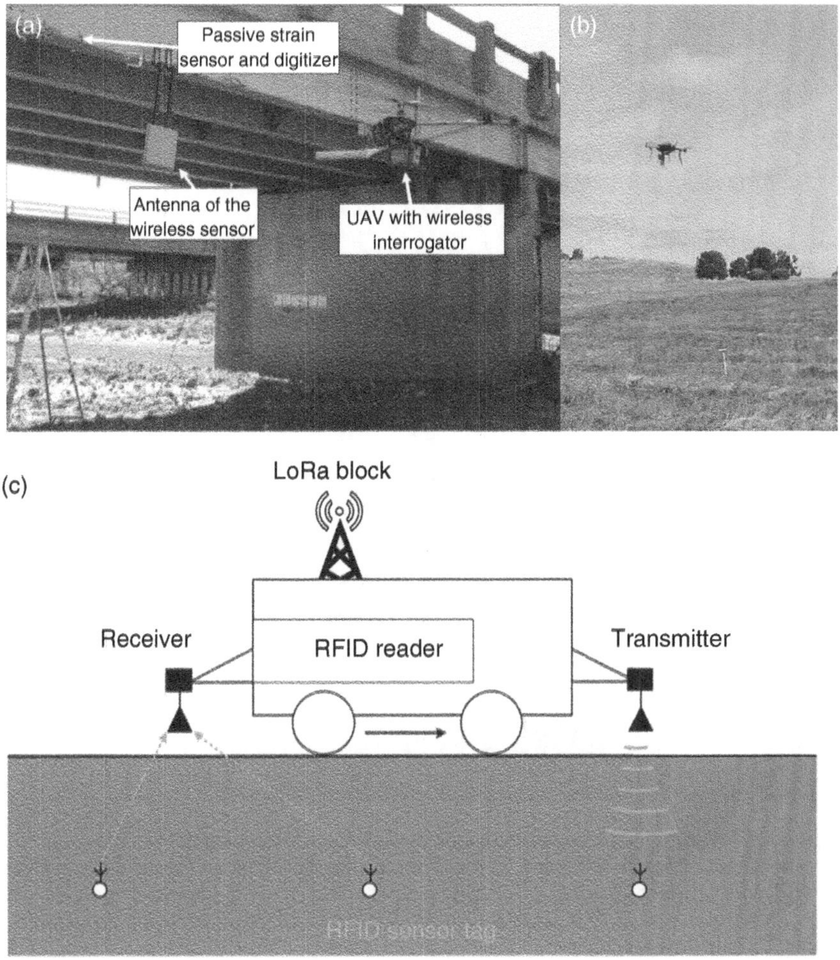

Figure 4.3 Mobile reader and fixed tags configuration, used here for: (a) strain sensing for structural health monitoring using a UAV, (b) sensing the moisture and temperature of the soil using a UAV, or (c) collecting data from a buried tag using a reader on a vehicle [6–8].

ruptures can be more reliably predicted (see Section 4.5). An RFID reader interrogates tags very rapidly—1–30 ms per tag depending on the communication protocol [11, 12]—however, its read range can cover an area of a few 1 000 m^2 at most. The RFID technologies, with tag cost of 0.1–€50 each and autonomous station cost of 1 000–€5 000, then appear competitive compared to wireless sensor networks (WSN) (500–€2 000 per node) in at least three situations. First, when many tags are deployed, their marginal cost (which includes the station cost)

(a)

(b)

Figure 4.4 Mobile reader and mobile tags configuration, used here to map the transportation of coarse sediments in rivers and on coasts. (a) Manual surveys over a river and (b) more efficient survey ($4\,m^2/s$) using a quad and an array of four readers on a coast. *Source:* (b) Dolphin et al. [9]/Coastal Education and Research Foundation, Inc. (a) Liébault et al. [10]/John Wiley & Sons.

decreases. For applications requiring more than 10 sensor nodes per station, an RFID reader with tags becomes an appealing option to consider. Second, when a very fine time resolution is required during years to decades, tags end up more energy efficient than WSNs, which usually require a trade-off between time resolution and monitoring duration to reduce energy consumption. Finally, when the monitored area presents the risk to destroy the sensors, it is much preferable to risk losing low-cost tags than WSN nodes, which are 10–1 000 times more expensive.

When the monitored objects move along a known trajectory, it is possible to install fixed detection portals to monitor access points along a linear route (Figure 4.2), for example, to detect pebbles or fishes passing in a river (see Section 4.4).

This configuration enlarges the area where tags can be deployed initially. It is adapted to detect timely events that occur at a very irregular pace, such as the increase of coarse sediment transportation during higher water debit, or the seasonal movements of fishes. However, the approach again requires a permanent reading infrastructure and can offer only a low spatial resolution which depends on the number of portals.

When data can be discontinuous in time, it can be collected on demand by moving the reader near fixed tags (Figure 4.3). This approach has been applied, for example, for soil monitoring in agriculture and for structural health monitoring of bridges (see Sections 4.6 and 4.7). It eliminates the cost of a permanent fixed reader and can survey much broader areas. Compared to traditional manually operated fixed sensors, the wireless interrogation allows much faster data collection: up to hundreds of RFID tags can be read almost simultaneously, without direct visibility. To improve the efficiency further, the reader can be moved with a land vehicle on transportation infrastructure (road, railway) and on accessible areas. Similarly, unmanned aerial vehicles (UAVs) can collect data efficiently in low-accessibility areas [6, 13]. In short, combining mobile readers with fixed tags is adapted to deploy very low-cost sensors on large areas, when discontinuous data is enough.

When tags move, they can also be relocated using the mobile reader approach. It is used, for example, to study the transport of coarse sediments or fish displacements along rivers and on coasts (see Section 4.4.3). The method has been used largely for research studies, using up to 1000 tags on up to $10^5\,m^2$ areas [9, 14]. However, the short read range of usually deployed tags requires a tight and time-consuming scanning of the entire area of interest. Therefore, this approach has the lowest time sampling, typically every 1–6 months for studies that last several years [15].

The challenges with fixed readers outdoors concern the infrastructure maintenance and the size of the reading area. Using fixed readers requires installing a permanent infrastructure, powering it, and providing maintenance. Outdoor installation might not be connected to the powering grid, and in this case, powering a fixed reader adds a challenge when powered on batteries, solar panels, or wind power (fixed readers on the market consume typically 10–30 W). When accounting the costs of the reader and wide range of telecommunication as well, each reading station ends up significantly more expensive than the cost of tags. That strengthens the second challenge, which is increasing the area read by one reader, allowing to monitor wider zones with fewer readers. The read range can increase with improved sensitivity of readers and tags, with higher-gain tags, or also, by better handling the zones shaded by multipathing to extend the range of reliable reading.

The challenges with mobile readers concern, in our opinion, mostly the survey automatization. Indeed, surveys require a human intervention to carry the reader, pilot the UAVs or drive the vehicle, therefore, preventing from high frequency (i.e. daily) surveys over long periods. The use of a vehicle also poses security issues, such as the risk of a crash during a UAV flight, or the data verification while driving a vehicle. Today, there is no commercial and widespread system that allows for safe and automatic surveying, using a mobile vector such as a UAV, a rover, a car, a quad, or a train. On unmanned systems, reading tag's ability to read tags despite the motion of the reader and of the RF interference of the vector (e.g. a UAV [16]), and the technical or legal limitation of its distance with the operator. As examples of applications, passive RFID tags have been used for maglev rail maintenance. Tags embedded in rail sections can report information such as maintenance history to RFID readers mounted on passing trains. The authors show that such data can be collected with high fidelity at speeds up to 59 km/h [17].

The Sections 4.3–4.6 discuss the challenges for RFID tag deployment in specific applications such as marine engineering, slope stability, agriculture, and infrastructure monitoring. All these applications are subject to climatic influences such as precipitation and obscuring due to vegetation, dust, and soil. Section 4.2, therefore, discusses the challenges posed by these common climatic influences and what work is being done in the literature to address them.

4.2 Weather and Environment Influence

Outdoor environments present many variable influences on RF signals, such as rainfall, dew, frost, snow, and vegetation, which can deteriorate the RF communication and the measurement accuracy. At the same time, each of these influences also represents an opportunity to develop new ways for sensing the tag's environment.

4.2.1 Effect of Rainfall and Dew

Liquid water on a tags couples with its antenna, which can decrease the Received Signal Strength Indicator (RSSI) and the phase up to −25 dB and −2 radians, depending on the tag design [18–20]. Outdoors, this occurs during rainfall, in the night when dew progressively accumulates on the tag, and in the morning when a layer of frost melts down. The coupling of water with the tag antenna has been largely exploited in laboratories for sensing the presence of water [18, 20, 21] and is exploited commercially.

The first challenge with wet tags is to keep good enough reading performances, with a minimal loss of RSSI. The second challenge is to reduce the influence of the water on the phase shift, in order to keep accurate phase-based sensing and localization outdoors [1, 19]. The potential solutions to reduce this influence come with appropriate tag design. For example, tag casings can be designed to keep the water film a few centimeters away from the tag antenna, where it has less influence [22]. Self-tuning tags [21, 23] might also compensate partially for the detuning of the tag, reducing the influence on the RSSI and allowing for a numerical correction on the phase.

4.2.2 Effect of Snow

Snow is very present in cold and mountainous areas and creates issues for most outdoor operations. One issue is to locate or identify objects (or persons) covered with snow. Another is to deliver reliable power to autonomous instruments, due to the damage of cold on the batteries and on the reduced effectiveness of solar panels and wind turbines in the presence of snow. RFID technologies have the potential to address these challenges, particularly when it comes to the identification of assets buried under snow. However, the interaction between snow and RFID, little documented today, encounters its own challenges [3].

Snow influences RFID installation by altering the RSSI and the phase [1, 3, 24] (Figure 4.5). The dielectric constant of snow is the main parameter of interest for RFID. It increases with the snow density (from 10 to 600 kg/m^3) and with its liquid water content (from 0% to 12%) [25]. The in-phase dielectric constant of snow ranges from slightly above 1 (for fresh, light, and dry snow) to up to 3.5 (for soaked

Figure 4.5 RFID tags deployed on a mountainous landslide during winter were used to monitor its displacement. The period of snowmelt was observed to increase, from 1 to 8 cm, the error of relative localization, which is attributed mostly to its alteration of the multipathing. *Source:* Le Breton et al. [1]/Reproduced with permission from Elsevier.

and dense snow) [26]. The quadrature dielectric constant, which depends mostly on the liquid water content, can reach 0.15 at 1 GHz [26]. The dielectric constant of snow is heterogeneous in time and in space: density fluctuates with compaction and percolation, and liquid water content fluctuates with daily temperature variations and with precipitations.

Snow influences the RFID communication in at least four ways: the coupling of the snow with the tag antenna, the partial reflection of the wave when it crosses interfaces in the snow layers, the slowness and attenuation of the RF wave propagating in the bulk snow, and the multipath interference [3]. The first effect—coupling—occurs with the deposit of frost or snow on the tag. It can alter the RSSI and the phase [3, 24, 27], depending on the aqueous phase of the snow composition. Tags loaded with dry ice may see a shift of the order of 200 MHz, 15 dB, and 0.2 rad of resonant frequency, RSSI and phase shift, and reach up to three times of these values when thawing [3, 24]. Frost can form progressively on a tag during the night, then thaw rapidly when the sun arrives, and finally evaporate, leading to strong fluctuations of phase and RSSI. The second effect—reflection loss—occurs when the signal partially reflects at the air–snow interface and between layers that have strong contrast of density. The reflection coefficient increases with higher incidence angles (with the exception of full transmission at the Brewster angle) and with the dielectric constant. The highest loss occurs when trying to read a tag with a high incidence angle, across a wet and dense snow surface. The third effect—transmission within the snowpack—induces attenuation and phase delay, depending on the path length within the snow. The attenuation is negligible in dry snow but increases strongly with wet snow, up to 10 dB/m [3, 26]. The propagation can be up to two times slower in soaked and dense snow than in air, severely influencing the phase. The last effect, multipath, is caused by the interference of waves reflected on the ground and on the snow surface. Snow modifies the interference patterns by adding new reflectors and by altering the reflected paths that transmit within the snowpack. It can result in strong variations of ± 20 dB and ± 2 rad in RSSI and phase. Because these four effects are all influenced mostly by wet snow, we think that dealing with wet snow is a main challenge in the presence of a snowpack.

The influence of snow on RFID is also an opportunity to use tags for snow sensing. At the regional scale, snow is a major water reservoir, which is essential to monitor, for hydropower management and flood anticipation. More locally, snow properties are also monitored for risk management, ski resorts, or loading on constructions. However, the properties of snow are extremely heterogeneous in space and in time; therefore, obtaining data on the snow is often time-consuming and difficult to operate. The RFID tags, being small, lightweight, battery less, and wireless, offer an opportunity for sensing the properties of snow at low cost, and with a small thermal disruption to the snowpack. Le Breton et al. [28] have demonstrated the ability to measure multiple properties of a dry snowpack using RFID

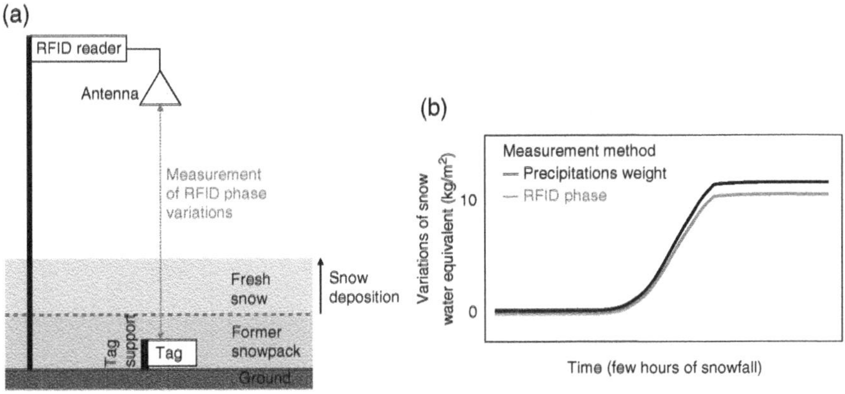

Figure 4.6 (a) Installation of a permanent RFID reader outdoors, interrogating a tag placed just above ground and covered by snow. (b) Schematic of the phase delay monitoring, which allows estimating changes in snow water equivalent over the complete layer of snow during a snow event. *Source:* Le Breton et al. [28]/The Cryopshere/ CC BY 4.0.

tags. First, the tags embedding a temperature sensor can measure the temperature. Then, the phase delay measured on tags placed below snow depends on the snow water equivalent (SWE, in kg/m^2) which increases with snowfall and decreases with snowmelt (see Figure 4.6). This offers the perspective to use tags for sensing the snow remotely.

4.2.3 Effect of Vegetation

The usage of RFID in natural areas will very often encounter the presence of vegetation. Vegetation causes issues to many measuring instruments, particularly the optical ones. It also increases the difficulty to find objects on the ground or to deploy and maintain wires. RFID offers the ability for wireless identification, localization, and sensing across vegetation (Figure 4.7) [3] with, however, a few challenges.

The vegetation near the tags increases the daily fluctuations of phase and RSSI, due to its influence on multipathing, on air humidity (which influences the formation of dew at night) and transmission (for tags below the vegetation). The loss due to communication across a dense grass coverage is rather small (0.2–0.5 dB/m, [3]), and depends on the vegetation density, its size, its humidity, and its ionic content [29].

The challenges of using RFID in vegetated areas are still to be identified and depend on the types of vegetated areas, such as wood, grass, or trees, and may differ between seasons. One hypothetical challenge might be to read and localize tags across high-moisture vegetation.

Figure 4.7 Picture of a reader outdoors in dense vegetation, in an experiment that studies the influence of dense grass on RFID communication. *Source:* Le Breton [3]/ theses.fr.

Again, this influence provides an opportunity for monitoring vegetation properties using tags. Le Breton [3] could estimate the density and the water content of a vegetal cover from the shift in the phase and signal strength, of a tag embedded in vegetative cover. Several applications may be envisioned, for example, to spot which area needs more watering in precision agriculture [30], or to know when and where the vegetation is the driest for wildfire prevention.

4.2.4 Effect of Mud and Tag Burial

RFID tags in outdoor environment can be covered by different types of materials such as soil, rocks, and vegetation. Each of these will impact the performance of the tags and several studies have attempted to either quantify these effects or factor them into RFID tag design.

Overlying materials, such as sand, wood particles, leaves, rocks, and water, on RFID can affect tag performance [31]. Tag count and read range decrease for tags buried under 0–2.5 m thick material coverings. The work is useful in quantifying the relative effect of common outdoor materials that may cover RFID tags. However, more detailed experiments are needed for some of the tested materials. For example, it would be useful to be able to understand how fresh versus dry leaves affect the tag performance. Similarly, sand is only one type of soil and in practice, there are many other soil types such as clays, loams, and mixtures of soil types. Understanding the effect of these on RFID tag performance is a potential area of research.

Dual-frequency RFID tags were designed for the detection of buried assets [32, 33]. The tag harvests energy at a fundamental frequency and backscatters ID response at a higher harmonic. Abdelnour's work harvests energy at fundamental frequencies in the UHF frequency range, while Ghazali's work harvests energy at a fundamental frequency of 2.5 GHz. The soil humidity also affects the tag's response [32]. The idea of separating the harvest and response frequency is interesting as it helps minimize the effect of environmental noise. However, this approach requires the design of a custom reader. Better understanding the trade-offs of this approach in terms of hardware cost versus sensor reliability might be a potential area of further research.

A broadband bowtie UHF RFID tag can give good performance for soil relative permittivity changes of 4–20 [34]. These changes are indicative of the variability of different soil types and moisture contents observed in the field. This improves the versatility of the tag and the environments in which it can be deployed. The dynamism of the environment, however, is also a challenge for this approach as the read range may change depending upon the soil type or soil moisture content. Combining this approach with other advancements in the field, such as self-tuning RFID ICs [23], might be an interesting avenue for further exploration.

4.3 Aquatic Environments

RFID tags have also been used in pure aquatic environments, such as rivers or coastal environments, to monitor the displacement of living beings, driftwood or coarse sediments.

4.3.1 Fish Population Estimation

Since the early 1980s [35], RFID passive integrated transponders (PIT tags) have been used to tag hundreds, thousands, and millions of individual fishes [36] (Figure 4.8). PIT tags of 8-, 12- or 23-mm length and 2 mm diameter are introduced by injection with a needle or by incision with a scalpel into the body cavity or musculature of the fish [38]. PIT tags are then sensed by fixed reading systems [39] or mobile antenna carried by walking operators or tracked by a boat [40, 41] (Figure 4.9). As the animals do not have to be captured, the detection rates are enhanced (compared to previous tagging method). Besides, it allows the

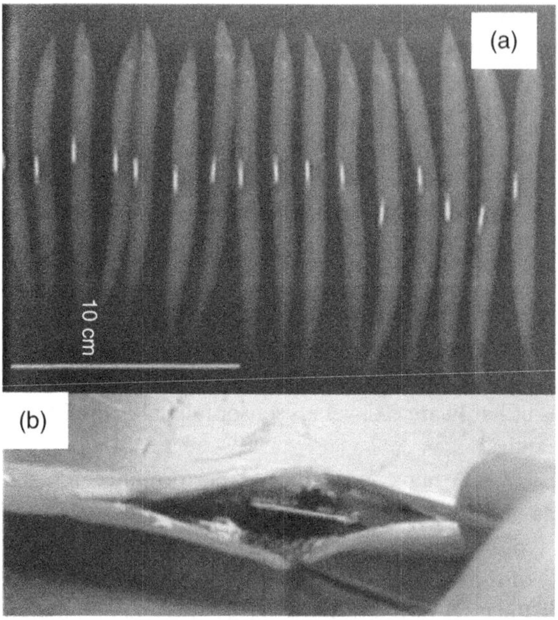

Figure 4.8 (a) Lateral X-rays and (b) dissection view of sandeels (Ammodytes tobianus). *Source:* Jørgensen et al. [37]/ Reproduced with permission from Elsevier.

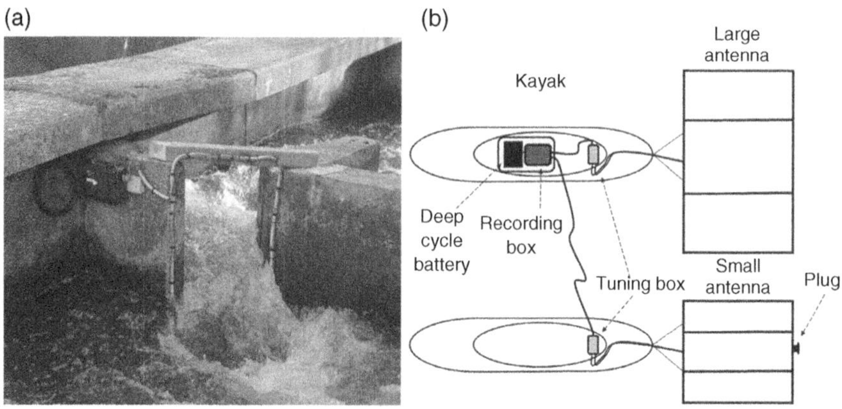

Figure 4.9 (a) Passive integrated transponder (PIT) tag array installed in a vertical slot fishway and (b) a mobile PIT antenna deployment from kayaks. *Source:* Zentner et al. [41]/ American Fisheries Society.

monitoring of habitat use, of movement patterns of individuals, population size, feeding behavior, predation rates, or effectiveness of fishways for restoring passage (see Ref. [42] for a review of tagging studies). One of the goals is to implant into the smallest fish possible [43]. Three main methodological points receive the researcher's attention as they can introduce bias into the data collected: tag retention, effect on the fishes, and ghost tags.

The retention of the tag over the life of the tagged individual is important as frequent PIT loss can lead to survivorship underestimation and abundance overestimation [44]; two metrics commonly used to assess the effectiveness of management actions [45]. PIT tags shedding rates have been investigated in captive settings with facilities in fish recapture [46, 47] and in free-swimming populations where a double-marking system is needed [48, 49]. The shed rates reported are generally low (<20%, in Ref. [50]) but exhibited variations depending on the fish species and age, tag types, tagging locations, and insertion techniques. One field of investigation consists in ensuring/improving the tag retention and reducing tagging effect of fish [51, 52].

A second key point is the effect of tagging on the physiology of the fishes, especially the youngest/smallest ones [53, 54]. The tag-mass:body-mass ratio (in air) and the tag-length:fish-length ratio are considered as controls on swimming performance, which eventually influence survival (predator avoidance) and growth (compete for food) rates and cumulative mortality across the time [55]. Yet, the abundance of the youngest fishes, age-0 abundance, is particularly investigated as it has been identified as a key driver of population dynamics by demographic models [56, 57]. The survival rate through the summer of the age-0 population would be the demographic model [58, 59]. Following operational recommendations [60], no significant effects were demonstrated on 40–49 mm FL Chinook Salmon *Oncorhynchus tshawytscha* [61], 35 mm FL Brook Trout (at least in the laboratory: food and water controlled, no predator; [53]), 42-mm *Oncorhynchus mykiss* [62].

Although the PIT tag retention is generally assumed to be permanent, the identification of ghost tags [63] is becoming a trendy investigated topic. The ghost tags are tags that have been shed by living fish or liberated when a tagged fish dies. These ghost tags can still be sensed and incorrectly interpreted as living fish which leads to biased estimates of population abundance [64], survival [65], or movement [66]. Furthermore, as the number and probability of ghost tags increase with time, it imperils long-term studies. Thus, some authors used intensively monitored watershed data to develop a method to differentiate between live fish and ghost tags based on random forest classification model [45] or generalized additive model [67].

The marine, coastal, and estuarine environments has been little investigated with RFID tags [42]. Technically, it is more difficult: the high conductivity of saline water reduces the read range, and the installation of a portal is more difficult. Although some technical solutions have been proposed [68] (Figure 4.10) and fish species successfully tagged and monitored [54]. The investigations in such an environment are more prone to rely on acoustic and radio telemetry solutions, which are larger and more expensive but that also provide larger sensing range [69].

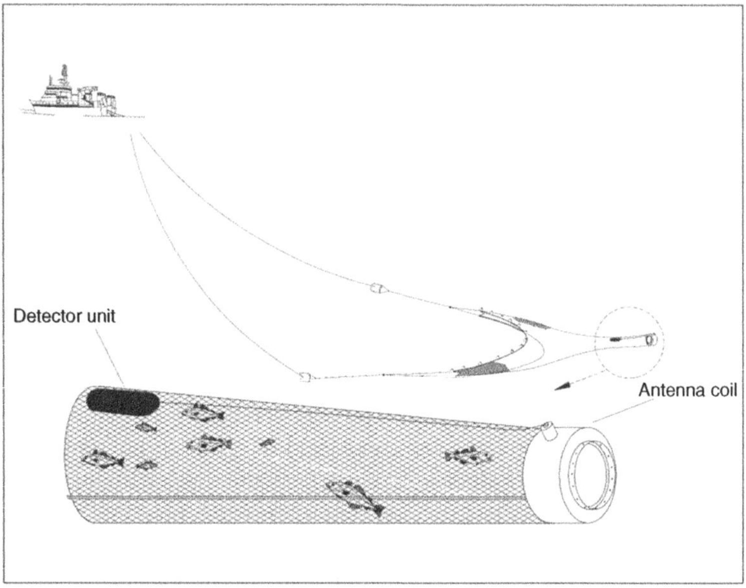

Figure 4.10 Sketch illustrating the mounting of the antenna at the end of an open codend. *Source:* Jørgensen et al. [68]/with permission of FAO/COISPA.

4.3.2 Driftwood Movement Monitoring

Large wood in rivers is considered as an important functional component of fluvial ecosystems, sediment transport, and riparian vegetation. But as it can also damage infrastructure, clog the channel, and induce flooding, it is also a potential hazard during high-magnitude flood [70]. Driftwood movement dynamics have been investigated with RFID tags by few studies, because imagery techniques were preferred. Due to the large distance of transport expected and observed (Table 4.1), the driftwood implementation has been realized with active—Ultra-High Frequency transponders (aUHF tags) coupled with GPS or PIT tags (see Table 4.1 and Figure 4.11). Tagged driftwoods have been sensed at their passage by automatic stations and or geolocated by mobile reading systems carried by boat or pedestrian operators. The use of automatic stations allows the monitoring during the high flow events and helps to determine which driftwoods must be relocated manually in the downstream section, once the flood is finished.

The very few studies published provided little information related to the mobile surveys, neither on method (by boat) nor on duration. Therefore, it is difficult to evaluate the efficiency of the different techniques used to tag and track driftwood

Table 4.1 Summary of the tagging technique and observations of driftwood monitoring studies.

Study	River (country)	Tagging system/detection mode	Detection range	Recovery rate Maximal distance
[71]	L'Ain (France)	PIT tags (23 mm) (a)/mobile reader	Inf. to 1 m	13–27% 2 km
		Active ultra-high frequency (b)/Station + mobile reader	Between 30 and 100 m (1)	
	Sacramento River (USA)	Radio transmitters (c)/Mobile reader on a boat	Approx. 800 m	74–93% 35.4 km
[72]	Lower Roanoke River (USA)	Active ultra-high frequency (b)/Station + mobile reader	Up to 300 m (2)	62% 101.1 km
[73]	Tagliamento (Italy)	Active ultra-high frequency (b)/Station + mobile reader (supposedly)	Approx. 200 m (2)	33–52% overall 43%
		GPS devices (d)	NA	20–81% overall 42% 51 km

(a): Tiris RI-TRP-WRHP; (b): RF Code 433 Hz; (c): advanced telemetry systems; (d): (3100-EXT purchased from LandAirSea Company) are composed by a GPS antenna receiver, and a case (80 × 100 × 40 mm) hosting 4 AA batteries and data logger; (1): tested on the field; (2): according to manufacturer.

other than by the recovery rate and the distance of transport comparison. As they are not battery powered, the PIT tags are suitable for very long-term studies but are limited by their short detection range which would prolong laborious surveying. The aUHF tags and radio transmitters, thanks to their larger detection range, appear to be very efficient (high recovery rate, long distance of monitoring), convenient (can be sensed by automatic station), and are suited for multiyear monitoring (≈ 5 years). The GPS devices are still expensive but provide the only way to monitor the multispatiotemporal-scale movements of driftwoods trajectory with the details of both moving and rest periods.

Thus, the RFID chips' ongoing challenges related to driftwood monitoring would be to combine (i) the multiyear monitoring (up to 10 years), (ii) over very long distance (~100 km) with (iii) the possibility of detailing (timestamping/geolocate) rest and movement periods. Such monitoring of individual logs would be very complimentary to time-lapse imagery and repeat aerial/satellite imagery.

Figure 4.11 (b and d) RFID tag and (a and c) GPS tracker devices as installed in logs in the Tagliamento River. *Source:* (a–d) Ravazzolo et al. [73]/Reproduced with permission from Elsevier. (e) The fixed RFID reader setup on a utility pole with the river in the background. (f) The reader required semi-annual downloads of the data. *Source:* (e, f) Schenk et al. [72]/John Wiley & Sons.

4.3.3 Sediment Tracing

Researchers, river managers, and design offices have tagged sediment to geolocate them before and after hydraulic forcing or to monitor their passages at fixed stations. Tagged sediment displacements (distance of transport or averaged velocities) are then analyzed depending on the particle properties (size and shape), the magnitude of hydro-meteorological events (floods or sea storms), and the morphologies of the investigated systems (slope and width). The effects of anthropic structures (dams, weirs, groynes, etc. see [74, 75]) have been more recently investigated. The first RFID transponders used were PIT tags (passive low-frequency (134.2 kHz) tags of 23 or 32 mm length) borrowed from fish studies [76, 77] (Figure 4.12). Those were deployed per hundreds of units (~200) in small channels (width <7 m) wherein they exhibited encouraging recovery rates (>85%). However, when PIT tags were deployed in very active or large rivers, recovery rates were found to be much lower (<50%). To compensate for this tracers' loss, the number of tracers was increased up to a thousand in such systems [78]. At the same time, the explanatory factors for low recovery rates have been identified [10, 76] and some of those were investigated by technical in lab/flume studies [79–81] but the short sensing range (<1 m) was pointed.

The first challenge is to improve recovery rates and survey efficiency/duration. Inspired by driftwood tracking results, a recent evaluation of active ultra-high frequency transponders (aUHF tags) assessed their potential in fluvial sediment medium [82]. Those tags exhibited a larger detection range (up to 80 m), an efficient anti-collision protocol allowing their grouped seeding and an RSSI that indicate the accuracy of their location and gives an order of their burial depth. These tags have, however, shorter lifespan (~5 years), are more expensive (40–60€, depending on the model and amount ordered), and are bigger (ø of 36 mm, thickness of 12 mm), which implies to mold them into artificial pebbles (Figure 4.12).

The aUHF tags have yet been deployed and successfully tracked on a very active system. They allowed to monitor distance of transport up to 3400 m during a single high-magnitude flood event [83] and virtual velocity between 13 and 27 m/h observed over a period of 2 days [84]. They are, however, less efficient in larger and deeper channels [85, 86] wherein RFID sediment tracing remains complicated. To investigate tracing protocols, an index (Es) was proposed to characterize and improve the infield survey efficiency. This efficiency score is based on (i) the surveying effort that is the time of surveying normalized by the study site surface (in hour/hectare) and (ii) the recovery rate [86].

$$\mathrm{Es} = R_r * A / T$$

with R_r, the tracer recovery rate, A, the prospecting area in hectares, and T, the prospecting time in h. Reducing the survey duration is also a challenge for beach sediment tracking experiments as the surface-to-prospect within the foreshore is daily submerged and cyclically evolves over time.

The second challenge is to estimate the tracer's burial depth. The tracers' burial depth is a means for the estimation of the river bed's active layer thickness during a flood event. Combined with distances of transport of the tracers, it allows the

Figure 4.12 (a) Natural clast equipped with PIT tags [10] and (b) artificial pebbles with aUHF tags. Surveying methods for PIT tags (c) and aUHF tags surveying (d) by pedestrians and (e) by UAV.

Table 4.2 Advantages, drawbacks, and limitations of the two methods for active layer thickness estimates.

	RFID column	**Wobblestone**
Study	[85, 90]	[91, 92]
Sensing range (m)	>20	<0.2
Vertical resolution (m)	0.5	0.1
Maximal burial depth (m)	1	0.15
Grain size	Between 35 and 45 mm of *b*-axis	Gravel to Cobble
Installation	Expert	None
Advantages	• Adapted to monitoring of long-distance of transport • Allow analysis of distance of transport as function of burial depth at the first survey • Measure the virtual velocity tracers constrained condition at first survey	• Adapted to long-term monitoring • Allows the repetitive measure spatially distributed of the burial depth • Allows analysis of distance of transport as function of burial depth between surveys. • Allows analysis of tracers displacement in constrained conditions at 2nd survey.

estimation of the time-averaged bedload flux [87–89]. Two methods to measure the active layer, based on RFID chips, have been recently assessed. Their respective limits still constitute challenges to address (advantages and drawbacks are summarized in Table 4.2).

The first method is called "Wobblestone" [91, 92] (Figure 4.13). It uses 12-mm long PIT tags introduced into two plastic trundle balls, separated by a lubricant layer. The inner ball has one half weighted so that the PIT tag always returns in a vertical position. This return in the vertical position maximizes and withdraws this variability in sensing range associated with the relative orientation between PIT tag and reading system antenna. It allows the use of the maximal height above the ground at which the PIT tag is sensed to calculate the trace burial depth. This system has shown encouraging results infield for tracers buried depth up to 15 cm after distance of transport up to 70 m. Although it allows spatially distributed measurements of the burial depth, it has the limits of the 12-mm PIT tag, especially its sensing range (0.2 m); and, when Wobblestone would be deployed in large/deep channels or very active systems, it is likely that either prospecting

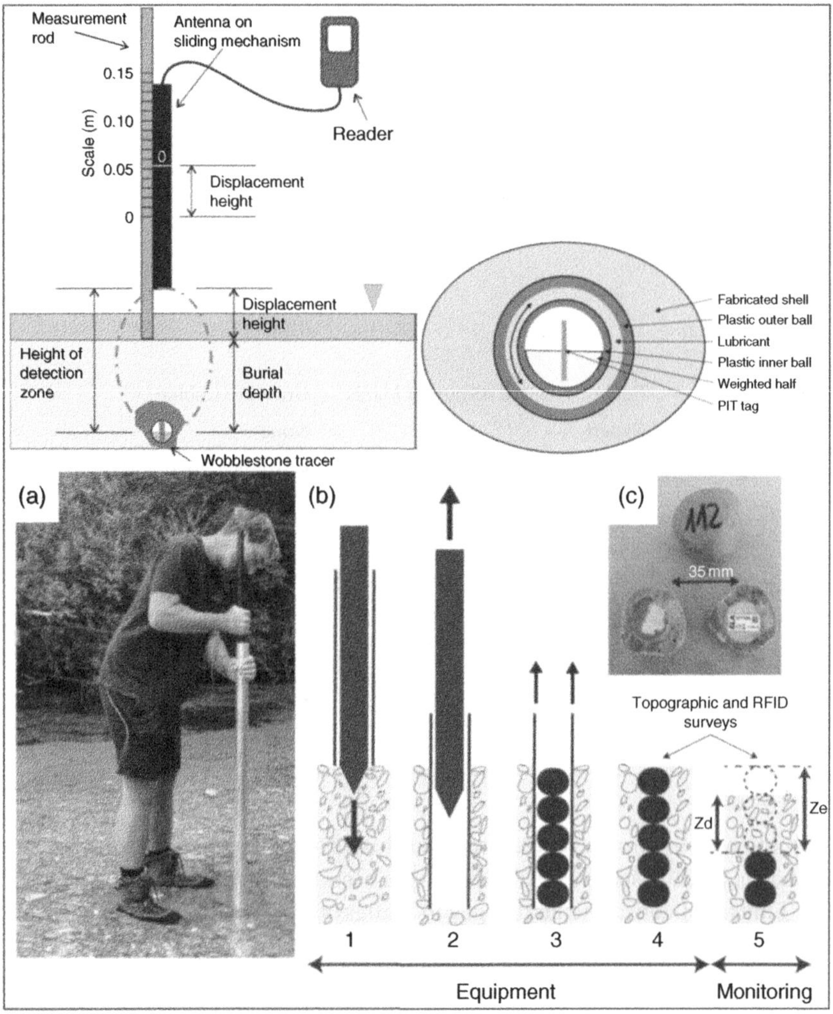

Figure 4.13 (Up–Left) Schematic of the burial depth measurement protocol using a straight RFID antenna and a Wobblestone tracer and (Up–Right) conceptual diagram of the Wobblestone. *Source:* Adapted from Papangelakis et al. [91]. (a and b) Installation and surveys of the RFID Column (Below with Ze: eroded layer thickness; Zd: deposited layer thickness) and (c) the artificial pebbles with aUHF tags. *Source:* Fabien Boutault [85]/theses.fr.

effort would increase and/or recovery rate would drop as tracers would be buried deeper or transported further.

The second method is called "RFID column" [85, 90] and is composed of several aUHF tag-equipped tracers (*b*-axis <45 mm) installed one above the others (Figure 4.13). To date, it has been installed up to depths of 0.25 and 1 m within shallow wet channels and gravel bars, respectively. Combined with topographic surveys at the point of the RFID column installation, it is possible to measure the thickness of the eroded and deposited layers. As they are incorporated within the sediment layer, the tracers of the RFID column geolocated downstream allow (i) the measure of transport distance in tracer constrained conditions at first survey and (ii) the analysis of the distance of transport as a function of the initial burial depth. However, to date, it is not possible to accurately estimate the burial depth of the tracers after they have been moved from the column and transported downstream. Indeed, the RSSI at the vertical location of the tracers exhibits too high variability caused by tag orientation [82], which cannot yet be corrected as it is with Wobblestone.

In addition to these works in progress, the RFID technology's ongoing challenges related to coarse sediment tracing are related to combine (i) multiyears monitoring (~10 years) and (ii) overlong distance of about 20 km (sediment tracing over 19.6 km have been performed on *the Severaisse* by *Envrionnement Ville Société laboratory* but not published yet). The (iii) ability to collect over long distance, detailed rest/movement periods, and hops length,is also a challenge: so far it has been performed only with radio transmitters over a distance of ≈200m [93].

4.4 Landslide and Rockfall Detection

Slope destabilization occurs in many regions of the world, imposing a heavy cost in life loss, property, and infrastructure. In the last decade, an average of 405 fatal landslides and 4 307 casualties were recorded per year [94], to add with the earthquake-induced landslides [95]. Their annual economic loss has been estimated regionally to $4, $2.6–5, and $3.6 billion in Japan, Italy, and the United States, respectively [96], representing a considerable cost worldwide. Furthermore, landslide occurrence is increasing due to human activities [94] and climate change in alpine regions [97].

RFID tags can be used to monitor a landslide, using phase-based ranging [98], to continuously monitor the displacement of tags [3]. The relative radial displacement (1D) was first measured using 19 tags deployed on the Pont-Bourquin landslide for 5 months (figure 4.14). The measurements were validated against a long-range wire extensometer, and surveys using a total station. The RFID technique appears little sensitive to most environmental effects, particularly with appropriate devices and calibration [19], as compared to the existing long-range wire extensometer. The accuracy of the RFID technique was 1 cm during normal weather and 8 cm during the hardest conditions like heavy snow events, for which the long-range wire extensometer was no longer reliable. Since then, the method has been also deployed on

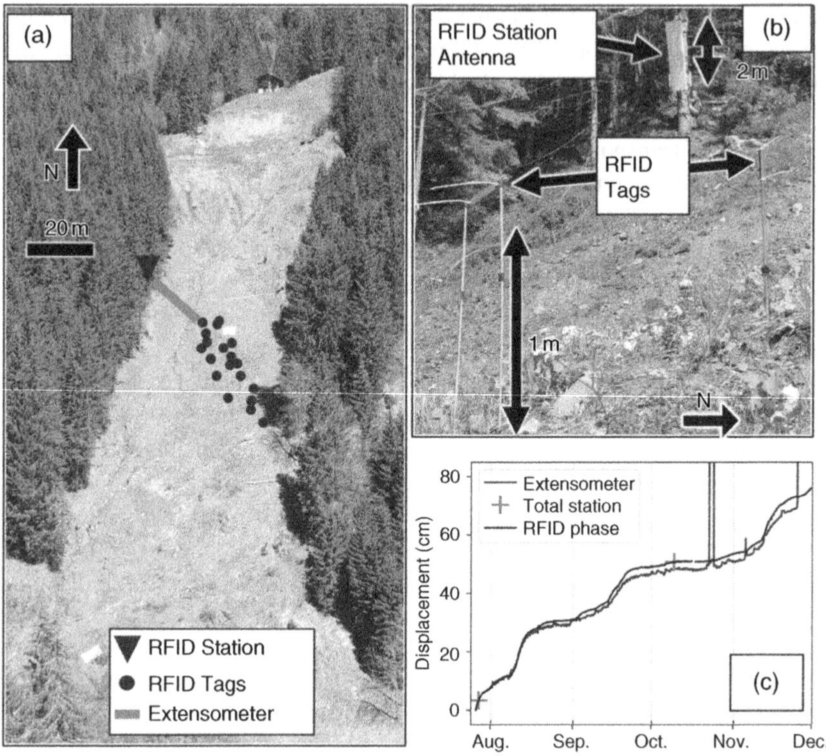

Figure 4.14 (a) General aerial view of the Pont-Bourquin landslide. *Source:* Le Breton et al. [1]/with permission of ELSEVIER. (b) Photograph showing tags no. 10 and no. 2 installed on the landslide, and the station which is placed on a stable zone, and (c) cumulative radial displacement measured over 4 months on the Pont-Bourquin landslide, using a wire extensometer, the phase of RFID tag no. 10, and a manual laser total station. *Source:* Le Breton et al. [1]/Reproduced with permission from Elsevier.

the landslides of Harmalière, France (2D tracking), [99] Valloire, France (strong snowfalls and steepness), and Bidart, France (coastal area) [100].

The first challenge of this application is to increase the read range. It is the major limit to address many landslides and deploy many tags, the condition for which the solution becomes really economical compared to other methods. Since the first installation in Ref. [1], the design of directive tags and the appearance of more sensitive readers on the market have multiplied the read range by 3.5 and further improvements are under preparation on the tags.

The second challenge is about the displacement accuracy, which is in practice deteriorated under some conditions. First, snowfall may create reversible drift up to 8 cm [1, 3, 19], which is likely due to changes in the multipath interference pattern. Then, the tags need to be installed on a support typically 0.5–1 m above the ground to increase the read range. However, the snow or ground motion can tilt this support, generating a rotational motion of the tag that does not represent the translational motion of the ground.

The last challenge is the ability to handle gaps of measurements, caused, for example, by RSSI loss due to multipathing interference, or by cuts of electrical power on the station. Such data gap occurring during a ground motion may result in an unwrapping error of the cumulative displacement, if the tag has moved by more than its ambiguity between two consecutive measurements ($\pm\pi/4$ in practice, equivalent to 4.3 cm at 866 MHz).

Another application consists of monitoring unstable cliffs with RFID tags, in order to detect its displacement preceding a potential rockfall [101]. The principle is to connect a wire to the anti-tampering function of tags, in order to inform if a crack has moved by more than a millimeter. In this situation, tag localization is not accurate enough; mechanical sensing, however, can detect such a motion. A continuous fixed reader (as in Figure 4.15) collects in real time the information on the state of the tampering sensor. This approach allows to avoid installing batteries or solar panels on the cliff, in order to reduce the need for maintenance, in such a difficult-to-access area.

One challenge is to reduce the cost of deploying the tags on a cliff. Indeed, in this context, the manpower needed for installation costs more than the tags. This is critical, because if only a few tags are installed because of economical reasons, then the cost per measured point (including the reader station) may become less competitive compared to WSNs. In this area, UAV deployment might, in the future, increase the cost efficiency of such deployments [6, 102].

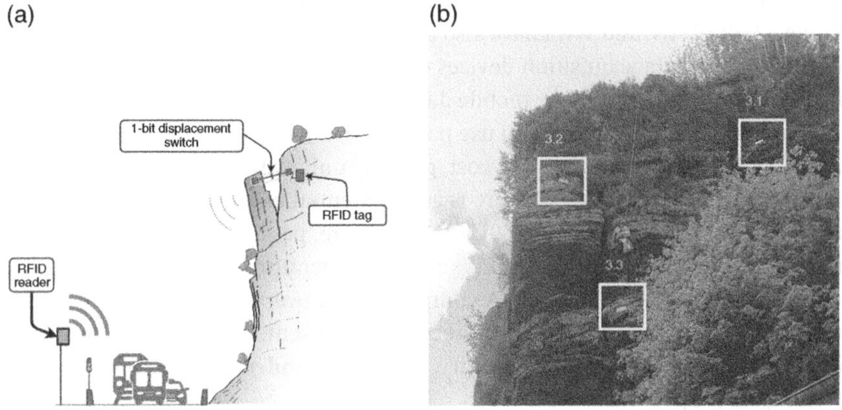

Figure 4.15 (a) Principle of measurement of fissures. *Source:* Le Breton et al. [101]/with permission of Géolithe company. (b) Sensing tags being deployed on a cliff by an operator. *Source:* Le Breton et al. [101]/Copernicus GmbH/CC BY 4.0.

Another challenge is to increase the long-term reliability of the anti-tampering sensor. Again, due to the difficult access, the maintenance intervention should be reduced. In practice, the anti-tampering input of the RFID chip can cause problems (some tags/ICs were defective and had to be discarded on this installation), and the sensing wire itself should be loose enough and protected, to avoid parasitic triggers of the sensor (a few false triggers occurred in the installation).

4.5 Agriculture

Precision agriculture, where water, fertilizer, and pest treatment are applied to crops as needed in order to better optimize resources and minimize pollution, has emerged as a prime area of focus in the agriculture industry [103]. Resource optimization is based on the data acquired from deployed sensors. Sensors used in agriculture have conflicting objectives. On the one hand, these sensors need to cover large geographical areas, have long communication range, and require minimal maintenance. As a result, satellite remote sensing is a popular technique currently utilized in agriculture [104].

On the other hand, the spatial resolution of these sensors cannot be too coarse and should be able to provide useful information for individual farms. Studies have shown that parameters such as soil moisture can vary significantly across a geographical landscape due to differences in soil types and topology [105] and can directly impact the yield of crops [106]. Unfortunately, spatial resolution can be a limitation for satellite remote sensing. In addition, cloud cover and the lack of on-demand data can also be a detraction [30].

To address resolution issues, WSNs based on many different protocols such as ZigBee, LoRaWAN, and WiFi have also been used and make use of a local bay station or mobile data acquisition devices called "data mules" [107]. UAVs are also increasingly being utilized for mobile data acquisition purposes [108].

There is also an opportunity to use passive UHF RFID technology for sensing. Passive UHF RFID tags offer low-cost, globally unique wireless identification in a broad range of industries so there is good potential to extend the transponders to serve as pervasive sensors [109]. In agriculture, passive and BAP UHF RFID tags have been used for applications such as temperature sensing [110, 111] and soil moisture sensing [112, 113]. A comprehensive overview of the various types of RFID soil moisture sensing approaches is presented in the literature that include antenna-based, reflectometry, humidity, and resistivity transduction techniques [75].

A monopole probe coupled with a passive UHF RFID tag is capable of monitoring the top 7.2 cm of soil for changes in moisture content and has been shown to work reliably in clay and sandy soils [114] (Figure 4.16).

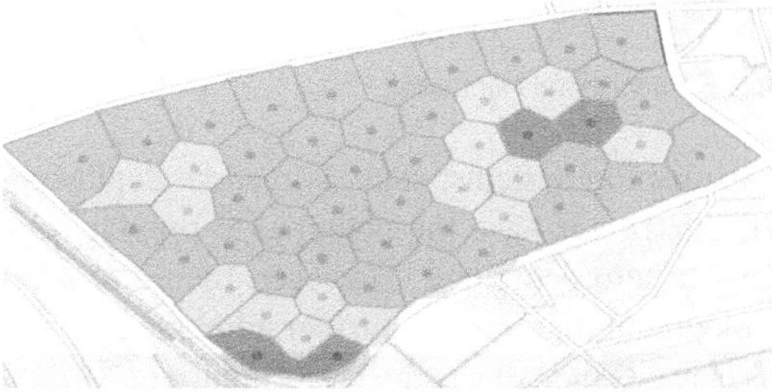

(a) Soil moisture detection at finer spatial granularity

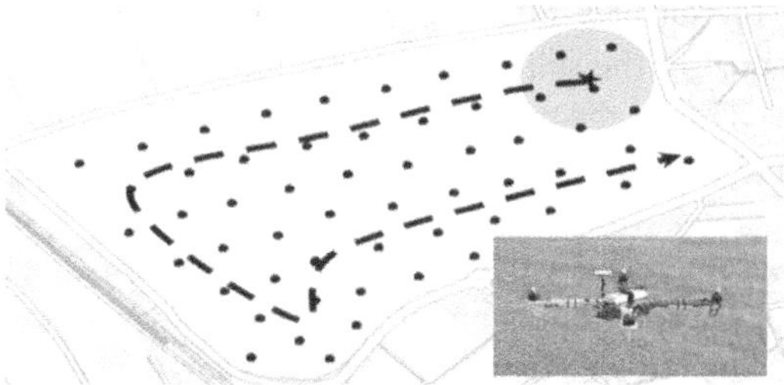

(b) Use of drones or data mules to monitor relatively short-range RFID tag sensors

(c) RFID tag attached to monopole probe comprises the soil sensor

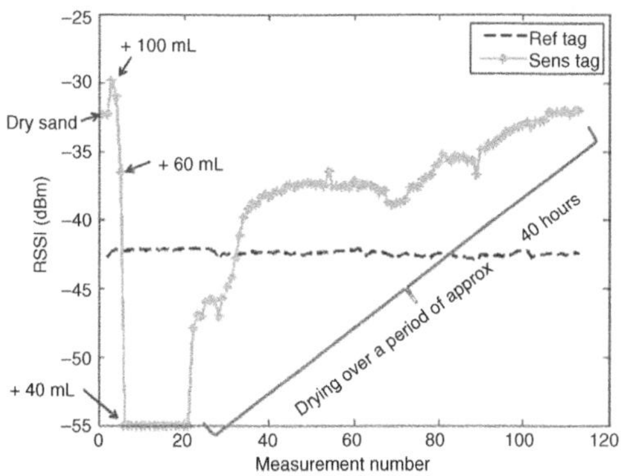

(d) Example of tag signal changing as soil dries out

Figure 4.16 Use of passive UHF RFID soil moisture sensors in agriculture: (a) due to low-cost the sensors can be deployed at increased spatial granularity; (b) the sensors can be monitored on demand with a mobile RFID reader; (c) soil moisture sensor developed by connecting a passive UHF RFID tag with a monopole probe. *Source:* (c) Azhar Hasan et al. [114]/IEEE. (d) Change in RSSI at the sensor as the soil moisture reduces. *Source:* (a, b, d) Adapted from Azhar Hasan et al. [114].

Together with mobile data acquisition systems, such as readers deployed on UAVs or farm equipment, UHF RFID tags have the potential to provide sensor data and fine spatial granularity at affordable cost. However, there are several challenges to their practical deployment that must be further addressed.

The ground material, such as clay or sand, affects the performance of RFID soil sensors [114], requiring extended studies to different types of soils. Furthermore, the distribution of moisture in the soil also influences the sensors. For example, if the water is uniformly distributed in the soil, the performance is different than when the water is applied directly near the probe. The effect of soil moisture distribution around the probe and soil compaction around the probe needs to be better examined.

Soil moisture is frequently accompanied with salinity, and studies have quantified the effect of both soil moisture and salinity on monopole probe performance at UHF RFID frequencies using a network analyzer [115]. However, the ability to relate both salinity and moisture levels to the wireless response of a passive UHF RFID tag has not been examined. There has been an effort to discriminate salinity and moisture levels using sensor attached to a tag that uses battery but this adds a layer of maintenance and cost to the sensor [116].

The deployment and subsequent removal of hundreds or thousands of these sensors across a farming landscape is also an issue that needs to be investigated further. Researchers have attempted to solve this by designing RFID tag sensors that can survive plowing by farm equipment and be deployed inside the ground [117], but this raises the complexity and costs of the sensors. A better understanding of the costs associated with sensor deployment, maintenance, and removal is needed.

Passive RFID sensors can be deployed in larger quantities, due to lower costs, but require more coverage with static or mobile data acquisition devices. Battery-assisted RFID sensors cost more but need sparser coverage due to their longer communication range. Some studies have attempted to bridge the gap between these two approaches by designing low-cost sensors that normally operate only in passive mode but switch to battery-assisted mode only in the event of an alarm or abnormal condition [118]. The trade-off in cost between the sensor and the data acquisition infrastructure needs to be better quantified.

4.6 Infrastructure

Civil infrastructure, particularly in industrial nations like the United States, is in a state of disrepair. The American Society of Civil Engineers estimates that $206B needs to be invested annually to improve the condition of infrastructure and maintain global competitiveness through [119]. Alavi and Buttlar also estimate that the poor condition of 32% of major US roads costs the nation $62B annually in 2019 [120]. As a result, there is a need and an interest in embedding monitoring solutions into new or existing infrastructure.

While wired sensors have been deployed in infrastructure since the start of the 21st century [121], the challenges in cable management and maintenance have led to the advent of wireless solutions. WSN dispense with the problems of cable management but suffer from their own issues such as cost [122], battery management, and transmission reliability [123]. With the advent of UAV technology, the ability to bring the monitoring infrastructure closer to the sensor presents itself and cost-effective, short-range sensor devices could be used more pervasively. As a result, infrastructure condition monitoring accounts for 45% of UAV usage [124].

Passive RFID technology is a popular option currently being used in the construction industry. While the technology has seen extensive use in the tracking of material, workers, and equipment in logistics [125], RFID data is increasingly being used for planning, building management purposes, and even structural health monitoring [126].

Accelerometer and strain sensor were also designed using an RFID tag as a front-end to a sensing system [127]. A microcontroller interrogates a three-axis accelerometer and strain sensor and encodes the data in the user memory bank of the tag's IC. The data is then conveyed to the reader using the RFID protocol. They demonstrate the ability to detect vibrational frequencies of up to 40 Hz and strains of at least 600 $\mu\varepsilon$ from a read range of 1.5 m [127]. The ability to convey strain and accelerometer data with no modifications to the Gen 2 protocol is a big advantage of the approach and one could make use of the same infrastructure used in building logistics for monitoring purposes. The disadvantage is that this sensor needs a power source and would, therefore, need periodic maintenance.

Humidity sensors interfaced with HF and UHF RFID tags were also designed to monitor relative humidity (RH) monitoring inside concrete [128]. However, they rely on a short read range of less than 10 cm to make the sensor completely passive and powered by inductive or radiative coupling from the reader. Such sensor can successfully monitor RH changes of 80–100% over a monitoring period of 172 days [128].

An antenna-based sensing approach may also be considered for health monitoring. Here, the parameter being sensed induces a controlled change in the phase, frequency, or amplitude of a tag's response that can be detected by the reader. RFID tag can be adapted as a position sensor by moving a metal plate within 30 mm of the tag [129]. Position can then be related to the backscatter signal strength of threshold reader transmitted power needed to just turn on the tag, while the corrosion of a steel plate near an RFID tag can be detected using signal changes at the tag [130].

Crack detection sensor can be created by observing the length and direction of propagation of a crack on a structural surface. That can be monitored by observing the crack influence on the relative signal strengths of a grid of RFID tags deployed on the surface [131]. The movement of metallic contacts next to two RFID tags can be used to monitor the width of an expansion joint in bridge or railway monitoring [132] (Figure 4.17).

While RFID is a promising technology for low-cost, pervasive sensor development in infrastructure health monitoring, there are some open challenges that need to be addressed. First, it is very likely that these sensors will be exposed to the element and buried under ice, snow, or mud of variable thickness. These materials will have a frequency dependence, with higher frequency (e.g. UHF or 2.4 GHz tags) being affected more strongly than HF tags. The impact and mitigation measures of material build-up on the tags will need to be considered (see Section 4.3).

Second, the spatial resolution of these tags may need to be improved in some situations. Take, for example, the crack or corrosion sensors. These tags will only be influenced by changes in proximity (i.e. within a few cm) of the RFID tag. Better spatial resolution may therefore be achieved by placing grids of tags close to one another. However, if the tags are too close together, they will couple with one another in unpredictable ways. In addition, a dense placement of too many tags will also drive up the costs. Therefore, the trade-off between the density of tag placement on the structural surface and the spatial resolution of the sensor will need to be better explored.

Finally, structural health monitoring needs to detect very small changes in the monitored parameters. For example, crack widths and displacements can be well below the mm level. Being able to measure parameters with such precision and at low cost is challenging and needs to be better addressed.

4.7 Conclusion on the Main Challenges

New usages of RFID tags are rising in harsh outdoor environments. We have presented specific challenges concerning monitoring transportation of sediments and fishes in rivers, monitoring unstable landslides and rock compartments, monitoring soil for agriculture, and structural health monitoring of infrastructure. Overall, these applications present common challenges and perspectives.

The first challenge is to read tags in complex mediums which contain liquid water, such as concrete, snow, ground, alluvial sediment, mud, vegetation, rivers, and coasts. These play a key role in the tags' performances, altering the RSSI and the phase delay. The medium couples with the tag antenna within its near field and modifies the wave propagation in the far field (multipathing, attenuation,

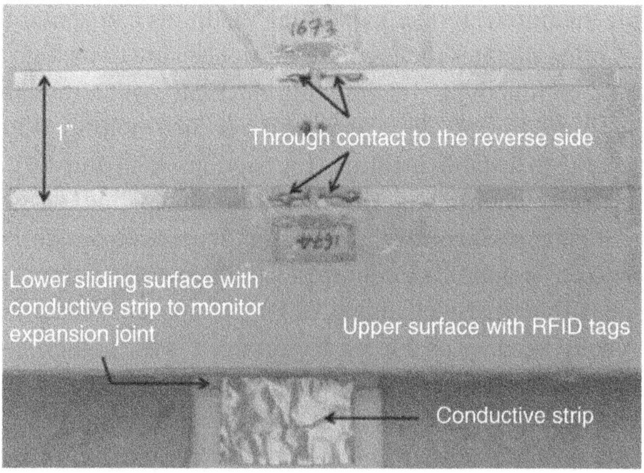

(a) Expansion joint monitor: The RFID tags are on one surface that moves relative to another. The other surface has a metal strip that moves below the RFID tags. Depending on the position of the strip, the tag signal changes that can be used to estimate the movement of the discontinuity.

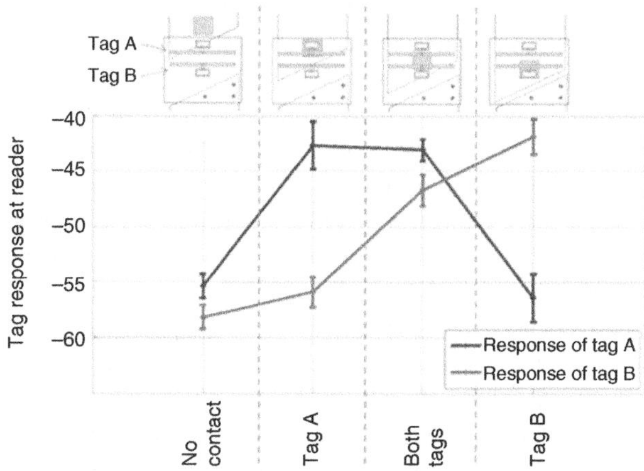

(b) The position of the strip influences the two RFID tags on the opposing surface. The position of the plate may be determined by the relative signal strengths of the two tags.

Figure 4.17 Passive RFID expansion joint monitor: (a) prototype of the sensor and (b) change in RSSI as a function of joint size [132].

and reflection at the air-medium interface), and such mediums can contain liquid water. Liquid water often dictates the medium influence, due to its high permittivity and to its ionic conductivity. Its amounts fluctuate over time due to many factors such as air humidity, precipitations, and freezing/melting. While it is possible to fine-tune a tag to a stable medium (see Sections 4.3.4 and 4.6), this operation is more difficult when its permittivity fluctuates due to water. Therefore, a challenge for the resilient use of RFID outdoors is to design RFID systems that can adapt to environments that have a variable water content.

The second challenge is the rapidly changing RFID technology landscape. As time passes, tags will likely become usable on larger areas, with higher spatial resolution and lower tag cost. This is possible due to improvements in read range, cost decreases, and to a larger extent in autonomous vehicles. One must keep in mind that RFID is a heterogeneous system and the set of underlying technologies are evolving rapidly. RFID is one of the main physical components of the Internet of Things, and the use of tags is increasing exponentially, with a yearly increase of tags volume produced >30% [133]. Automatization in general is also increasing, for example in agriculture. The combination of RFID with autonomous vehicles and UAVs—which also presents an exponential growth [134]—offers perspectives for new ways to collect data using arrays of low-cost, long-duration, and sustainable sensors. Additionally, industrial readers are becoming more sensitive every year, and tags become more sensitive or have more capabilities. Applications that require battery-assisted tags today may fit well with battery-less tags in the future due to lower energy requirements or implementation of energy harvesters.

Another challenge is the adoption of RFID sensing in large businesses, which will improve the technologies available for outdoor usage. Today RFID is adding value in many businesses using only identification [135], but using tags as sensors remain in its early stages. Nontechnical challenges for adoption of RFID sensing can relate to concerns such as privacy, the will to obtain and exploit data on existing processes, or the concerns of responsibility in case of warnings raised by the data. Such concern is out of the technical scope of this chapter but must be kept in mind. On the technical side, tags have strongly evolved in 20 years and are far more reliable today. Tags that offer sensing functionality are also improving continuously, with manufacturers like Farsens, EM Microelectronic, ASYGN, or Axzon offering solutions to many sensing modalities.

To conclude, we envision a positive feedback loop for RFID technology development in outdoor applications. Advancements in RFID technology unlock new applications in outdoor or large-area monitoring and these in turn help drive the development of improved solutions in said applications. In particular, RFID appears as a complementary monitoring technique that can bridge a gap between classical in situ sensors and large-scale WAN sensing. It is hoped that the challenges identified in this chapter can help better channelize development efforts to make RFID deployment more compelling outdoors.

References

1 Le Breton, M., Baillet, L., Larose, E. et al. (2019). Passive radio-frequency identification ranging, a dense and weather-robust technique for landslide displacement monitoring. *Engineering Geology* 250: 1–10.

2 Mao, L., Dell'Agnese, A., and Comiti, F. (2017). Sediment motion and velocity in a glacier-fed stream. *Geomorphology* 291: 69–79.

3 Le Breton, M., Baillet, L., Larose, É., 2023. Tomography of the quantity of grass using RFID propagation-based sensing, in: IEEE 13th Int. Conf. on RFID Tech. and App. (RFID-TA). Presented at the IEEE 13th Int. Conf. on RFID Tech. and App. (RFID-TA), pp. 142–145. https://doi.org/10.1109/RFID-TA58140.2023.10290481

4 Amato, F., Torun, H.M., and Durgin, G.D. (2018). RFID backscattering in long-range scenarios. *IEEE Transactions on Wireless Communications* 17 (4): 2718–2725.

5 Qi, C., Amato, F., Alhassoun, M., and Durgin, G.D. (2021). A phase-based ranging method for long-range RFID positioning with quantum tunneling tags. *IEEE Journal of Radio Frequency Identification* 5 (2): 163–173.

6 Wang, J., Schluntz, E., Otis, B., and Deyle, T. A new vision for smart objects and the internet of things: mobile robots and long-range UHF RFID sensor tags. *arXiv* 1507.02373. http://arxiv.org/abs/1507.02373.

7 Mascareñas, D., Flynn, E., Todd, M. et al. (2008). Wireless sensor technologies for monitoring civil structures. *Sound and Vibration Magazine* 5: 16–20.

8 Deng, F., Zuo, P., Wen, K., and Wu, X. (2020). Novel soil environment monitoring system based on RFID sensor and LoRa. *Computers and Electronics in Agriculture* 169: 105169.

9 Dolphin, T., Lee, J., Phillips, R. et al. (2016). Velocity of RFID tagged gravel in a non-uniform longshore transport system. *Journal of Coastal Research* 75 (sp1): 363–367.

10 Liébault, F., Bellot, H., Chapuis, M. et al. (2012). Bedload tracing in a high-sediment-load mountain stream. *Earth Surface Processes and Landforms* 37 (4): 385–399.

11 EPC-Gen2 (2015). EPC (tm) Radio-Frequency Identity Protocols Generation-2 UHF RFID. EPCglobal Inc. Report No.: 2.0.1. http://www.gs1.org/epcrfid/epc-rfid-uhf-air-interface-protocol/2-0-1.

12 Klair, D.K., Chin, K.-W., and Raad, R. (2010). A survey and tutorial of RFID anti-collision protocols. *IEEE Communications Surveys & Tutorials* 12 (3): 400–421.

13 Buffi, A., Motroni, A., Nepa, P.,Tellini, B., and Cioni, R. A SAR-based measurement method for passive-tag positioning with a flying UHF-RFID reader. *IEEE Transactions on Instrumentation and Measurement* 68(3) (March 2019): 845-53. https://doi.org/10.1109/TIM.2018.2857045.

14 Bradley, D.N. and Tucker, G.E. (2012). Measuring gravel transport and dispersion in a mountain river using passive radio tracers. *Earth Surface Processes and Landforms* 37 (10): 1034–1045.

15 Le Breton, M., Liébault, F., Baillet, L. et al. (2021). Dense and long-term monitoring of Earth surface processes with passive RFID – a review. *arXiv*:2112.11965. https://arxiv.org/abs/2112.11965.
Le Breton, M., Liébault, F., Baillet, L. et al. (2022). Dense and long-term monitoring of earth surface processes with passive RFID —a review. *Earth Science Reviews* 234: 104225.

16 Casati, G., Longhi, M., Latini, D. et al. (2017). The interrogation footprint of RFID-UAV: electromagnetic modeling and experimentations. *IEEE Journal of Radio Frequency Identification* 1 (2): 155–162.

17 Tanaka, M., Takahashi, N., Ikeda, R., and Suzuki, M.2012). RFID based maintenance management system for ground coils of superconducting maglev. *2012 IEEE International Conference on RFID-Technologies and Applications (RFID-TA)*, Nice, France (5–7 November 2012), 330–334.

18 Caccami, M.C., Manzari, S., and Marrocco, G. (2015). Phase-oriented sensing by means of loaded UHF RFID tags. *IEEE Transactions on Antennas and Propagation* 63(10): 4512-20.

19 Le Breton, M., Baillet, L., Larose, E. et al. (2017). Outdoor UHF RFID: phase stabilization for real-world applications. *IEEE Journal of Radio Frequency Identification* 1 (4): 279–290.

20 Occhiuzzi, C., Caizzone, S., and Marrocco, G. (2013). Passive UHF RFID antennas for sensing applications: principles, methods, and classifcations. *IEEE Antennas and Propagation Magazine* 55 (6): 14–34.

21 Caccami, M.C. and Marrocco, G. (2018). Electromagnetic modeling of self-tuning RFID sensor antennas in linear and nonlinear regimes. *IEEE Transactions on Antennas and Propagation* 66 (6): 2779–2787.

22 Dobkin, D. and Weigand, S. (2005). Environmental effects on RFID tag antennas. *MTT-S International Microwave Symposium Digest*, Long Beach, CA, USA (17 June 2005), 135–138.

23 Zannas, K., Matbouly, H.E., Duroc, Y., and Tedjini, S. (2018). Self-tuning RFID tag: a new approach for temperature sensing. *IEEE Transactions on Microwave Theory and Techniques* 66 (12): 5885–5893.

24 Wagih, M. and Shi, J. (2021). Wireless ice detection and monitoring using flexible UHF RFID tags. *IEEE Sensors Journal* 21 (17): 18715–18724.

25 Fierz, C., Amstrong, R.L., Durand, Y. et al. (2009). *The International Classification for Seasonal Snow on the Ground*. Paris: UNESCO-IHP Report No.: IHP-VII Technical Documents in Hydrology N°83, IACS Contribution N°1.

26 Tiuri, M., Sihvola, A., Nyfors, E., and Hallikaiken, M. (1984). The complex dielectric constant of snow at microwave frequencies. *IEEE Journal of Oceanic Engineering* 9 (5): 377–382.

27 Nummela, J., Ukkonen, L., and Sydänheimo, L. (2008). Passive UHF RFID tags in arctic environment. *International Journal of Communication* 2 (1): 135–142.

28 Le Breton, M., Larose, É., Baillet, L. et al. (2023). Monitoring snow water equivalent using the phase of RFID signals. The Cryosphere 17(8): 3137-56. doi: https://doi.org/10.5194/tc-17-3137-2023.

29 El-rayes, M.A. and Ulaby, F.T. (1987). Microwave dielectric spectrum of vegetation-Part I: experimental observations. *Transactions on Geoscience and Remote Sensing* GE-25 (5): 541–549.

30 Mulla, D.J. (2013). Twenty five years of remote sensing in precision agriculture: key advances and remaining knowledge gaps. *Biosystems Engineering* 114 (4): 358–371.

31 Li, C.H., Lao, K.W., and Tam, K.W. (2018). A flooding warning system based on RFID tag array for energy facility. *2018 IEEE International Conference on RFID Technology Application (RFID-TA)*, Macau, Macao (26–28 September 2018), 1–4.

32 Abdelnour, A., Lazaro, A., Villarino, R. et al. (2018). Passive harmonic RFID system for buried assets localization. *Sensors* 18 (11): 3635.

33 Ghazali, M.I.M., Karuppuswami, S., and Chahal, P. (2016). Embedded passive RF tags towards intrinsically locatable buried plastic materials. *2016 IEEE 66th Electronic Components and Technology Conference (ECTC)*, Las Vegas, Nevada, USA (31 May–3 June 2016), 2575–2580.

34 Mondal, S., Rothwell, E.J., and Chahal, P. (2018). A wideband antenna for buried RFID applications. *2018 IEEE International Symposium on Antennas and Propagation USNC/URSI National Radio Science Meeting*, Boston, Massachusetts, USA (8–13 July 2018), 327–328.

35 Prentice, E.F., Flagg, T.A., and McCutcheon, C.S. et al. (1990). Equipment, methods, and an automated data-entry station for PIT-tagging. *American Fisheries Society Symposium*, 335–340.

36 Feldheim, K., Gruber, S., de Marignac, J., and Ashley, M. (2002). Genetic tagging to determine passive integrated transponder tag loss in lemon sharks. *Journal of Fish Biology* 61: 1309–1313.

37 Jørgensen, M.G., Deurs, M., Butts, I. et al. (2017). PIT-tagging method for small fishes: a case study using sandeel (*Ammodytes tobianus*). *Fisheries Research* 193: 95–103.

38 Pine, W., Hightower, J., Coggins, L. et al. (2012). Design and analysis of tagging studies. In: *Fisheries Techniques*, 3ee (ed. A.V. Zale, D.L. Parrish, and T.M. Sutton), 521–572. Bethesda, MD: American Fisheries Society.

39 MacLeod, I. and Gagen, C. (2018). New applications of radio frequency identification stations for monitoring fish passage through headwater road crossings and natural reaches. *Journal of the Arkansas Academy of Science* 72: 109.

40 Saboret, G., Dermond, P., and Brodersen, J. (2021). Using PIT-tags and portable antennas for quantification of fish movement and survival in streams under different environmental conditions. *Journal of Fish Biology* 99 (2): 581–595.

41 Zentner, D.L., Wolf, S.L., Brewer, S.K., and Shoup, D.E. (2021). A review of factors affecting PIT tag detection using mobile arrays and use of mobile antennas to detect PIT-tagged suckers in a wadeable ozark stream. *North American Journal of Fisheries Management* 41 (3): 697–710.

42 Cooke, S.J., Woodley, C.M., Brad Eppard, M. et al. (2011). Advancing the surgical implantation of electronic tags in fish: a gap analysis and research agenda based on a review of trends in intracoelomic tagging effects studies. *Reviews in Fish Biology and Fisheries* 21 (1): 127–151.

43 Fischer, E.S., Blackburn, S.E., Liss, S.A. et al. (2019). How small can we go? Evaluating survival, tag retention, and growth of juvenile chinook salmon implanted with a new acoustic microtag. *North American Journal of Fisheries Management* 39 (6): 1329–1336.

44 Schwarz, C. and Stobo, W. (2011). Estimation and effects of tag-misread rates in capture-recapture studies. *Canadian Journal of Fisheries and Aquatic Sciences* 56: 551–559.

45 Stout, J.B., Conner, M., Budy, P. et al. (2020). Keeping it classy: classification of live fish and ghost PIT tags detected with a mobile PIT tag interrogation system using an innovative analytical approach. *Canadian Journal of Fisheries and Aquatic Sciences* 77 (9): 1564–1573.

46 Moore, A. (1992). Technical notes: passive integrated transponder tagging of channel catfish. *Progressive Fish-Culturist* 54 (2): 125–127.

47 Onders, R.J., Mims, S.D., Wang, C., and Pearson, W.D. (2001). Reservoir ranching of paddlefish. *North American Journal of Aquaculture* 63 (3): 179–190.

48 Stobo, W.T. and Horne, J.K. (1994). Tag loss in grey seals (*Halichoerus grypus*) and potential effects on population estimates. *Canadian Journal of Zoology* 72 (3): 555–561.

49 Bruyndoncx, L., Knaepkens, G., Meeus, W. et al. (2005). The evaluation of passive integrated transponder (PIT) tags and visible implant elastomer (VIE) marks as new marking techniques for the bullhead. *Journal of Fish Biology* 60: 260–262.

50 Clark, S. (2016). Effects of passive integrated transponder tags on the physiology and swimming performance of a small-bodied stream fish. *Transactions of the American Fisheries Society* 145: 1179–1192.

51 Buzby, K. and Deegan, L. (1999). Retention of anchor and passive integrated transponder tags by arctic grayling. *North American Journal of Fisheries Management* 19: 1147–1150.

52 Hamel, M., Hammen, J., and Pegg, M. (2012). Tag retention of T-bar anchor tags and passive integrated transponder tags in Shovelnose sturgeon. *North American Journal of Fisheries Management* 32: 533–538.

53 O'Donnell, M.J. and Letcher, B.H. (2017). Implanting 8-mm passive integrated transponder tags into small brook trout: effects on growth and survival in the laboratory. *North American Journal of Fisheries Management* 37 (3): 605–611.

54 Kimball, M.E. and Mace, M.M. (2020). Survival, growth, and tag retention in estuarine fishes implanted with passive integrated transponder (PIT) tags. *Estuaries and Coasts* 43 (1): 151–160.

55 Vollset, K.W., Lennox, R.J., Thorstad, E.B. et al. (2020). Systematic review and meta-analysis of PIT tagging effects on mortality and growth of juvenile salmonids. *Reviews in Fish Biology and Fisheries* 30 (4): 553–568.

56 Bassar, R.D., Letcher, B.H., Nislow, K.H., and Whiteley, A.R. (2016). Changes in seasonal climate outpace compensatory density-dependence in eastern brook trout. *Global Change Biology* 22 (2): 577–593.

57 Kanno, Y., Pregler, K., Hitt, N.(T.) et al. (2015). Seasonal temperature and precipitation regulate brook trout young-of-the-year abundance and population dynamics. *Freshwater Biology* 61: 88–99.

58 Letcher, B.H., Schueller, P., Bassar, R.D. et al. (2015). Robust estimates of environmental effects on population vital rates: an integrated capture–recapture model of seasonal brook trout growth, survival and movement in a stream network. *The Journal of Animal Ecology* 84 (2): 337–352.

59 Vincenzi, S., Mangel, M., Jesenšek, D. et al. (2016). Within- and among-population variation in vital rates and population dynamics in a variable environment. *Ecological Applications* 26 (7): 2086–2102.

60 Jepsen N. (2005). A brief discussion of the 2% tag/bodymass rule. Aquatic Telemetry: Advances and Applications, *Proceedings of the Fifth Conference on Fish Telemetry*, 255–259.

61 Tiffan, K., Perry, R., Connor, W. et al. (2015). Survival, growth, and tag retention in age-0 chinook salmon implanted with 8-, 9-, and 12-mm PIT tags. *North American Journal of Fisheries Management* 354: 845–852.

62 Tiffan, K.F., Jezorek, I.G., and Perry, R.W. (2019). A field evaluation of the growth and survival of age-0 *Oncorhynchus mykiss* tagged with 8-mm passive integrated transponder (PIT) tags. *Animal Biotelemetry* 7 (1): 9.

63 Bond, R.M., Nicol, C.L., Kiernan, J.D., and Spence, B.C. (2019). Occurrence, fate, and confounding influence of ghost passive integrated transponder tags in an intensively monitored watershed. *Canadian Journal of Fisheries and Aquatic Sciences* 76 (2): 286–298.

64 O'Donnell, M., Horton, G., and Letcher, B. (2010). Use of portable antennas to estimate abundance of PIT-tagged fish in small streams: factors affecting detection probability. *North American Journal of Fisheries Management* 30: 323–336.

65 Fetherman, E., Avila, B., and Winkelman, D. (2014). Raft and floating radio frequency identification (RFID) antenna systems for detecting and estimating abundance of PIT-tagged fish in rivers. *North American Journal of Fisheries Management* 34: 1065–1077.

66 Sloat, M.R., Baker, P.F., and Ligon, F.K. (2011). Estimating habitat-specific abundances of PIT-tagged juvenile salmonids using mobile antennas: a comparison with standard electrofishing techniques in a small stream. *North American Journal of Fisheries Management* 31 (5): 986–993.

67 Šmejkal, M., Bartoň, D., Děd, V. et al. (2020). Negative feedback concept in tagging: ghost tags imperil the long-term monitoring of fishes. *PLoS One* 15 (3): e0229350.

68 Jørgensen, T., Ovredal, J.T., and Huse, I. (2005). A new pass-through PIT tag detection system for marine use. *FI, Conference on Fish Telemetry in Europe S 5.* FAO/COISPA, Ustica, Italy (9–13 June 2003).

69 Cooke, S.J., Midwood, J.D., Thiem, J.D. et al. (2013). Tracking animals in freshwater with electronic tags: past, present and future. *Animal Biotelemetry* 1 (1): 5.

70 Ruiz-Villanueva, V., Piégay, H., Gurnell, A.M. et al. (2016). Recent advances quantifying the large wood dynamics in river basins: new methods and remaining challenges. *Reviews of Geophysics* 54 (3): 611–652.

71 MacVicar, B.J., Piégay, H., Henderson, A. et al. (2009). Quantifying the temporal dynamics of wood in large rivers: field trials of wood surveying, dating, tracking, and monitoring techniques. *Earth Surface Processes and Landforms* 34 (15): 2031–2046.

72 Schenk, E.R., Moulin, B., Hupp, C.R., and Richter, J.M. (2014). Large wood budget and transport dynamics on a large river using radio telemetry. *Earth Surface Processes and Landforms* 39 (4): 487–498.

73 Ravazzolo, D., Mao, L., Picco, L., and Lenzi, M.A. (2015). Tracking log displacement during floods in the Tagliamento River using RFID and GPS tracker devices. *Geomorphology* 228: 226–233.

74 Hassan, M.A. and Roy, A.G. (2016). Editor(s):G. Mathias Kondolf, Hervé Piégay. Coarse particle tracing in fluvial geomorphology. In: *Tools in Fluvial Geomorphology*, 306–323. Wiley-Blackwell http://onlinelibrary.wiley.com/doi/abs/10.1002/9781118648551.ch14.

75 Le Breton, M., Liébault, F., Baillet, L. et al. (2022). Dense and long-term monitoring of earth surface processes with passive RFID — a review. *Earth Science Reviews* 234: 104225.

76 Lamarre, H., MacVicar, B., and Roy, A.G. (2005). Using passive integrated transponder (PIT) tags to investigate sediment transport in gravel-bed rivers. *Journal of Sedimentary Research* 75 (4): 736–741.

77 Nichols, M.H. (2004). A radio frequency identification system for monitoring coarse sediment particle displacement. *Applied Engineering in Agriculture* 20 (6): 783–787.

78 Piégay, H., Arnaud, F., Grégoire, N., et al. (2016). Caractérisation physique et thermique des habitats aquatiques de l'Ain. EVS – UMR 5600; IRSTEA; Ecole des Mines de Saint Etienne; Laboratoire de Géologie de Lyon. p. 66. https://hal.archives-ouvertes.fr/hal-01313825.

79 Chapuis, M., Bright, C.J., Hufnagel, J., and MacVicar, B. (2014). Detection ranges and uncertainty of passive Radio Frequency Identification (RFID) transponders for sediment tracking in gravel rivers and coastal environments: detection ranges and uncertainty of passive RFID transponders. *Earth Surface Processes and Landforms* 39 (15): 2109–2120.

80 Arnaud, F., Piégay, H., Vaudor, L. et al. (2015). Technical specifications of low-frequency radio identification bedload tracking from field experiments: differences in antennas, tags and operators. *Geomorphology* 238: 37–46.

81 Cassel, M., Piégay, H., and Lavé, J. (2017). Effects of transport and insertion of radio frequency identification (RFID) transponders on resistance and shape of natural and synthetic pebbles: applications for riverine and coastal bedload tracking. *Earth Surface Processes and Landforms* 42 (3): 399–413.

82 Cassel, M., Dépret, T., and Piégay, H. (2017). Assessment of a new solution for tracking pebbles in rivers based on active RFID. *Earth Surface Processes and Landforms* 42 (13): 1938–1951.

83 Brousse, G., Arnaud-Fassetta, G., Liébault, F. et al. (2020). Channel response to sediment replenishment in a large gravel-bed river: the case of the Saint-Sauveur dam in the Buëch River (Southern Alps, France). *River Research and Applications* 36 (6): 880–893.

84 Misset, C., Recking, A., Legout, C. et al. (2020). Combining multi-physical measurements to quantify bedload transport and morphodynamics interactions in an Alpine braiding river reach. *Geomorphology* 351: 106877.

85 Boutault, F. (2020), Etude de l'impact cumulé des facteurs d'anthropisation sur la Dordogne moyenne et préconisations en vue d'une restauration écologique du cours d'eau [These de doctorat]. Lyon. https://www.theses.fr/2020LYSE3066.

86 Cassel, M., Piégay, H., Fantino, G. et al. (2020). Comparison of ground-based and UAV aUHF artificial tracer mobility monitoring methods on a braided river. *Earth Surface Processes and Landforms* 45 (5): 1123–1140.

87 Haschenburger, J. and Church, M. (1998). Bed material transport estimated from the virtual velocity of sediment. *Earth Surface Processes and Landforms* 23: 791–808.

88 Liébault, F. and Laronne, J.B. (2008). Evaluation of bedload yield in gravel-bed rivers using scour chains and painted tracers: the case of the Esconavette Torrent (Southern French Prealps). *Geodinamica Acta* 21 (1–2): 23–34.

89 Houbrechts, G., Campenhout, J.V., Levecq, Y. et al. (2012). Comparison of methods for quantifying active layer dynamics and bedload discharge in armoured gravel-bed rivers. *Earth Surface Processes and Landforms* 37 (14): 1501–1517.

90 Brousse, G., Liébault, F., Arnaud-Fassetta, G. et al. (2020). Gravel replenishment and active-channel widening for braided-river restoration: The case of the Upper Drac River (France). *Science of the Total Environment* 766: 142517.

91 Papangelakis, E., Muirhead, C., Schneider, A., and Macvicar, B. (2019). Synthetic radio frequency identification tracer stones with weighted inner ball for burial depth estimation. *Journal of Hydraulic Engineering* 145 (12): 06019014.

92 Cain, A., Iannetta, M., Muirhead, C. et al. (2020). Tracking the mean annual velocity and vertical mixing of bedload tracers. In: *River Flow 2020*, 1ee, 10pp. CRC Press. https://www.taylorfrancis.com/chapters/edit/10.1201/b22619-53/tracking-mean-annual-velocity-vertical-mixing-bedload-tracers-cain-iannetta-muirhead-papangelakis-raso-macvicar-ashmore.

93 Habersack, H.M. (2003). Use of radio-tracking techniques in bed load transport investigations. In: *Erosion and Sediment Transport Measurement in Rivers: Technological and Methodological Advances*, vol. 283, 172–180. IAHS Publication. (Proceedings of the Oslo Workshop, June 2002).

94 Froude, M.J. and Petley, D.N. (2018). Global fatal landslide occurrence from 2004 to 2016. *Natural Hazards and Earth System Sciences* 18 (8): 2161–2181.

95 Marano, K.D., Wald, D.J., and Allen, T.I. (2010). Global earthquake casualties due to secondary effects: a quantitative analysis for improving rapid loss analyses. *Natural Hazards* 52 (2): 319–328.

96 Alimohammadlou, Y., Najafi, A., and Yalcin, A. (2013). Landslide process and impacts: a proposed classification method. *Catena* 104: 219–232.

97 Gariano, S.L. and Guzzetti, F. (2016). Landslides in a changing climate. *Earth Science Reviews* (162): 227–252.

98 Nikitin, P.V., Martinez, R., Ramamurthy, S., et al. (2010) Phase based spatial identification of UHF RFID tags. *IEEE International Conference on RFID*, Orlando, FL, USA. IEEE, 102–109. http://ieeexplore.ieee.org/xpls/abs_all.jsp?arnumber=5467253.

99 Charléty, A., Le Breton, M., Larose, E., Baillet, L., 2022. 2D Phase-Based RFID Localization for On-Site Landslide Monitoring. Remote Sensing 14, 3577. https://doi.org/10.3390/rs14153577

100 Charléty, A., Le Breton, M., Baillet, L., Larose, E., 2023. RFID Landslide Monitoring: Long-Term Outdoor Signal Processing and Phase Unwrapping. IEEE Journal of Radio Frequency Identification 7, 319–329. https://doi.org/10.1109/JRFID.2023.3256560.

101 Le Breton, M., Grunbaum, N., Baillet, L, and Larose, É. (2021). Monitoring rock displacement threshold with 1-bit sensing passive RFID tag. *Copernicus Meetings*. Report No.: EGU21-15305. https://meetingorganizer.copernicus.org/EGU21/EGU21-15305.html EGU21-15305, updated on 04 Mar 2021 https://doi.org/10.5194/egusphere-egu21-15305.

102 Longhi, M. and Marrocco, G. (2017). Ubiquitous flying sensor antennas: radiofrequency identification meets micro drones. *IEEE Journal of Radio Frequency Identification* 1 (4): 291–299.

103 Mulla, D. and Khosla, R. (2016). Historical evolution and recent advances in precision farming. *Soil-Specific Farming Precision Agriculture [Internet].* 10.1201/b18759-2.

104 Pinter Jr Paul, J., Hatfield, J.L., Schepers, J.S. et al. (2003). Remote sensing for crop management. *Photogrammetric Engineering and Remote Sensing* 69 (6): 647–664.

105 Korres, W., Koyama, C.N., Fiener, P., and Schneider, K. (2010). Analysis of surface soil moisture patterns in agricultural landscapes using Empirical Orthogonal Functions. *Hydrology and Earth System Sciences* 14 (5): 751–764.

106 Jaynes, D.B., Kaspar, T.C., Colvin, T.S., and James, D.E. (2003). Cluster analysis of spatiotemporal corn yield patterns in an iowa field. *Agronomy Journal* 95 (3): 574–586.

107 Touseau, L. and Le Sommer, N. (2019). Contribution of the web of things and of the opportunistic computing to the smart agriculture: a practical experiment. *Future Internet* 11 (2): 33.

108 Hunt, E.R. and Daughtry, C.S.T. (2018). What good are unmanned aircraft systems for agricultural remote sensing and precision agriculture? *International Journal of Remote Sensing* 39 (15–16): 5345–5376.

109 Bhattacharyya, R., Floerkemeier, C., and Sarma, S. (2010). Low-cost, ubiquitous RFID-tag-antenna-based sensing. *Proceedings of the IEEE* 98 (9): 1593–1600.

110 Luvisi, A., Panattoni, A., and Materazzi, A. (2016;123). RFID temperature sensors for monitoring soil solarization with biodegradable films. *Computers and Electronics in Agriculture* 135–141.

111 Palazzi, V., Gelati, F., Vaglioni, U. et al. (2019). Leaf-compatible autonomous RFID-based wireless temperature sensors for precision agriculture. *2019 IEEE Topical Conference on Wireless Sensors and Sensor Networks (WiSNet)*, Orlando, FL, USA (20–23 January 2019), 1–4.

112 Kim, S., Le, T., Tentzeris, M.M. et al. (2014) An RFID-enabled inkjet-printed soil moisture sensor on paper for 'smart' agricultural applications. *IEEE SENSORS Proceedings*, Valencia, Spain, (2–5 November 2014), 1507–1510.

113 Wang, J., Chang, L., Aggarwal, S. et al. (2020). Soil moisture sensing with commodity RFID systems. *Proceedings of the 18th International Conference on Mobile Systems, Applications, and Services*. Toronto Ontario Canada: ACM, 273–285. https://dl.acm.org/doi/10.1145/3386901.3388940.

114 Hasan, A., Bhattacharyya, R., and Sarma, S. (2015). Towards pervasive soil moisture sensing using RFID tag antenna-based sensors. *IEEE International Conference RFID Technologies and Applications*, Tokyo, Japan (16–18 September 2015), 165–170.

115 Dey, S., Karmakar, N., Bhattacharyya, R., and Sarma, S. (2016). Electromagnetic characterization of soil moisture and salinity for UHF RFID applications in precision agriculture. *2016 46th European Microwave Conference (EuMC)*, London, United Kingdom: IEEE, 616–619. http://ieeexplore.ieee.org/document/7824418/.

116 Fonseca, N., Freire, R., Fontgalland, G. et al. (2018). A fully passive UHF RFID soil moisture time-domain transmissometry based sensor. *2018 3rd International Symposium on Instrumentation Systems, Circuits and Transducers (INSCIT)*, Bento Goncalves, Brazil (27–31 August 2018), 1–6.

117 Wang, C., George, D., and Green, P.R. (2014). Development of plough-able RFID sensor network systems for precision agriculture. *2014 IEEE Topical Conference on Wireless Sensors and Sensor Networks (WiSNet)*, Newport Beach, California, USA (19–22 January 2014), 64–66.

118 Kantareddy, S.N.R., Bhattacharyya, R., and Sarma, S. (2018). UHF RFID tag IC power mode switching for wireless sensing of resistive and electrochemical transduction modalities. *2018 IEEE International Conference on RFID (RFID)*. Orlando, FL: IEEE, 1–8. https://ieeexplore.ieee.org/document/8376201/.

119 American Society of Civil Engineers (2017). *A comprehensive assessment of America's infrastructure: 2017 Report Card for America's Infrastructure*, 20. American Society of Civil Engineers https://www.infrastructurereportcard.org/wp-content/uploads/2016/10/2017-Infrastructure-Report-Card.pdf.

120 Alavi, A.H. and Buttlar, W.G. (2019). An overview of smartphone technology for citizen-centered, real-time and scalable civil infrastructure monitoring. *Future Generation Computer Systems* 93: 651–672.

121 Wang, Y., Lynch, J.P., and Law, K.H. (2007). A wireless structural health monitoring system with multithreaded sensing devices: design and validation. *Structure and Infrastructure Engineering* 3 (2): 103–120.

122 Spencer, B.F., Park, J.W., Mechitov, K.A. et al. (2017). Next generation wireless smart sensors toward sustainable civil infrastructure. *Procedia Engineering* 171: 5–13.

123 Yang, L., Fu, C., Li, Y., and Su, L. (2019). Survey and study on intelligent monitoring and health management for large civil structure. *International Journal of Intelligent Robotics and Applications* 3 (3): 239–254.

124 Shakhatreh, H., Sawalmeh, A.H., Al-Fuqaha, A. et al. (2019). Unmanned Aerial Vehicles (UAVs): a survey on civil applications and key research challenges. *IEEE Access* 7: 48572–48634.

125 Valero, E., Adán, A., and Cerrada, C. (2015). Evolution of RFID applications in construction: a literature review. *Sensors* 15 (7): 15988–16008.

126 Duan, K.K. and Cao, S.Y. (2020). Emerging RFID technology in structural engineering – a review. *Structure* 28: 2404–2414.

127 Jayawardana, D., Liyanapathirana, R., and Zhu, X. (2019). RFID-based wireless multi-sensory system for simultaneous dynamic acceleration and strain measurements of civil infrastructure. *IEEE Sensors Journal* 19 (24): 12389–12397.

128 Strangfeld, C., Johann, S., Müller, M., and Bartholmai, M. (2017). Embedded passive RFID-based sensors for moisture monitoring in concrete. *2017 IEEE Sensors*, Glasgow, Scotland, UK (29 October–1 November 2017), 1–3.

129 Bhattacharyya, R., Floerkemeier, C., and Sarma, S. (2009). Towards tag antenna based sensing – An RFID displacement sensor. *2009 IEEE International Conference on RFID*, Orlando, FL, USA (27–28 April 2009), 95–102.

130 Bouzaffour, K., Lescop, B., Talbot, P. et al. (2021). Development of an embedded UHF-RFID corrosion sensor for monitoring corrosion of steel in concrete. *IEEE Sensors Journal* 21: 12306–123121.

131 Kalansuriya, P., Bhattacharyya, R., and Sarma, S. (2013). RFID tag antenna-based sensing for pervasive surface crack detection. *IEEE Sensors Journal* 13 (5): 1564–1570.

132 Bhattacharyya, R., Kalansuriya, P., and Sarma, S. (2013). An antenna-based RFID expansion joint monitor. *2013 IEEE Antennas and Propagation Society International Symposium (APSURSI)*, 2305–2036.

133 Halliday, S. (2022). Of Hamburgers . . . and RAIN RFID IC Tags. RAIN RFID Alliance, Steve Halliday, President of RAIN Alliance. https://rainrfid.org/blog/of-hamburgers-and-rain-rfid-ic-tags/.

134 Precedence Research (2022). UAV drones market size to worth around US$ 102.38 Bn by 2030. GlobeNewswire News Room. https://www.globenewswire.com/news-release/2022/03/01/2394915/0/en/UAV-Drones-Market-Size-to-Worth-Around-US-102-38-Bn-by-2030.html.

135 Tzeng, S.F., Chen, W.H., and Pai, F.Y. (2008). Evaluating the business value of RFID: evidence from five case studies. *International Journal of Production Economics* 112 (2): 601–613.

5

Harmonic Transponders for Tracking and Sensing

Valentina Palazzi

Department of Engineering, University of Perugia, Perugia, Italy

5.1 Introduction

Wireless sensors are the backbone of Internet of Things (IoT) solutions and provide support to many 5G applications [1]. In applications such as precision farming and industrial IoT, where the access to the grid can be difficult and large areas need to be monitored, wireless solutions make real-time data concerning key parameters (such as position, temperature, vibration, and so forth) available for centralized decision-making algorithms with limited installation costs. Wireless sensors can be also used to monitor mobile platforms and mobile equipment, where they enable essential condition monitoring and preventive maintenance services [2].

However, outdoor and industrial environments are particularly challenging for electronic devices. First of all, the high temperatures and humidity of these environments can challenge the reliability of electronics. This is especially true for active components, which can be seriously damaged by harsh environmental conditions [3]. Second, industrial environments are characterized by the presence of severe reflections due to metal equipment and the sensor placement close to the ground (or even below it). These strong reflections can deteriorate wireless links and can ultimately make the receiver blind to the sensor data. Additionally, sensors can be inaccessible or placed in remote areas, which is a problem for battery-operated devices. The latter aspect, in particular, is a bottleneck for a pervasive diffusion of wireless sensors. Indeed, wireless sensor devices must be able to acquire the sensor data and transmit this information

Wireless Identification and Sensing Systems for Harsh and Severe Environments,
First Edition. Edited by Smail Tedjini and Valentina Palazzi.

Published 2025 by John Wiley & Sons, Inc.

wirelessly, which means that they are usually equipped with an active circuitry to perform signal conditioning and to process and transmit data. Whenever possible, batteries are replaced by energy-harvesting solutions, used to either prolong the lifetime of batteries or to fully replace batteries [4]. However, alternative energy sources are variable and unpredictable, and they are not always available.

Among others, harmonic backscattering has been analyzed as a possible technological solution to develop passive wireless sensing in harsh and severe environments.

Harmonic transponders are a special class of RFID devices, which base their operation on the harmonic generation [5]. According to this approach, harmonic transponders are interrogated by a continuous-wave (CW) sinusoidal signal at a given frequency, f_0. The transponders perform a frequency conversion of the interrogating RF signal and backscatter the signal at higher harmonic frequencies. Their basic block diagram consists of a harmonic antenna system and a nonlinearity (usually a diode or a frequency doubler). Due to their simple operation, these transponders do not require any additional power supply and can be very compact and robust [6]. Additionally, since the tag response occurs at a frequency which is far from the interrogating frequency, harmonic systems are intrinsically insensitive to the transmitter phase noise and to clutter, which makes them perfect candidates to operate in harsh and remote areas. Originally proposed for automotive applications (under the name of harmonic radar) [7], in the last couple of decades, harmonic transponders have found application in a plethora of fields, ranging from insect tracking to victim rescue and electronic article surveillance [8]. Indeed, the capability of harmonic systems to clearly separate the tag response from unwanted reflections caused by the environment is particularly favorable for localization and tracking applications. Among others, harmonic tags have proven to be especially beneficial for buried asset localization, where traditional RFID systems are impaired by the strong reflections due to ground proximity.

Recently, research has focused on integrating harmonic transponders with various transducer types (for instance, temperature, crack, and humidity transducers), so as to realize zero-power wireless harmonic transponders suitable for sensing. Adding sensing capabilities to these devices is particularly challenging since the adopted transducers must be able to encode the sensing information in the backscatter signal by only leveraging the interrogating radio-frequency (RF) signal, which is both used as power source and as a carrier for communication. Basic harmonic transponders transmit one bit of information (presence/absence of the transponders). Therefore, a hot research topic consists of investigating solutions to encode multi-bit information in harmonic transponders without relying on external energy sources.

This chapter is aimed at giving a comprehensive overview of the state of the art for harmonic transponders. The first part will be dedicated to providing the reader with the theoretical background of their operating principle and the main features of their basic building blocks. Then, examples of harmonic systems for tracking and sensing will be reported. Particular attention will be paid to describing their performance in harsh and severe environments, and their features will be compared to more traditional RFID technologies.

5.2 Harmonic Backscattering

The basic architecture of a wireless system based on harmonic backscattering is illustrated in Figure 5.1 [5]. The harmonic reader consists of a transmitter, which generates and transmits a signal at the fundamental frequency, f_0, and a receiver, tuned to the second harmonic $2f_0$. The signal detected by the harmonic reader at $2f_0$ is generated by harmonic tags only, and eventual reflections from the environment do not impact it, which justifies the high immunity to clutter of harmonic systems. The harmonic tag acquires the interrogating signal, converts it to the second harmonic, and transmits back the converted signal. It consists of a harmonic antenna system (either composed of two separate antennas or by a single dual-frequency antenna) and a passive frequency doubler ("× 2" block in Figure 5.1).

Therefore, the performance of the tag depends on the antenna system, on the nonlinearity, and on the impedance matching between them.

The frequency doubler is a critical building block since its conversion loss C_l, defined as the ratio of the available RF input power at f_0 to the power delivered to the load at $2f_0$, is usually high (generally larger than 10 dB for

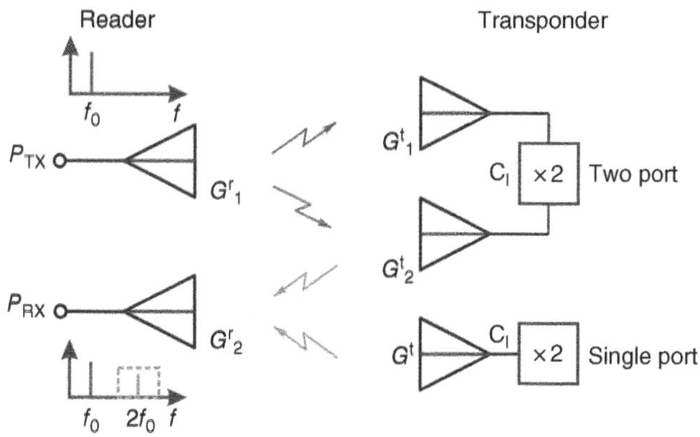

Figure 5.1 Basic architecture of a wireless system based on harmonic backscattering.

passive frequency doublers), and significantly affects the read range of the system.

Assuming the signals propagate in free space, the power received by the reader at $2f_0$, P_{RX} can be estimated as follows:

$$P_{RX} = \frac{1}{4} \frac{P_{TX} G_1^r G_1^t G_2^t G_2^r}{C_1(P_{in}) L_{sys}} \left(\frac{\lambda_0}{4\pi d} \right)^4 \tag{5.1}$$

where P_{TX} is the transmitted power at f_0, d is the tag-to-reader distance, G_i^j the specific antenna gain (i = 1 fundamental, i = 2 2nd harmonic, j = t tag, j = r reader), C_1 the overall frequency conversion loss of the tag, and L_{sys} includes all additional losses of the system (such as polarization loss, insertion loss, and so forth). Generally, the conversion loss C_1 is a function of the available input power P_{in}. Therefore, first, the available input power is estimated using the link budget analysis, the corresponding C_1 is determined, and then P_{RX} is estimated using (5.1).

Unlike UHF RFID systems, where the read range of the tag is generally limited by the minimum power required to switch on the chip, the read range of harmonic systems is mostly limited by the sensitivity of the receiver, that is, by its capability to detect low-power signals. For similar operating conditions (i.e. fundamental frequency, transmitted power, distance, and antenna gains), the power received by the reader of harmonic systems is generally lower than that of UHF RFID systems. This is due to the high conversion loss in the harmonic generation and the higher path loss experienced by the backscattered higher-order harmonic components with respect to the fundamental tone. On the other hand, the harmonic receiver is not affected by the self-jamming problems of UHF RFID readers, and the read range is not limited by the minimum power required to switch to the UHF RFID chip. The high immunity to clutter makes it possible to have receivers with a higher sensitivity than UHF RFID reader since the receiver noise floor is set by thermal noise rather than the phase noise of strong leakage signals [9].

Depending on the application, stringent geometric requirements can be imposed on harmonic transponders. For instance, in insect tracking applications, small, conformal and lightweight transponders are needed not to alter the insect behavior. To achieve compact transponders, dual-band single antennas are generally used. The harmonic tags used for insect tracking are composed of just a dipole connected to a diode with a tuning inductor. Fractal antennas have been also explored, to generate compact antenna layouts [10]. In Ref. [11], two nested annular slot antennas are used, so that the whole transponder area corresponds to the area of the slot working at f_0.

5.3 Frequency Doubler for Harmonic Transponders

The device responsible for the harmonic frequency generation can be a one-port [12] or a two-port network [11]. As already mentioned, the C_l of the tags varies with the available RF input power. Therefore, the design of the zero-bias frequency doublers must be optimized for low available input powers to maximize the read range of harmonic systems [12].

The reported one-port frequency doublers are based on a single-diode impedance matched to the dual-frequency antenna. A lumped inductor is connected in parallel with the diode to both compensate for the capacitive impedance contribution of the diode and to provide a DC return for the self-generated DC component. A circuit schematic is shown in Figure 5.2. R_r is the radiation resistance of the antenna, R_l describes the loss, and X_a is the antenna reactance. The diode is described using the Shockley model valid for small signals, where R_s is the series resistance, R_j is the junction resistance, and C_{j0} is the junction capacitance of the diode. L_s and C_p are the parasitics of the diode package.

Studies are performed to identify the diode parameters, which minimize the conversion loss for low available input powers. In Ref. [12], a theoretical study is performed, which demonstrates that at high frequencies ($f_0 > 1\,\text{GHz}$) the capacitive nonlinearity of the diode (either varactor or detector diode) is dominant with respect to the resistive nonlinearity. A proof-of-concept prototype is manufactured and tested, based on a varactor diode (SMV2019 model from Skyworks) and a folded dipole antenna. The prototype is shown in Figure 5.3. The antenna is designed to be matched to the diode impedance both at $f_0 = 1\,\text{GHz}$ and at $2f_0 = 2\,\text{GHz}$. The tag,

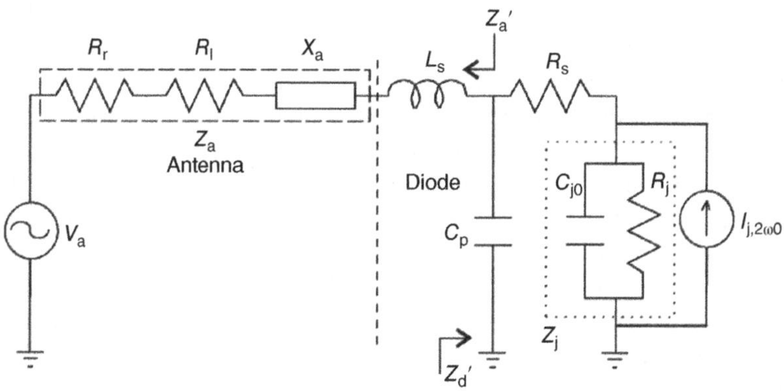

Figure 5.2 Circuit schematic of a passive one-port frequency doubler. *Source:* Rasilainen et al. [12]/with permission of IEEE.

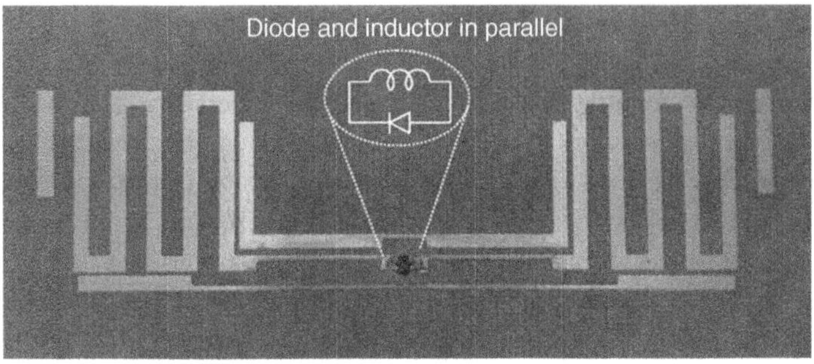

Figure 5.3 Prototype of a harmonic tag based on a one-port frequency doubler [12].

tested in an anechoic environment, features a conversion loss of about 20 dB at 5 m distance from the reader for a transmitted power of 20.4 dBm EIRP.

This approach leads to very compact layouts, although the performance of the frequency doubler can be only estimated from the performance of the overall harmonic transponder. In fact, the frequency doubler is not separately measured, so conversion loss estimations are based on link budget analysis involving the whole transponder, and the impact of impedance mismatch and other nonidealities cannot be easily identified. On the other hand, with this approach, losses can be minimized avoiding the presence of input and output matching networks.

Two-port frequency doublers are generally based on a series-connected low-barrier Schottky diode (model HSMS-2850) and harmonic filters [5]. The schematic of a two-port frequency doubler is shown in Figure 5.4. Schottky diodes, which are commonly used in frequency multipliers [13], are chosen for their capability to operate without any bias signals and for small available RF input powers. Two quarter-wave short-circuited and open-circuited stubs, connected in parallel at the two sides of the diode, operate as harmonic filters. The short-circuited stub connected at the anode of the diode is an open circuit for the f_0 tone and a short circuit for the $2f_0$ tone, while the open-circuited stub connected at the cathode of the diode is a short circuit for the f_0 tone and an open circuit for the $2f_0$ tone. This way, all the voltage drop of the input tone at f_0 occurs across the diode (i.e. it is used to activate the diode). The $2f_0$ tone generated by the diode, instead, can flow through the output load, while the dynamic ground at the diode anode prevents this component from flowing toward the source. The input matching network is used to match the large-signal input impedance of the circuit to the source impedance at the specific design input power and frequency of the doubler. The output matching network, consisting of an LC network (L_m and C_m in Figure 5.4), is used to transform the load impedance of the doubler into the optimum impedance that minimizes the conversion loss of the circuit. The LC network also plays the role of a bias tee (i.e. the self-generated DC component of

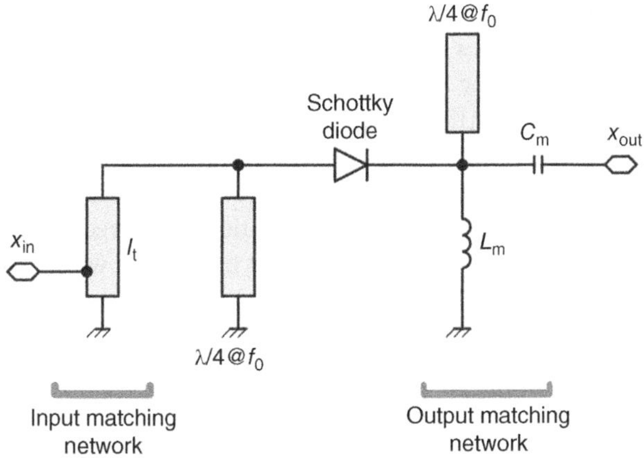

Figure 5.4 Circuit schematic of a passive two-port frequency doubler. *Source:* Palazzi et al. [5]/with permission of IEEE.

the diode has a return through the inductor L_m). This approach makes it possible to arbitrarily set the input and output impedance of the doubler, thereby making it easy to connect the frequency doubler with antennas, additional RF circuits and eventual sensing components.

An example of a frequency doubler proof-of-concept prototype is shown in Figure 5.5. The doubler is manufactured on a flexible paper substrate using copper adhesive laminate and photolithography, as described in Ref. [14]. Input and output matching networks are designed to make the frequency doubler optimally operate with input and output impedances of 50 Ω. Two surface-mount lumped components are used, namely the low-barrier Schottky diode HSMS-2850 and a ceramic capacitor. The operating frequency is set to $f_0 = 1.04$ GHz. A photo of the manufactured prototype is shown in Figure 5.5a. The conversion loss of the

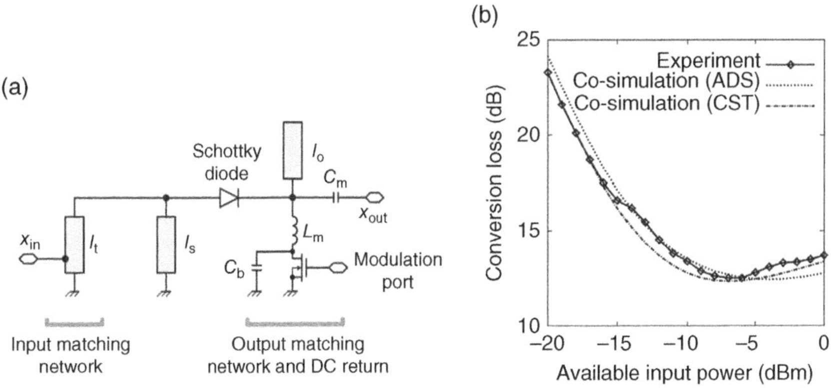

Figure 5.5 Prototype of a two-port frequency doubler. (a) Photo of the prototype and (b) conversion loss. *Source:* Palazzi et al, [5]/IEEE.

frequency doubler versus its available input power is shown in Figure 5.5b. The doubler shows a minimum conversion loss of 13 dB for an available input power of −7 dBm. For $P_{\text{in}} < -10$ dBm, the conversion loss increases linearly by about 10 dB/decade (the diode operates in the so-called linear region, where the quadratic nonlinearity dominates).

In Ref. [6], a variant is proposed, where the open-circuited quarter-wave stub is removed to reduce the area occupation of the frequency doubler. The LC network is designed to both provide a load impedance with a small resistive component (about 1 Ω) at f_0 and to provide a load impedance equal to the complex conjugate of the diode impedance at $2f_0$. To further reduce the number of lumped components used, the ceramic capacitor is replaced by an interdigital capacitor with a floating ground plane. A prototype designed for $f_0 = 1.2$ GHz is shown in Figure 5.6a. The circuit is manufactured on paper, and the diode model is HSMS-2850. The overall area occupation of the doubler is $2.8 \times 0.8\,\text{cm}^2$. The conversion loss of the doubler is shown in Figure 5.6b. The doubler is optimized for a lower input power than the previous case. Indeed, the minimum conversion loss of 15.7 dB is measured for $P_{\text{in}} = -13$ dBm.

Besides diode-based frequency multipliers, we would also like to mention nonlinear transmission-line-based frequency multipliers and transistor-based frequency multipliers. The latter ones are used to implement frequency doublers at millimeter-waves and beyond. However, they are usually impractical for harmonic transponders due to their high-power consumption.

Nonlinear transmission lines (NLTL) are transmission lines periodically loaded with nonlinear capacitances. The capacitances are usually implemented using junction diodes or varactor diodes [15]. NLTLs can be also modeled as the periodic combination of series inductors and shunt varactor diodes [16]. The conversion loss of NLTL-based harmonic transponders is generally higher than single

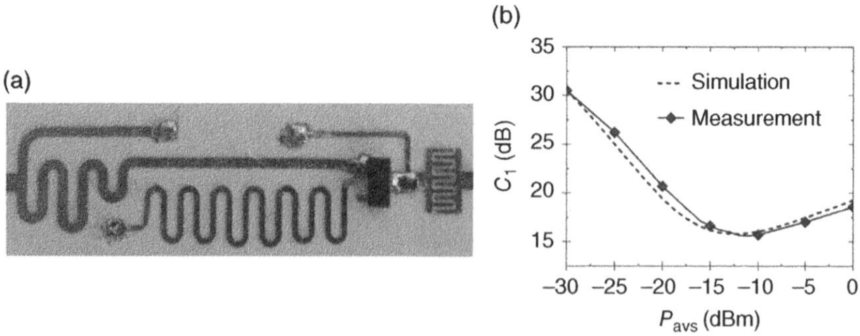

Figure 5.6 Prototype of a two-port frequency doubler without open-circuited stub. (a) Photo and (b) conversion loss. *Source:* Palazzi et al. [11]/IEEE.

diode-based frequency doublers for $P_{in} < -10$ dBm in absence of any DC bias signals. On the other hand, NLTLs are conveniently used to generate amplitude-modulated second harmonic signals and to implement broadband frequency doublers [9].

5.4 One-Bit Harmonic Transponders

Besides the example shown in Figure 5.3, based on a single-port frequency doubler and a dipole antenna, two approaches are here reported, describing harmonic tags based on two-port frequency doublers. In the first example, illustrated in Figure 5.7, the two-port frequency doubler is connected to two nested slot antennas [11]. The metal between the slot antennas is used as the ground plane for the frequency doubler. This is possible because the currents at the slot borders decrease exponentially and do not interfere with the doubler operation. The slots are excited through proximity-coupled microstrip feed lines. Circular stubs, placed at the end of each feed line, are used to match the antennas to 50 Ω. The two slots are orthogonal to reduce their mutual coupling. Tapered slot profiles are adopted to increase the impedance bandwidth of the antennas and improve the tag robustness to eventual detuning caused by the environment. The two antennas are linearly polarized, feature an almost omnidirectional radiation pattern (which is beneficial for the system coverage) and feature a measured gain of 3.3 and 3 dBi, respectively. The harmonic generation is performed with the frequency doubler illustrated in Figure 5.6. The whole tag is in paper substrate. The read range of the tag is measured in a laboratory

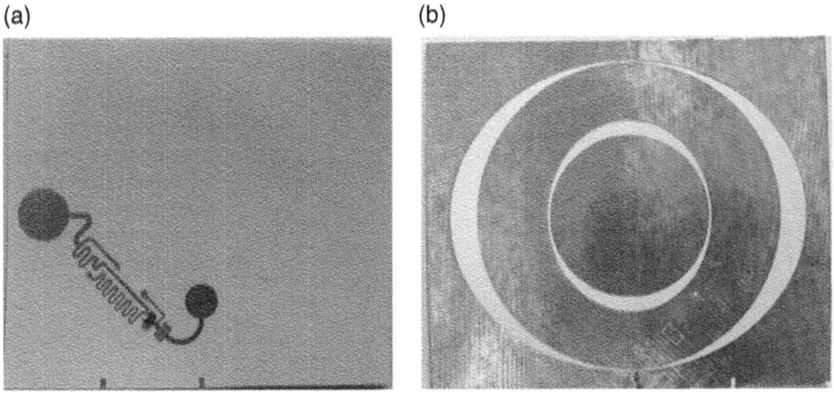

Figure 5.7 Prototype of a harmonic tag based on nested slot antennas. (a) Top side and (b) bottom side. *Source:* Palazzi et al. [11]/IEEE.

environment, utilizing a reader, consisting of a signal generator and a spectrum analyzer, equipped with circularly polarized antennas. A maximum read range of 4 m is observed for a transmitted power of 17 dm EIRP in good agreement with the link budget analysis based on Ref. [1]. The tag is flexible and lightweight (tag weight: 3 g).

In the second example, the frequency doubler is connected to a single dual-frequency spiral slot using a frequency duplexer [19]. The antenna consists of two spiral slots excited by the same proximity-coupled feedline. The antenna is circularly polarized and features a gain of 4 and 3.8 dBi at f_0 and $2f_0$, respectively. The diplexer is based on a U-shaped half-wave open-circuited microwave resonator at f_0 and the ports are capacitively coupled at specific locations of the resonator to maximize the coupling at the desired frequencies while suppressing unwanted signals. The frequency doubler, based on the schematic shown in Figure 5.8, is optimized for input power in the range [−25, −35] dBm. The tag is designed for $f_0 = 3.5\,\text{GHz}$, and the adopted substrate is Rogers RT/duroid 6002 with 20-mil thickness. The read range of the tag is tested in a laboratory environment, and a maximum read-out distance of 7 m is observed for a transmitted power of 41.7 dBm EIRP.

Table 5.1 shows the performance of harmonic transponders in the state of the art. The figure of merit labeled "tag C_1" is defined as

$$\text{tag } C_1 = \frac{C_1}{G_1^t G_2^t}. \tag{5.2}$$

This metric quantifies the capability of SoA harmonic transponder to effectively convert the incident RF power density to a second harmonic.

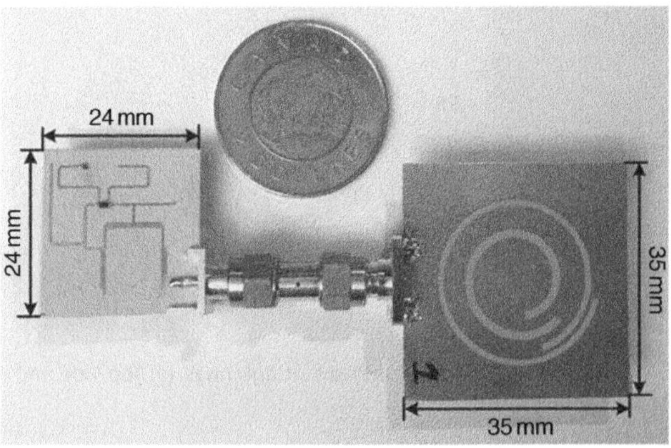

Figure 5.8 Prototype of a harmonic tag based on spiral slot antennas and a frequency diplexer. *Source:* Gu et al. [19]/IEEE.

Table 5.1 State of the art one-bit harmonic transponders.

Ref.	No. of antennas	Antenna type	f_0 (GHz)	Area/λ_0^2	Tag C_L (dB)	P_{in} (dBm)	Substrate
Palazzi et al. [11]	2	Slot	1.2	0.3 × 0.26	9.4	−13	Paper
Psychoudakis et al. [10]	1	Loop	5.91	0.19 × 0.19	7	/	/
Rasilainen et al. [12]	1	Dipole	1	0.27 × 0.12	9	−16.3	Rogers 5870
Lazaro et al. [17]	2	Patch	0.868	0.58 × 0.37	13.8	−7.5	FR4
Kubina et al. [18]	2	Patch and dielectric resonator	3–3.1	1.02 × 1.02	8.4	−20	Rogers TMM6 and 5870
Gu et al. [19]	1	Slot	3.5	0.7 × 0.41	9.2	−25	Rogers RT7duroid 6002

5.5 Harmonic Tracking Systems

Tracking for harmonic transponders can be based on RSSI measurements or on phase-recovery measurements.

RSSI measurements can be performed using an incoherent reader, which can consist of a signal generator connected to the transmitting antenna and a spectrum analyzer connected to a $2f_0$ antenna. However, RSSI measurements are prone to multipath, misalignments, and propagation nonidealities, which makes this measurement approach inaccurate.

Phase-recovery measurements, on the other hand, allow us to achieve high accuracy. However, a coherent reader is needed. Additionally, traditionally multi-frequency backscattering schemes must be adopted to resolve phase integer ambiguity. In [20], a tracking system based on harmonic tags is developed, where the reader consists of a two-port software-defined radio. The same reference digital clock signal is fed into two frequency synthesizers, one working at f_0 for the transmitter and the other at $2f_0$ for the receiver (LO port of the mixer). A ranging accuracy of around 50 μm is obtained for $f_0 = 1$ GHz, a transmitted power of 27 dBm EIRP, and a tag-to-reader distance of 2.5 m, with the tag in air, and immersed in construction rand and water. The tag is based on an NLTL and an RF switch, which is used to modulate the backscattered second harmonic (see Section 5.6).

Alternative approaches for localization based on harmonic transponders include pulsed pseudorandom code harmonic radars [21] and frequency-modulated continuous-wave radars [22].

In Ref. [21], for example, the distance from the reader of bees equipped with miniaturized harmonic transponders is estimated by correlation of pseudorandom codes. The transmitted carrier is modulated by a pseudorandom sequence with 1023 chips. The correlation between the received signal at $2f_0$ and the transmitted signal is calculated to determine the delay. This technique is also beneficial to improve the sensitivity of the receiver, and a detection range of 60 m with a maximum measurement error of 75 cm is demonstrated for a transmitted power of 1.75 kW at $f_0 = 9.4$ GHz.

5.6 Multi-Bit Harmonic Transponders

In UHF RFID transponders, multi-bit information is conveyed through the amplitude modulation or phase modulation of the backscattered signals. The same approach can be used for harmonic tags, by modulating the generated second harmonic signal. Generally, a digital modulation unit or an oscillator is used to periodically change the DC bias point of the Schottky diode, or the DC bias point of varactor diodes along NLTLs.

In Ref. [16], an amplitude-modulated harmonic transponder is proposed. The block diagram of the transponder is shown in Figure 5.9a. The input antenna of the tag captures the interrogating signal. The signal is then divided into two parts

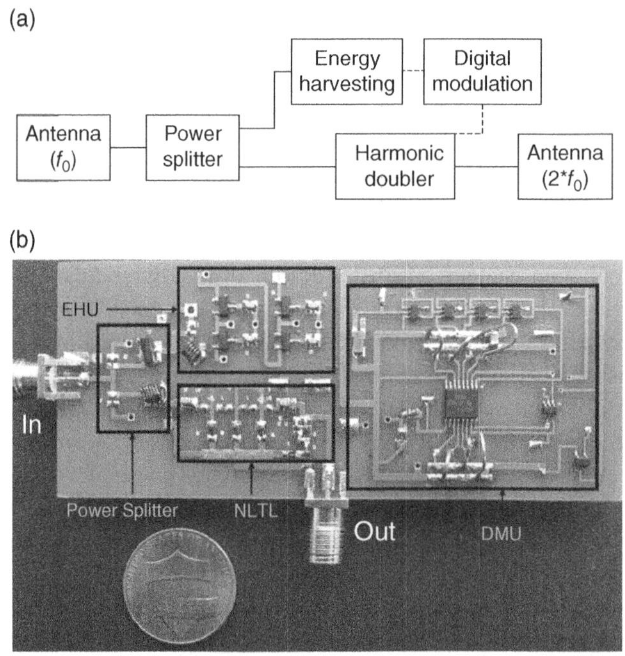

Figure 5.9 Multi-bit harmonic transponder based on NLTL loaded with varactor diodes. (a) Block diagram and (b) photo of a proof-of-concept prototype. *Source:* Mondal et al. [16]/IEEE.

using a power divider: one part represents the input signal for the RF-DC converter in the energy-harvesting unit, while the other part represents the fundamental carrier for the harmonic generation in the harmonic doubler. Once the energy-harvesting unit stores the required energy, it activates the digital modulation unit, which switches the harmonic doubler on-and-off based on the information to be transferred. The frequency doubler consists of a three-stage NLTL loaded with varactor diodes. The on-and-off states are obtained by varying the DC bias voltage of the varactors. Therefore, the communication is based on amplitude shift keying. An unequal power splitting is performed to maximize the read range based on the different power requirements of the digital modulation unit and the harmonic generation units.

A proof-of-concept prototype working at $f_0 = 434\,\text{MHz}$ was manufactured and tested (a photo, excluding the tag antennas, is shown in Figure 5.9b). Two meander dipole antennas are adopted to reduce the area occupation of the tag. The energy-harvesting unit is based on an 8-stage Dickson charge pump rectifier and a 1.8 V Linear Drop Output voltage regulator. The bias point of the NLTL is varied between 0.6 and 0 V, corresponding to a dynamic range between the "on" and "off" states of the frequency doubler of 26 dB for $P_{in} = -10\,\text{dBm}$. The tag demonstrated to be able to transmit an 8-bit ID with a minimum RF input power of −6 dBm.

The harmonic WISP system presented in Ref. [9] is based on a similar principle, where however the energy harvesting and the communication phases occur in two different moments.

A recent work from my group shows how to implement an on-off keying modulation in Schottky-based frequency doublers by varying the DC impedance termination of the doubler [23]. The circuit of the doubler is shown in Figure 5.10a switch, consisting of an NMOS transistor is connected between the shunt inductor in the output matching network and ground. As anticipated in Section 5.3, the LC network performs both the function of output matching network at $2f_0$ and of bias tee for the DC component generated by the diode. When a voltage larger than the threshold voltage of the MOSFET is applied to the gate, the MOSFET is in a triode region: the channel between drain and source is formed and the inductor is connected to ground. This guarantees that the diode is zero-biased (DC voltage across the diode is zero). On the other hand, if no voltage is applied to the gate, the MOSFET is off. The DC current is zero, and it can be demonstrated that a positive voltage is developed at the diode cathode [23]. Therefore, the diode becomes reverse biased, which causes a reduction in the generated output second harmonic (on/off keying modulation). A bypass capacitor is connected in parallel with the MOSFET for the RF signal components. Since, the switch only affects the DC component, the doubler behavior is not affected by the switch parasitics, and both low-frequency NMOS transistors and electromechanical switches can be used. Additionally, since the modulation

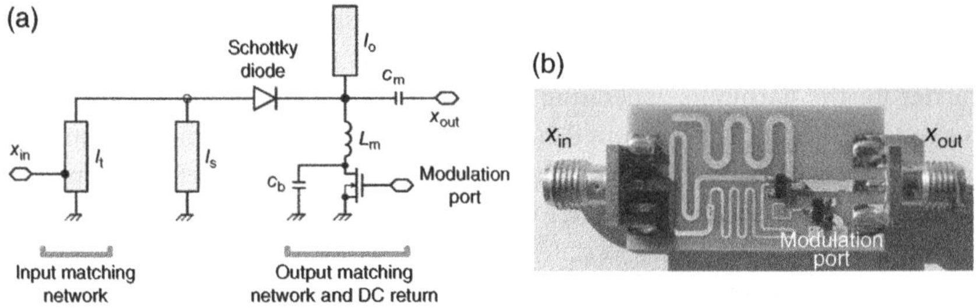

Figure 5.10 Multi-bit harmonic transponder based on DC impedance termination. (a) Circuit schematic and (b) photo of a proof-of-concept prototype. *Source:* Palazzi et al. [23]/MDPI/CC BY 4.0.

signal is applied to the gate, no static power consumption is experienced, which is highly beneficial for low-power applications.

A proof-of-concept prototype is manufactured, as shown in Figure 5.10b. The frequency doubler is designed for $f_0 = 1.04\,\text{GHz}$, and the MOSFET model BSS1234 is used to control the DC impedance termination. For demonstration purposes, the modulation port is controlled by a waveform generator. Square waves with a 3 V 0-to-peak voltage and frequencies up to 15 MHz were applied to the MOSFET. The doubler was tested in a wireless experiment (input and output ports were connected to patch antennas resonating at f_0 and $2f_0$, respectively), and the spectrum acquired by the spectrum analyzer for an interrogation signal of 16 dBm EIPR, at a tag-to-reader distance of 50 cm, and for modulation signals of 500 kHz and 1 MHz are reported in Figure 5.11. The received spectrum is in good agreement

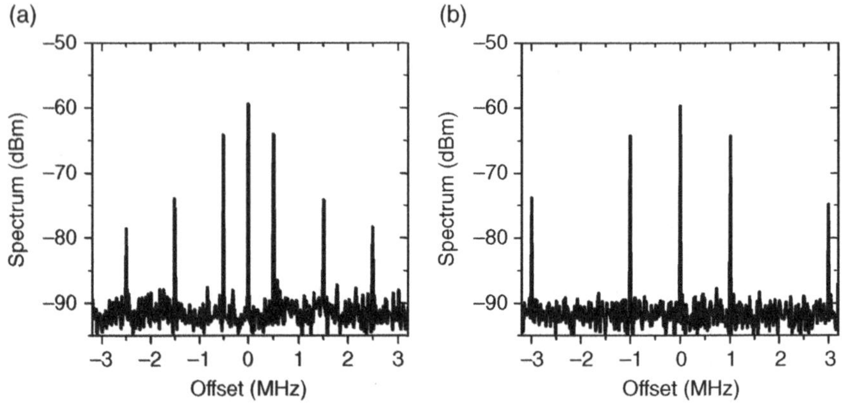

Figure 5.11 Spectrum received of the wireless multi-bit harmonic transponder based on DC impedance termination. *Source:* Palazzi et al. [23]/MDPI/CC BY 4.0. (a) Modulation frequency of 500 kHz and (b) modulation frequency of 1 MHz.

with the spectrum of a sine wave multiplied by an ideal square wave, demonstrating the capability of the proposed harmonic transponder to support high-speed data transfer.

5.7 Harmonic Tag for Rotation Sensing

In industrial machinery, the measurement of the rotational speed of wheels, blades, and motors is of high interest. This measurement is usually performed with optical sensors, which, however, can be inaccurate in environments characterized by dust and by the presence of liquids. A harmonic rotational speed sensor has been conceived based on the circuit schematic shown in Figure 5.10. In the latter case, though, the MOSFET switch is replaced by an electromechanical switch, which is actuated by magnets (see Figure 5.12) [23]. Electromechanical switches are characterized by large parasitics, which generally hinder their usage in RF circuits. However, in the proposed circuit, the electromechanical switch only controls the DC impedance termination. Therefore, the switch parasitics do not impact the circuit operation.

To prove the concept, the circuit is used to measure the rotational speed of a small wheel. The adopted experimental setup is shown in Figure 5.13a. Four magnets are placed at equal intervals on the border of a rotating wheel, while the harmonic sensor including the electromechanical (reed) switch is placed in close proximity. The wheel rotates at a speed of 10 Hz (this value is measured with a photoelectric sensor). When the switch is in front of a magnet, the switch is closed, while for the rest of the time, the switch is open. The spectrum of the received signal is shown in Figure 5.13b. The two sidebands at 40 Hz from the carrier, due to the modulation of the second harmonic output signal, are clearly visible, demonstrating the correct operation of the sensor.

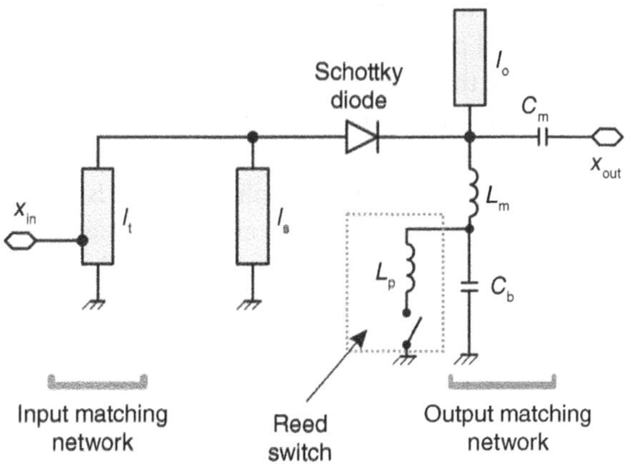

Figure 5.12 Schematic diagram of a rotational speed harmonic sensor based on an electromechanical switch. *Source:* Palazzi et al. [23]/ MDPI/ CC BY 4.0.

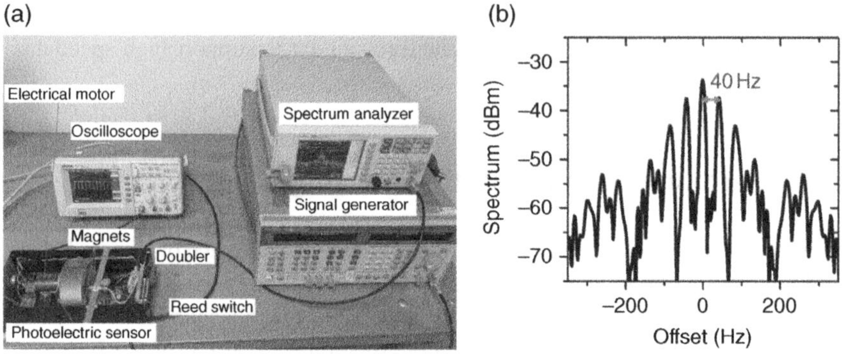

Figure 5.13 Rotational speed harmonic sensor. (a) Experimental setup and (b) measured spectrum. *Source:* Palazzi et al. [23]/MDPI/CC BY 4.0.

5.8 Harmonic Tag for Temperature Sensing

Temperature is a key parameter not only in industrial environments, but in a plethora of different applications, ranging from smart homes to precision agriculture and assisted living. In all of the aforementioned cases, passive wireless transponders are essential due to the difficulties to deploy wired sensors and to periodically change the battery of the transponders. For these reasons, research has focused on developing passive harmonic transponders for temperature sensing.

In Ref. [24], a harmonic transponder is described, which transduces temperature variations in phase variations. The transponder schematic is shown in Figure 5.14.

The input antenna is connected to the frequency doubler, which generates the second harmonic. A power divider separates the generated second harmonic signal into two parts. One part is backscattered in vertical polarization, while the other part is fed into a temperature-controlled phase shifter, and it is then backscattered in horizontal polarization. This way, the signal in vertical polarization is a phase reference for the other signal, and the sensor information can be recovered without accurate knowledge of the tag-to-reader distance. The reader consists of a CW transmitter and an I/Q receiver to retrieve the tag information.

The temperature-controlled phase shifter is implemented using a 90° hybrid junction (i.e. a branch coupler) and a bimorph cantilever (see Figure 5.15a). The bimorph cantilever is composed of two metal layers stuck together which have different thermal expansion coefficients. The bimorph cantilever is used to implement variable reactive impedances at the direct and coupled ports of the cantilever. The cantilever is anchored at one end, while the other end is free to move, and forms a parallel plate capacitance with three metal pads: one pad is connected to ground with vias, while the other two pads are connected to the direct and coupled

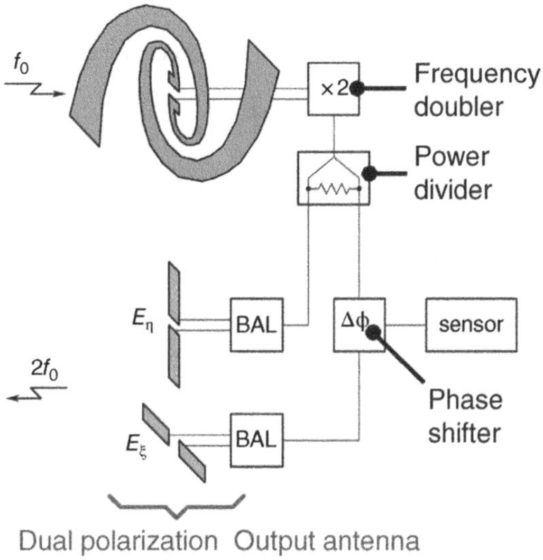

Figure 5.14 Schematic diagram of a harmonic transponder for temperature sensing. *Source:* Alimenti et al. [24]/with permission of IEEE.

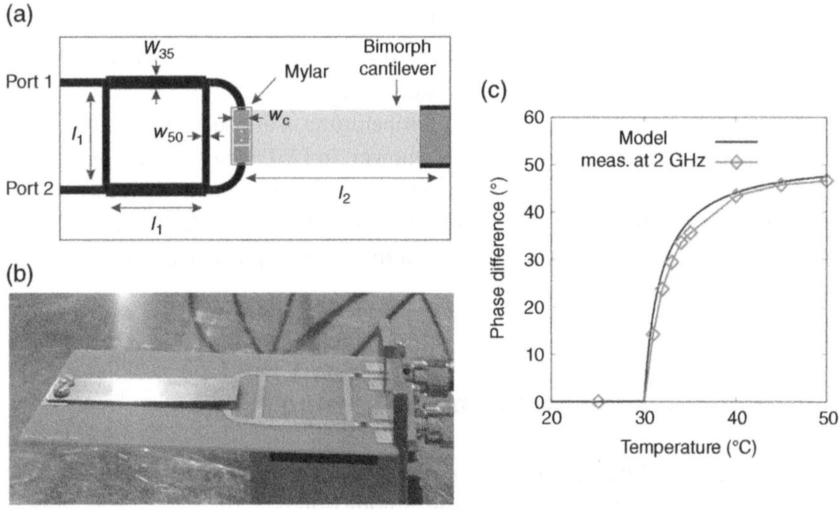

Figure 5.15 Temperature-controlled phase shifter. (a) Layout, (b) photo of the proof-of-concept prototype, and (c) phase difference versus temperature. *Source:* Alimenti et al. [24]/IEEE.

ports of the branch line coupler. A mylar layer is attached to the pads to avoid short circuits with the cantilever. When the temperature increases, the cantilever bends. Therefore, the capacitance decreases, causing a variation in the phase shift between port 1 and port 2 of the reflection-type phase shifter.

The phase shift between port 1 and port 2 of the reflection-type cantilever depends on the capacitance value C at the direct and couple ports of the branch line as follows:

$$\angle S_{21}(T) = \varphi - 2\tan^{-1}\left[2\pi f R_0 C(T)\right] \tag{5.3}$$

where φ is a constant phase value, due to connecting lines and to the branch line geometry, f is the operating frequency, R_0 is the reference impedance, and T is the temperature. A model that describes the phase variation with the bending of the bimorph cantilever is illustrated in detail in Ref. [24].

A proof-of-concept prototype was manufactured on FR4 (see Figure 5.15b), working at $f_0 = 1\,\text{GHz}$, and the phase change versus temperature is analyzed, as shown in Figure 5.15c. There are no significant changes in the measured phase for $T < 30°$; since for these temperatures, the cantilever would bend in the direction of the PCB. On the other hand, the phase changes significantly for $30° < T < 40°$, where we observe a phase variation of 47 dB. For $T > 40°$, the phase values tend to saturate (the capacitance value is almost negligible and further bending does not impact significantly on the phase variation).

Alternative approaches have been also adopted. For instance, in Ref. [18] a temperature-sensitive antenna resonator is used. The interrogating signal is swept from 3 to 3.1 GHz. A temperature variation causes a frequency shift in the peak of the received spectrum. Therefore, the temperature is encoded in the frequency corresponding to the maximum received power. In [25], a harmonic temperature sensor for railway applications is proposed, based on a single Schottky diode. Since the conversion loss of the doubler varies with temperature, a temperature variation can be encoded in the variation of the received power (this approach can be used in stationary environments).

5.9 Harmonic Tag for Vibration Sensing

Vibration sensors find application in structural health monitoring, preventive maintenance, and condition monitoring, among others.

Harmonic transducers have been considered for vibration sensing in conjunction with piezoelectric cantilevers.

The circuit architecture in Ref. [26] is based on an input antenna, a frequency doubler, a vibration-controlled phase shifter, and a $2f_0$ antenna. The block diagram of the sensing system is shown in Figure 5.16.

The vibration-controlled phase shifter is based on a reflection-type phase shifter, where the direct and coupled ports of the branch line coupler are closed on two identical varactor diodes. The piezoelectric cantilever is used to control the bias

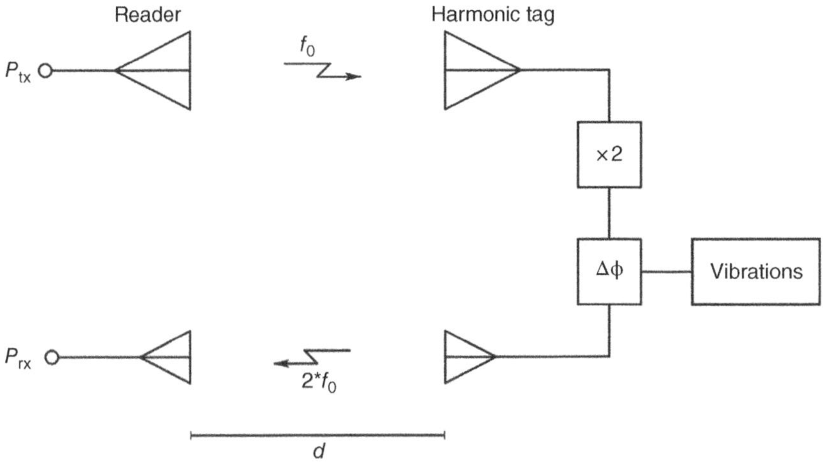

Figure 5.16 Schematic diagram of a harmonic transponder for vibration sensing. *Source:* Simoncini et al. [26]/with permission of IEEE.

point of the varactors, thereby periodically changing their equivalent capacitance. The capacitance variation causes a phase variation in the transmission coefficient between port 1 and port 2. This way, the second harmonic signal generated by the frequency doubler is modulated in phase, and the vibration information can be recovered from the sidebands of the received signal.

A proof-of-concept prototype was manufactured in an FR4 substrate (see Figure 5.17a). The branch line is connected to a piezoelectric transducer with two wires. The piezoelectric transducer is a commercial piezoelectric sensor from Mide. One end of the cantilever is attached to a shaker with a screw. A commercial accelerometer is placed on the shaker as well to check the vibration gain.

The sensor is tested with a vibration frequency at the natural resonant frequency of the cantilever, at 130 Hz. The measure is repeated for different acceleration values. The results are shown in Figure 5.17b. The vibration frequency is encoded in the frequency difference between the carrier (corresponding to 0 in Figure 5.17b) and the sidebands; the acceleration is encoded in the modulation index (i.e. in the ratio of the first sideband to the carrier).

5.10 Harmonic Tag for Crack Sensing

Crack sensors find application in structural health monitoring of buildings and infrastructures and in anti-tampering systems. In both scenarios, battery-less wireless sensors are of great interest: in the first case, wired sensors are

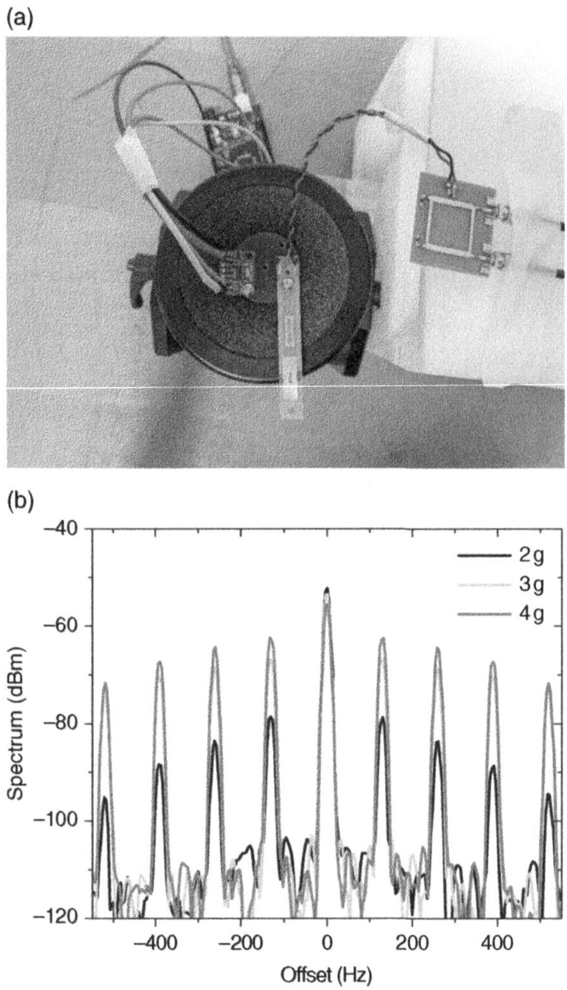

Figure 5.17 Phase-based vibration sensor. *Source:* Simoncini et al. [26]/IEEE. (a) Photo of the prototype and (b) experimental results.

impractical since the transponders must be deployed over large surfaces, while battery-operated wireless transducers are also impractical, since the transponders cannot be easily accessed for battery replacement. The same considerations apply to anti-tampering systems, where wired sensors cannot be used to monitor mobile items, and batteries would make transponders bulky, expensive, and would add maintenance issues. In both scenarios long read ranges are also required to guarantee good system coverage.

Harmonic transponders can satisfy the above requirements and are therefore of great interest for crack sensing.

In [27], a harmonic strain and crack sensor is developed, based on antenna sensing. The strain/crack information is encoded in the resonance frequency of the input antenna. When a deformation occurs, also the antenna deforms. In particular, the antenna length increases, which causes a down-shift in the antenna resonant frequency. This causes a variation in the frequency of the interrogating signal corresponding to the maximum power for the second harmonic. A prototype manufactured on a Rogers substrate is shown in Figure 5.18. The tag, tested at a distance of 50 cm with a transmitted power of 15 dBm EIRP, demonstrated a strain sensitivity of −5.232 kHz/με for crack widths below 5.49 mm. For broader cracks, the crack grows through the entire antenna width, and no more response is obtained from the tag.

The harmonic crack sensor proposed in [6] instead is based on an open-circuited quarter-wave stub at f_0 placed in a shunt at the input of the frequency doubler. The block diagram of the system is shown in Figure 5.19. When the tag is intact, the input signal is short-circuited by the stub (which operates as a notch filter). Therefore, although the tag is interrogated by a CW signal at f_0, virtually no second harmonic is generated. On the other hand, when a crack occurs, the stub is removed, and the transponder operates as a common harmonic tag. Therefore, the tag produces an alarm in presence of a crack. This approach is compatible with a system including multiple tags, where an alarm is generated if one of them is broken [28].

The stub is placed at the input of the doubler to maximize the second harmonic attenuation in intact condition. In fact, the attenuation introduced by the stub at

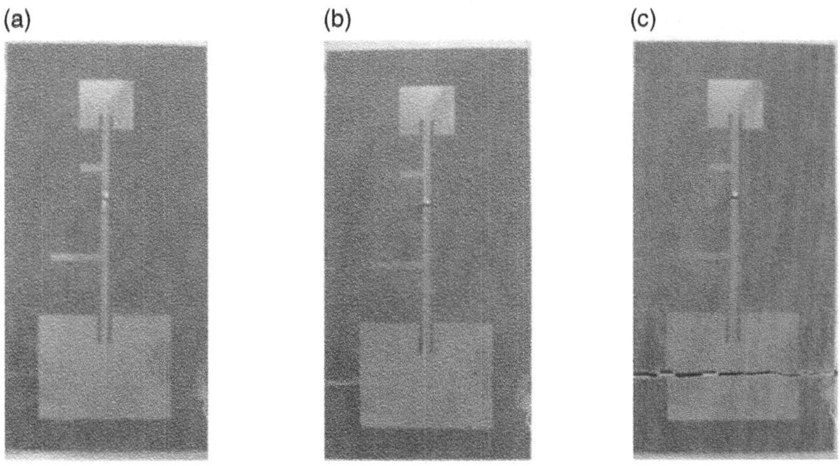

Figure 5.18 Photos of deformed harmonic antenna sensor at different crack opening sizes. *Source:* Cho et al. [27]/IEEE. (a) Crack width 460 μm, (b) crack width 1.16 mm, and (c) crack width 5.49 mm.

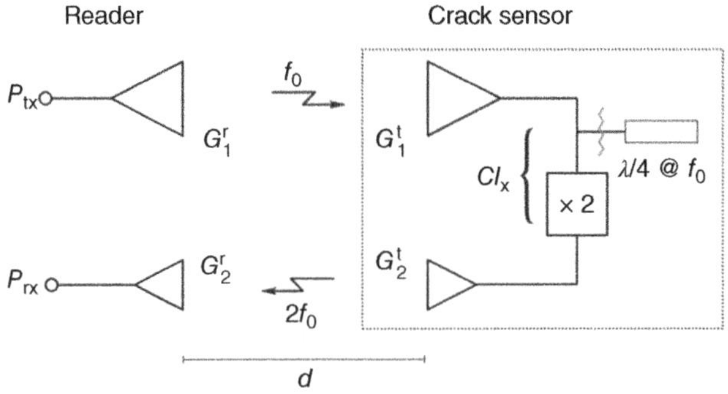

Figure 5.19 Block diagram of a harmonic crack sensor based on a notch filter. *Source:* Adapted from Palazzi et al. [6].

the fundamental frequency is usually limited (this is especially true when the stub is manufactured on a high-loss substrate). However, the conversion loss of the doubler increases significantly as the input power decreases (C_l decreases by 10 dB/decade in the linear region). This way, an attenuation of 25 dB of the stub causes a reduction in the output power at $2f_0$ of about 50 dB.

A proof-of-concept prototype is manufactured on a paper substrate, and it is illustrated in Figure 5.20. The transponder is designed for $f_0 = 2.45$ GHz. A system of nested annular slots is adopted, like the ones described in Figure 5.7. In this

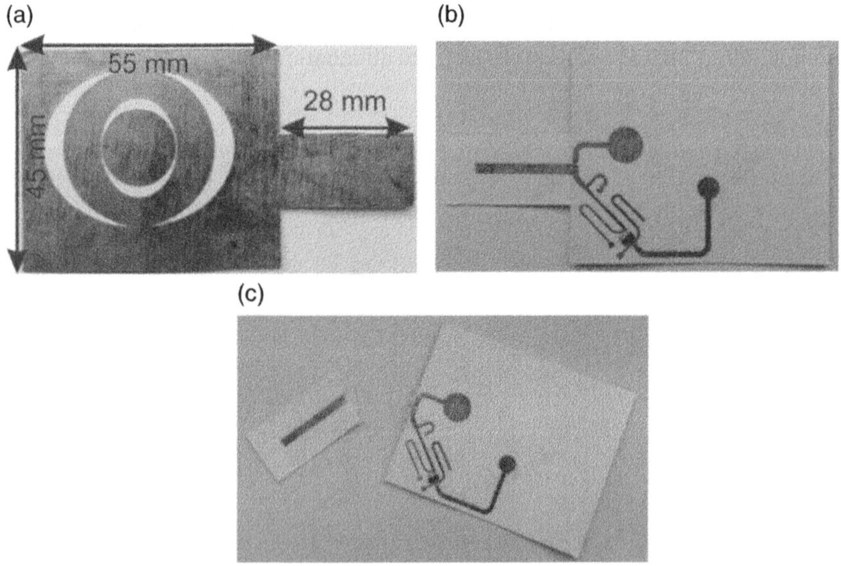

Figure 5.20 Photo of the harmonic crack sensor based on a notch filter. (a) Bottom side, (b) top side of the intact tag, and (c) top side of the cracked tag. *Source:* Palazzi et al. [6]/IEEE.

transponder, though, a physical short is implemented in correspondence with a null of the electric field in the outer slot. This way, the microstrip line, which feeds the interior slot, can cross the output slot to be connected to the output port of the frequency doubler. A read range from 1 to 5 m is experimentally demonstrated for a transmitted power of 25 dBm EIRP. The minimum read range is defined by the second harmonic leakage when the tag is intact, while the maximum read range is determined by the sensitivity of the receiver (equal to -100 dBm in the reported experiment). A maximum read range of 9 m is foreseen for a transmitted power of 36 dBm EIRP [28]. The read range could be extended by using an LNA at the receiver side.

When comparing the read range of harmonic crack sensors with other passive wireless transponders, in addition to the inherent robustness to clutter, a significant advantage in terms of read range per transmitted power can be noticed, as shown in Table 5.2.

5.11 Harmonic Tags for Buried Items Localization

Due to their intrinsic robustness to strong reflections from the environment, harmonic tags are particularly appealing for the detection and localization of buried items.

For example, the commercial RECCO® system is developed to localize victims buried by the snow [35]. The system includes an active harmonic reader and a passive harmonic tag (i.e. 1-bit tag) for wearable applications, and has a penetration depth of 80 m in the air and 20 m in the packed snow.

In [36], passive harmonic tags are used as a nondestructive technique to both detect and identify buried pipelines. The block diagram of the tag is shown in Figure 5.21. The reader interrogates the tag using a CW signal. The second harmonic generated by the two-port frequency doubler is modulated using a low-power, low-frequency oscillator (modulation frequency f_m). The oscillator is based on two NAND gates and a quartz crystal. The modulation frequency is used to identify different pipe types. Additionally, the adoption of a frequency modulation technique makes it possible to relax the requirement on the filters needed to remove the leakage of the spurious second harmonic self-generated by the transmitter to the receiver.

The reader transmits the CW f_0 signal using a circularly polarized antenna. The tag is equipped with a dual-polarized patch antenna. One port of the antenna is connected to the frequency doubler, while the other one is connected to an RF-DC converter used to power the low-frequency oscillator. The oscillator is connected to the quarter-wave stub at the input of the Schottky diode with a bias tee. The generated low-frequency voltage signal is used to vary the bias point of the

Table 5.2 State of the art wireless crack sensors.

Ref.	Transponder type	Sensing strategy	Substrate	Freq. (GHz)	P_t EIRP (dBm)	Max tag-to-reader distance (m)
Palazzi et al. [6]	Harmonic	Stub removal	Paper	2.45; 4.9	25	5
Cho et al. [27]	Harmonic	Resonator frequency shift	Rogers RT/duroid 5880	2.78–2.89; 5.56–5.78	15	0.5
Marindra and Tian [29]	Chipless	Resonator frequency shift	Taconic CER-10-0500	2–6	0[a]	0.3
Javed et al. [30]	Chipless	Resonator frequency shift	Rogers RT/Duroid® 6010.2LM	2–8	/	0.06533
Kalansuriya et al. [31]	UHF RFID	RSSI	No substrate	0.915	36	0.75
Dey et al. [32]	Chipless	Time domain reflectometry	Taconic TLX-0	2–20	/	0.3
Caizzone et al. [33]	UHF RFID	Phase difference	FR4	0.867	/	0.9
Zhang et al. [34]	UHF RFID	Resonant frequency shift	FR4	0.90275–0.92725	36	1

State of the art wireless crack sensors.
[a] the antenna gain of the transmitter is not reported.

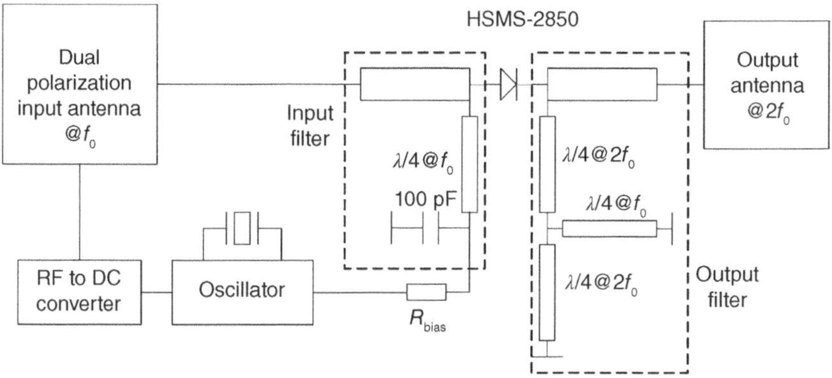

Figure 5.21 Circuit schematic of a harmonic tag for buried asset localization. *Source:* Abdelnour et al. [36]/MDPI/CC BY 4.0.

Schottky diode, which causes a periodic amplitude modulation of the generated second harmonic, causing the appearance of sidebands in the spectrum of the received signal. Since the tag detection is based on the power received at the sidebands, the conversion loss of the tag is defined as the ratio of the available input power at f_0 to the power at $2f_0+f_m$.

A proof-of-concept prototype on a 1.6-mm thick FR4 substrate is shown in Figure 5.22. The transmitted CW f_0 signal lies in the European UHF RFID band (i.e. 865.868 MHz), and the oscillator works at the modulation frequency of 32 kHz. The tag is placed inside a plastic box, with the aim to protect it from moisture and to reduce the antenna detuning caused by the proximity of the patch

Figure 5.22 Photo of a harmonic tag for buried asset localization. *Source:* Abdelnour et al. [36]/MDPI/CC BY 4.0.

antennas with high-permittivity materials (i.e. with wet soil), and it is buried in the soil. The tag was detected up to a depth of 60 cm in the soil, for volumetric water contents up to 25%.

5.12 Conclusion

An overview of some of the most recent advancements in harmonic transponders and harmonic backscattering has been reported. Starting from the analysis of harmonic generation, the main performance of basic one-bit harmonic tags has been described, which highlights the potential of harmonic transponders for medium-range communication. Strategies to implement low-power multi-bit harmonic transponders are reported as well.

Finally, a series of examples of passive wireless sensors based on harmonic transponders are reported, which demonstrates the interest in this subject, and clarifies the potential of harmonic transponders to perform sensing with good read ranges in highly cluttered environments and in harsh environmental conditions.

References

1 Williams, J.M., Khanna, R., Ruiz-Rosero, J.P. et al. (2017). Weaving the wireless web: toward a low-power, dense wireless sensor network for the industrial IoT. *IEEE Microwave Magazine* 18 (7): 40–63.

2 Koene, I., Viitala, R., and Kuosmanen, P. (2019). Internet of things based monitoring of large rotor vibration with a microelectromechanical systems accelerometer. *IEEE Access* 7: 92210–92219.

3 Grosinger, J. and Michalowska-Forsyth, A. (2022). Space tags: ultra-low-power operation and radiation hardness for passive wireless sensor tags. *IEEE Microwave Magazine* 23 (3): 55–71.

4 Costanzo, A., Benassi, F., and Monti, G. (2022). Wearable, energy-autonomous RF microwave systems: chipless and energy-harvesting-based wireless systems for low-power, low-cost localization and sensing. *IEEE Microwave Magazine* 23 (3): 24–38.

5 Palazzi, V., Alimenti, F., Mezzanotte, P. et al. (2014). Low-power frequency doubler in cellulose-based materials for harmonic RFID applications. *IEEE Microwave and Wireless Components Letters* 24 (12): 896–898.

6 Palazzi, V., Alimenti, F., Mezzanotte, P. et al. (2017). Zero-power, long-range, ultra low-cost harmonic wireless sensors for massively distributed monitoring of

cracked walls. *2017 IEEE MTT-S International Microwave Symposium (IMS)*, Honolulu, HI, USA, pp. 1335–1338. doi: 10.1109/MWSYM.2017.8058859.

7 Shefer, J. and Klensch, R.J. (1973). Harmonic radar helps autos avoid collisions. *IEEE Spectrum* 10 (5): 38–45.

8 Colpitts, B.G. and Boiteau, G. (2004). Harmonic radar transceiver design: miniature tags for insect tracking. *IEEE Transactions on Antennas and Propagation* 52 (11): 2825–2832.

9 Ma, Y., Hui, X., Kan, E.C. (2016). Harmonic-WISP: a passive broadband harmonic RFID platform. *2016 IEEE MTT-S International Microwave Symposium (IMS)*, San Francisco, CA, USA, pp. 1–4. doi: 10.1109/MWSYM.2016.7540224.

10 Psychoudakis, D., Moulder, W., Chen, C.-C. et al. (2008). A portable low-power harmonic radar system and conformal tag for insect tracking. *IEEE Antennas and Wireless Propagation Letters* 7: 444–447.

11 Palazzi, V., Alimenti, F., Kalialakis, C. et al. (2017). Highly integrable paper-based harmonic transponder for low-power and long-range IoT applications. *IEEE Antennas and Wireless Propagation Letters* 16: 3196–3199.

12 K. Rasilainen, J. Ilvonen, A. Lehtovuori, J.-M. Hannula e V. Viikari, On design and evaluation of harmonic transponders, *IEEE Transactions on Antennas and Propagation*, vol. 63, n. 1, pp. 15–23, 2015.

13 Maas, S. (2003). *Nonlinear Microwave and RF Circuits*. Artech.

14 Alimenti, F., Mezzanotte, P., Dionigi, M. et al. (2012). Microwave circuits in paper substrates exploiting conductive adhesive tapes. *IEEE Microwave and Wireless Components Letters* 22 (12): 660–662.

15 Yu, F., Lyon, K.G., and Kan, E.C. (2010). A novel passive RFID transponder using harmonic generation of nonlinear transmission lines. *IEEE Transactions on Microwave Theory and Techniques* 58 (12): 4121–4127.

16 Mondal, S. and Chahal, P. (2019). A passive harmonic RFID tag and interrogator development. *IEEE Journal of Radio Frequency Identification* 3 (2): 98–107.

17 Lazaro, A., Villarino, R., and Girbau, D. (2014). A Passive Harmonic Tag For Humidity Sensing. *International Journal of Antennas and Propagation* 2014: 1–11.

18 Kubina, B., Mandel, C., Schüßler, M., and Jakoby, R. (2015). Compact quasi-chipless harmonic radar sensor with a dielectric resonator antenna. *2015 IEEE MTT-S International Microwave Symposium*, Phoenix, AZ, USA, pp. 1–3. doi: 10.1109/MWSYM.2015.7166820.

19 Gu, X., Srinaga, N.N., Guo, L. et al. (2019). Diplexer-based fully passive harmonic transponder for sub-6-GHz 5G-compatible IoT applications. *IEEE Transactions on Microwave Theory and Techniques* 67 (5): 1675–1687.

20 Hui, X. and Kan, E.C. (2019). Harmonic UHF RFID ranging with 50-micron accuracy and 1-kHz sampling rate. *2019 IEEE International Conference on RFID (RFID)*, Phoenix, AZ, USA.

21 Tsai, Z.-M., Jau, P.-H., Kuo, N.-C. et al. (2013). A high-range-accuracy and high-sensitivity harmonic radar using pulse pseudorandom code for bee searching. *IEEE Transactions on Microwave Theory and Techniques* 61 (1): 666–675.

22 Storz, G. and Lavrenko, A. (2020). Compact low-cost FMCW harmonic radar for short range insect tracking. *2020 IEEE International Radar Conference (RADAR)*, Washington, DC, USA, pp. 642–647, doi: 10.1109/RADAR42522.2020.9114612.

23 Palazzi, V., Roselli, L., Tentzeris, M.M. et al. (2022). Energy-efficient harmonic transponder based on on-off keying modulation for both identification and sensing. *Sensors* 22 (2): 620.

24 Alimenti, F., Palazzi, V., Mezzanotte, P., and Roselli, L. (2020). Zero-power temperature sensor for wireless harmonic systems based on a reflection-type phase shifter and a bimorph cantilever. *2020 IEEE Topical Conference on Wireless Sensors and Sensor Networks (WiSNeT)*, San Antonio, TX, USA, pp. 5–8. doi: 10.1109/WiSNeT46826.2020.9037619.

25 Silveira, T.M., Pinho, P., and Carvalho, N.B. (2022). Harmonic RFID Temperature Sensor Design for Harsh Environments. *IEEE Microwave and Wireless Components Letters* 32 (10): 1239–1242.

26 Simoncini, G., Salvati, R., Palazzi, V. et al. (2022). Zero-power vibration sensor for wireless harmonic systems based on a reflection-type phase shifter and a piezoelectric transducer. *2022 IEEE Topical Conference on Wireless Sensors and Sensor Networks (WiSNeT)*, Las Vegas, NV, USA.

27 Cho, C., Yi, X., Li, D. et al. (2016). Passive wireless frequency doubling antenna sensor for strain and crack sensing. *IEEE Sensors Journal* 16 (14): 5725–5733.

28 Palazzi, V., Alimenti, F., Mezzanotte, P. et al. (2018). Analysis of a multi-node system for crack monitoring based on zero-power wireless harmonic transponders on paper. *2018 IEEE Topical Conference on Wireless Sensors and Sensor Networks (WiSNet)*, Anaheim, CA, USA, pp. 92–95. doi: 10.1109/WISNET.2018.8311573.

29 Marindra, A.M.J. and Tian, G.Y. (2018). Chipless RFID sensor tag for metal crack detection and characterization. *IEEE Transactions on Microwave Theory and Techniques* 66 (5): 2452–2462.

30 Javed, N., Azam, M.A., and Amin, Y. (2021). Chipless RFID multisensor for temperature sensing and crack monitoring in an IoT environment. *IEEE Sensors Letters* 5 (6): 1–4.

31 Kalansuriya, P., Bhattacharyya, R., and Sarma, S. (2013). RFID tag antenna-based sensing for pervasive surface crack detection. *IEEE Sensors Journal* 13 (5): 1564–1570.

32 Dey, S., Kalansuriya, P., and Karmakar, N.C. (2014). Chipless RFID based high resolution crack sensing through SWB technology. *2014 IEEE International Microwave and RF Conference (IMaRC)*, Bangalore, India.

33 Caizzone, S., Di Giampaolo, E., and Marrocco, G. (2014). Wireless crack monitoring by stationary phase measurements from coupled RFID Tags. *IEEE Transactions on Antennas and Propagation* 62 (12): 6412–6419.

34 Zhang, J., Huang, B., Zhang, G., and Tian, G.Y. (2018). Wireless passive ultra high frequency RFID antenna sensor for surface crack monitoring and quantitative analysis. *Sensors* 18 (7): 2130.

35 Recco(R) Technology (2023). Online. https://recco.com/technology/ (accessed 24 July 2024).

36 Abdelnour, A., Lazaro, A., Villarino, R. et al. (2018). Passive harmonic RFID system for buried assets localization. *Sensors* 18 (11): 3635.

6

Passive Wireless Sensors in Radiation Environments

Jasmin Grosinger[1] *and Alicja Michalowska-Forsyth*[2]

[1] *Institute of Microwave and Photonic Engineering, Graz University of Technology, Graz, Austria*
[2] *Institute of Electronics, Graz University of Technology, Graz, Austria*

6.1 Introduction

The technological advancement in wireless sensor technology to batteryless systems enables new applications, like proximity communication, identification, or environmental monitoring. Consequently, wireless communication and sensing devices have become more widespread and easily accessible. The functional capabilities of batteryless wireless sensor technology would have a lot to offer, particularly in harsh environments, with restricted human access. Interesting settings would be the broad-sense industrial, medical, and space environments, where monitoring instruments are exposed to extreme temperatures, vibrations, humidity, and—of particular interest in this work—ionizing radiation, as presented in Figure 6.1.

There are several large sectors and applications that operate with the ionizing radiation. These include the medical sector, energy sector, homeland security, experimental science, and a multitude of industrial applications. As a consequence of many of those above, nuclear waste management comes on top. In medicine, X-ray imaging is the most popular branch where ionizing radiation is employed. Further, radionuclides in positron emission tomography (PET) imaging and heavy ion therapy become more and more accessible. There are also facilities of immense intensity and energy that are used for the sterilization of medical equipment. Similarly, X-ray generators, gamma rays emitted from radioisotopes, and electron beams are broadly used in food processing to inhibit the sprouting of

Wireless Identification and Sensing Systems for Harsh and Severe Environments,
First Edition. Edited by Smail Tedjini and Valentina Palazzi.

Published 2025 by John Wiley & Sons, Inc.

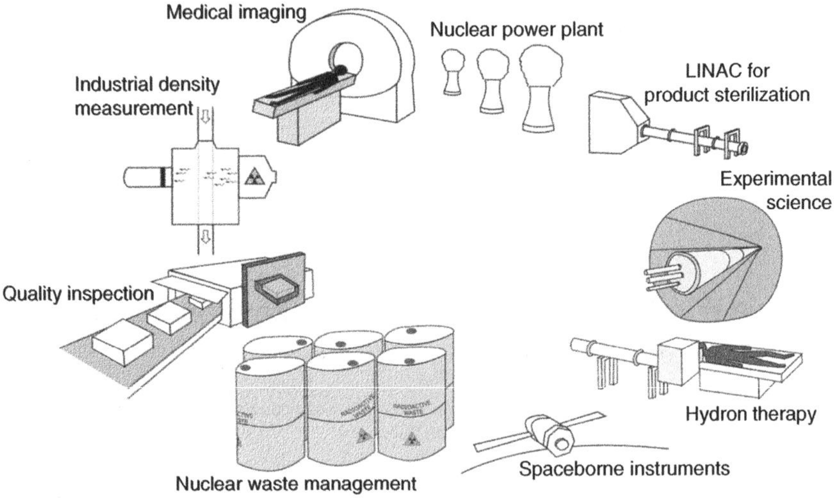

Figure 6.1 Batteryless wireless sensor technology in harsh application environments experiencing ionizing radiation.

vegetables, destroying insects, microorganisms, and bacteria, for example, salmonella in food. Another application broadly observed in the industry is quality inspection, used possibly for any product: packaged food, processed materials, composites, and assembled components. In all the above examples, there is a need for sensing and measurement instruments.

In nuclear reactors and accelerators, in the presence of high-energy radiation fields, operation-induced radioactivity is observed. This leads to the necessity of designated larger control zones, where human presence is limited to short periods of time for safety reasons. Restrictions with access control are imposed over the nuclear waste management facilities for treatment and storage. Therefore, the use of technology such as automation and robotics has led to higher safety standards. This is a clear trend evident in many industries. Actually, all the applications with X-rays, radionuclides, or natural radiation environments, limit human access and rely on sensing and measurement instruments. However, in many cases, due to harsh conditions, in addition to the radiation environment, there are other environmental stress factors. In consequence, the application has to rely on highly robust electronic instruments. The obvious example is spaceborne satellites. Further in the industry, there are corrosive chemical environments, extreme temperatures, or extreme pressure, where for example, the production parameters are monitored with methods based on ionizing radiation, such as density measurement using gamma rays. These applications are seeking new radiation-hard technologies that enable sensing of the environmental parameters, tracking the process, and facilitating data transmission. Passive wireless sensor technologies for environmental sensing and communication could offer their potential.

Nevertheless, state-of-the-art batteryless or passive wireless sensors are currently unsuitable for harsh environments. The main reliability issues must be addressed to enable passive wireless sensor technology for such demanding environments. The challenges include tight power constraints due to the batteryless operation of sensors and the hardening of the associated electronic circuits to survive in harsh environments with ionizing radiation. Passive wireless sensors typically rely on microchips realized in a low-cost complementary metal-oxide-semiconductor (CMOS) technology which do not provide hardness for ionizing radiation. Looking at different semiconductor materials, CMOS devices based on silicon (Si) material are not as radiation-hard as, for example, Gallium Arsenide (GaAs) devices. GaAs devices rely on the wide bandgap of this material, resulting in a better resistance to radiation damage. However, CMOS is superior in some significant aspects [1]. There is no alternative to silicon in passive wireless sensors as they rely on ultra-low-power integrated circuits (ICs) realized in silicon-based CMOS technologies.

One favorable possibility to realize passive wireless sensors operating without batteries at ultra-low power is to use the passive backscatter radio frequency (RF) technology, which exploits far-field wireless power transfer and communication through reflected power [1]. Today, for example, a backscatter RF technology is used for inventory procedures at the International Space Station (ISS) [2]. The backscatter RF technology is particularly interesting for applications in harsh environments like space, as the interrogator device locates almost entirely the system's complexity and power consumption. The batteryless sensor devices require no wires or power supply and can be added in harsh environments with restricted human access.

Backscatter RF systems are useful for access control and tracking. This is, in particular, valuable in large facilities and areas with restricted and controlled human access. Especially the installations in the proximity of ionizing radiation are strictly controlled for safety reasons. Batteryless sensor devices embedded in the tracked products and parts of the equipment coming in and out of the facility can improve transparency and efficiency and, most importantly, minimize the likelihood of human error in the installations, repairs, or decommissioning of facilities. Obviously, in order to be applicable, they must be robust to the harsh environment and remain reliable. This was never on the agenda for commercially used passive tags.

6.2 Passive Wireless RFID Sensors

With the availability of inexpensive, ultra-low-power integrated circuits in the early 1990s, considerable research and development have been invested in backscatter RF systems for identification (ID) purposes, also known as radio frequency identification (RFID) systems [3]. RFID systems rely on the wireless communication between an RFID reader and a multitude of batteryless/passive RFID transponders (tags). Since 2009, researchers have been working consistently toward reaching beyond the ID in RFID [4], that is, by integrating sensing capabilities in passive RFID tags to additionally monitor the tag environment (e.g. temperature, curvature, liquid level). An RFID tag consists of an antenna and a CMOS microchip (chip). Researchers have developed batteryless sensor tags in two ways – either by incorporating sensor circuitry into the tag chip itself, which includes a sensor interface and an external sensor, or by creating tags without the need for additional sensor circuitry. The latter type of RFID sensor tags uses passive transducers that change their RF parameters when the tag environment changes.

Figure 6.2 shows examples of sensor tags with and without additional sensor circuitry. The backscatter RF sensor tags with and without additional sensor circuitry

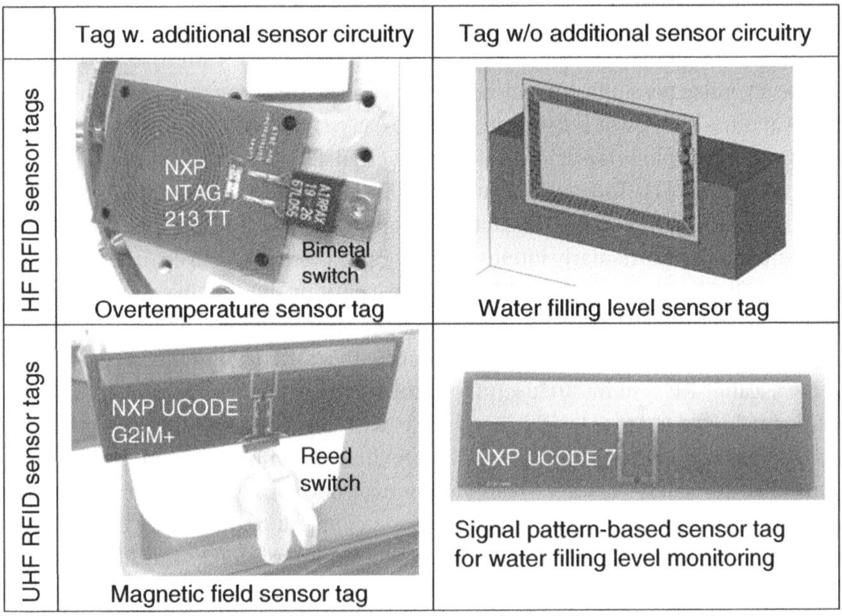

Figure 6.2 Overview of RFID sensor tags: the two UHF RFID sensor tags are discussed in detail in the main text. For completeness, the overview also shows two sensor tags based on the high frequency (HF) RFID technology using load modulation in near-field magnetic resonant coupling links. The overtemperature sensor HF RFID tag with additional sensor circuitry consists of a custom-built ferrite-backed coil, an NXP NTAG 213 TT, and a bimetal switch to monitor the overtemperature of automotive battery packs. The shown electromagnetic simulation model of an HF RFID sensor tag without (w/o) additional sensor circuitry consists of a sensing coil and an HF RFID tag chip [6] to monitor water filling levels. *Source:* Grosinger and Michalowska-Forsyth [1] /IEEE; Grosinger [5] /IEEE.

rely, in particular, on the ultra-high frequency (UHF) RFID technology based on the ISO/IEC 18000-6 standard of the International Organization for Standardization (ISO) and the International Electrotechnical Commission (IEC). The UHF RFID magnetic field sensor tag with sensor circuitry consists of a custom-built dipole antenna and an NXP UCODE G2iM+ chip. The chip has a one-bit sensor interface that is connected to a reed switch. The switch is used to monitor the open or close status of a valve that has a permanent magnet attached to it. When the valve is closed, the magnetic field of the permanent magnet triggers the reed switch to become conductive, thus creating a closed loop. As a result, the chip changes the state of the specific sensor bit in its memory, which can be read by the reader [7]. The UHF RFID sensor tag without additional sensor circuitry uses the antenna as a sensing device and consists of a custom-built dipole antenna and an NXP UCODE 7 chip to monitor a water canister's full or empty status [8, 9]. This sensor concept is discussed in more detail in Section 6.2.4. In the following, however, UHF RFID systems are described in more detail.

6.2.1 UHF RFID Systems

The UHF RFID system relies on wireless communication between the RFID reader and a multitude of passive RFID tags based on the ISO/IEC 18000-6 standard [10]. The wireless communication link between the RFID reader and the RFID tag can be divided into a forward link channel and a backward link channel, which can be related to downlink and uplink channels in wireless communication systems [11]. The read range of a UHF RFID system is defined as the maximum communication range between the reader and the tag. Worldwide UHF RFID systems operate in the frequency range of 860–960 MHz, covering reading distances of up to 10 m [10]. In Europe, UHF RFID systems operate specifically in the frequency range of 865.6–867.6 MHz.

In the forward link, the RFID reader transmits RF power and data wirelessly to the batteryless tag. To achieve robust wireless power transfer from reader to tag, the absorbed power by the tag chip must be higher than the chip sensitivity P_{min}, which is the minimum input power required for the tag chip circuitry to operate (e.g. −19 dBm for NXP UCODE 7 chip in reading mode). If the power is less than P_{min}, the forward link of the communication system is limited [12]. In the backward link, the reader triggers the tag to initiate its data transfer by modulating the backscattered signal. Then, depending on the transmitted tag data, the tag reflects the reader signal at the tag chip. The reader detects the backscattered tag signal and demodulates it. For robust backscatter communication, the received power at the reader receiver (RX) must be higher than the RX sensitivity. The RX sensitivity is the minimum amount of input power at the reader required to ensure that the tag data is received successfully (e.g. −80 dBm for a custom-built reader [13]).

If the power received is below the reader RX sensitivity threshold, the backward link of the communication system is limited [12].

As mentioned, a conventional RFID tag consists of an antenna and a chip. Figure 6.3 shows an equivalent tag circuit. Their impedances characterize the tag antenna and chip, that is, by the antenna impedance $Z_{Ant} = R_{Ant} + jX_{Ant}$ and by the chip impedances at the absorbing and reflecting modes, $Z_{Abs} = R_{Abs} + jX_{Abs}$ and $Z_{Ref} = R_{Ref} + jX_{Ref}$, respectively. R denotes the real part, and X denotes the imaginary part of the impedance. The tag generally modulates the backscattered signal amplitude, communicating its ID number using on–off keying modulation [14]. To do this, the tag chip switches between the chip impedance in the absorbing mode Z_{Abs} and the chip impedance in the reflecting mode Z_{Ref} as denoted in Figure 6.3. The reflected tag signal at the chip input can be related to the reflection coefficient in the absorbing mode S_{Abs} (representing a logical "0") and to the reflection coefficient in the reflecting mode S_{Ref} (representing a logical "1"), defined by the antenna and chip impedances [15],

$$S_{Abs} = \frac{Z_{Abs} - Z_{Ant}^*}{Z_{Abs} + Z_{Ant}} \quad \text{and} \quad S_{Ref} = \frac{Z_{Ref} - Z_{Ant}^*}{Z_{Ref} + Z_{Ant}} \tag{6.1}$$

6.2.2 RFID Reader

The UHF RFID reader comprises a baseband processor, a transmitter (TX), an RX, and one or more antennas [14]. The baseband processor generates the command sequences, demodulates the received signals, and controls the protocol by the ISO/IEC 18000-6 standard. The TX produces the RF signal to power the tag and sends commands, while the RX amplifies and purifies the received signal. During the communication process, the transmit signal leaks into the receive path,

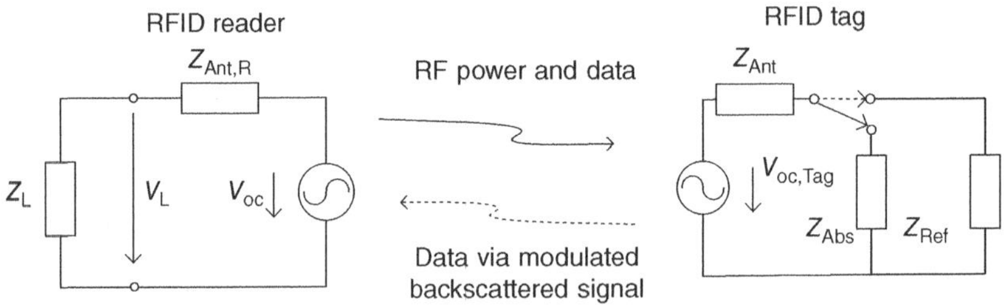

Figure 6.3 Thevenin equivalent circuits for the reader and tag in receiving modes [8, 16]: the tag antenna is modeled by an open-circuit voltage $V_{oc,Tag}$ and the tag antenna impedance Z_{Ant}. The backscattered tag signal is modulated by switching between the two chip impedances, that is, the absorbing impedance Z_{Abs} and the reflecting impedance Z_{Ref}. The reader RX antenna is modeled by an open-circuit voltage V_{oc} induced by the backscattered tag signal and the reader antenna impedance $Z_{Ant,R}$. The voltage V_L at the RX input depends on V_{oc}, $Z_{Ant,R}$, and the reader RX input impedance Z_L.

denoted as carrier leakage. In a practical system, the carrier leakage can be 65–90 dB higher than the backscattered tag signal [17]. Thus, a carrier cancellation is a must and is discussed in the next section.

Concerning the reader antennas, a single antenna is used for transmission and reception in the case of a monostatic antenna configuration. A circulator or a directional coupler then separates the TX and RX paths. A bistatic antenna configuration uses two antennas for transmission and reception [14]. UHF RFID readers with a multiple-input multiple-output (MIMO) antenna configuration exploit many TX and RX antennas, which can be favorably used for tag localization and tracking purposes [17].

6.2.2.1 Reader Receiver

As mentioned above, the baseband processor demodulates the received signals at the reader. At the baseband RX, the tag absorb state *State* (*A*) is detected when the tag is in its absorbing mode, and the tag reflect state *State* (*R*) is detected when the tag starts to transmit its ID [18]. The tag signal h can then be calculated by $h = State\ (R) - State\ (A)$, which equals a carrier cancellation, as the tag absorb state is essentially equal to the carrier leakage [18].

Figure 6.4 shows an exemplary tag signal h after carrier compensation as measured in the RX baseband of a custom-built reader [13, 17]. The tag signal is plotted in a signal constellation diagram [11]. As shown in Figure 6.4, the custom-built reader RX is capable of measuring the tag signal in amplitude and phase, which is important for detecting antenna-based sensor tags (see Section 6.2.4). To methodologically design antenna-based sensor systems, it is important to relate the tag signal detected at the reader receiver to the differential scattered electric field of the sensor tag antenna [8, 9]. In the following, a general signal model is deduced to create this relation. At the reader receiver, the baseband and passband signals can be clearly related to each other if the reader RX parameters are known [8]. Thus, the baseband voltage level *State* (*A*) can be clearly related to the passband voltage level V_{abs}, and *State* (*R*) can be related to V_{ref}. Then, the tag signal at the reader receiver can be calculated in the reader RX passband, that is,

$$h_{RF} = V_{ref} - V_{abs} \tag{6.2}$$

The voltages at the receiver input V_{abs} and V_{ref} can be related to the scattered electric field of the tag antenna by examining the equivalent circuit of the reader antenna in receiving mode in Figure 6.3. Referring to this circuitry, the RX input voltage V_L (i.e. V_{abs} and V_{ref}, respectively) can be calculated at the RX input impedance Z_L with this equation:

$$V_L = V_{oc} \frac{Z_L}{Z_{Ant,R} + Z_L} = \vec{\ell}_e \cdot \vec{E}_{inc} \frac{Z_L}{Z_{Ant,R} + Z_L} \tag{6.3}$$

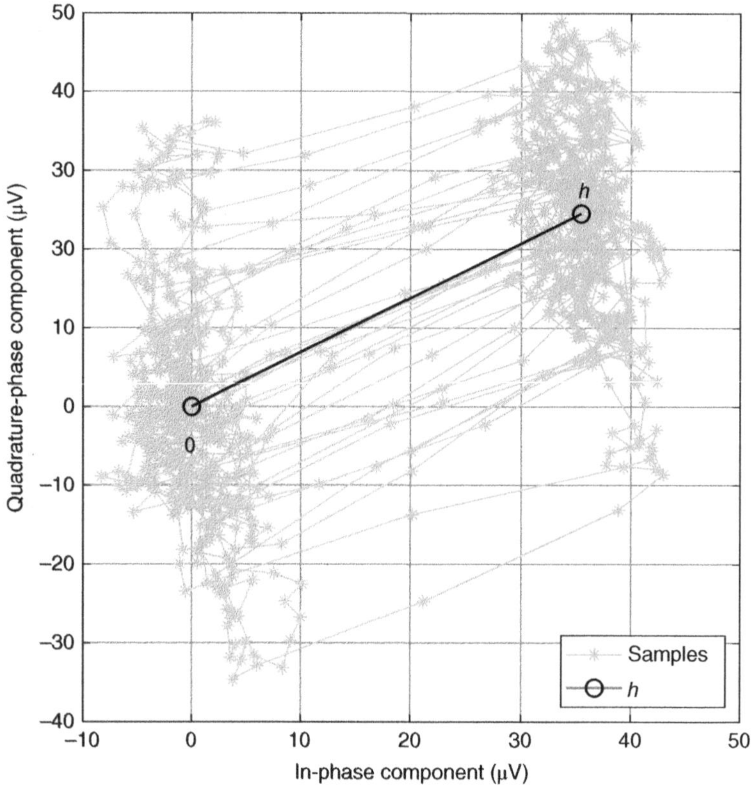

Figure 6.4 Measured signal constellation diagram of an exemplary tag signal h after carrier compensation: the tag signal was measured over time in an anechoic chamber using a custom-built RFID reader with an RX sensitivity of about −80 dBm [13, 17]. In the tag absorbing mode, the reader baseband receiver estimates the tag absorb state *State* (*A*) as the temporal mean of the received symbol. The tag reflect state *State* (*R*) is estimated when the tag initiates backscatter modulation. The tag signal h is then determined as $h = State\ (R) - State\ (A)$.

where $Z_{\text{Ant,R}}$ is the impedance of the reader RX antenna. The scattered electric field of the tag antenna induces an open-circuit voltage V_{oc}, which depends on the vector effective antenna length $\vec{\ell}_{\text{e}}$ of the reader antenna in the tag direction [8, 19], and the incident electric field $\vec{E}_{\text{inc}}$, which is the scattered electric field of the tag antenna depending on the respective chip impedance, $\vec{E}_{\text{Scat}}\left(Z_{\text{Abs}}\right)$ and $\vec{E}_{\text{Scat}}\left(Z_{\text{Ref}}\right)$, respectively. Inserting (6.3) in (6.2) leads to the tag signal in the reader RX passband:

$$h_{\text{RF}} = \vec{\ell}_{\text{e}}\ \Delta\vec{E}_{\text{Scat}} \frac{Z_{\text{L}}}{Z_{\text{Ant,R}} + Z_{\text{L}}} \tag{6.4}$$

with the differential scattered electric field defined as $\Delta\vec{E}_{\text{Scat}} = \vec{E}_{\text{Scat}}(Z_{\text{Ref}}) - \vec{E}_{\text{Scat}}(Z_{\text{Abs}})$. Equation (6.4) shows the tag signal h_{RF} is proportional to the differential scattered electric field of the tag antenna $\Delta\vec{E}_{\text{Scat}}$, multiplied by a factor that depends on the reader RX equivalent circuit.

A free-space-system scenario is considered to derive the differential scattered electric field, $\Delta\vec{E}_{\text{Scat}}$ (i.e. the reader and tag antennas are located in each others' far fields). At a specific observation point in the far field, the load-dependent scattered electric field $\vec{E}_{\text{Scat}}(Z_{\text{Abs|Ref}})$ of the receiving tag antenna can be represented by [20, 21]: A free-space-system scenario is considered to derive the differential scattered electric field, $\Delta\vec{E}_{\text{Scat}}$ (i.e. the reader and tag antennas are located in each others' far fields). At a specific observation point in the far field, the load-dependent scattered electric field $\vec{E}_{\text{Scat}}(Z_{\text{Abs|Ref}})$ of the receiving tag antenna can be represented by [8, 20, 21]:

$$\vec{E}_{\text{Scat}}(Z_{\text{Abs|Ref}}) = \vec{E}_{\text{Scat}}(Z^*_{\text{Ant}}) - S_{\text{Abs|Ref}} \frac{I_0 Z_{\text{Ant}}}{2R_{\text{Ant}}} \frac{\vec{E}_{\text{Ant}}}{I_{\text{Ant}}} \tag{6.5}$$

The first term of (6.5) represents the structural mode component, that is, the field scattered in the case of a conjugate-matched load. The second term of (6.5) represents the antenna mode component, which depends on the tag reflection coefficients S_{Abs} and S_{Ref}, respectively, the tag antenna impedance Z_{Ant} and its respective resistance R_{Ant}, the antenna short-circuit current I_0, and the tag antenna transmitting parameters $\vec{E}_{\text{Ant}}$ and I_{Ant} [8, 20, 21]. Assuming an antenna driving current I_{Ant} at the tag antenna input, $\vec{E}_{\text{Ant}}$ is the transmitted electric field of the tag antenna at the specific observation point. Once (6.5) has been calculated, the differential scattered electric field $\Delta\vec{E}_{\text{Scat}}$ can be obtained as:

$$\Delta\vec{E}_{\text{Scat}} = \vec{E}_{\text{Scat}}(Z_{\text{Abs}}) - \vec{E}_{\text{Scat}}(Z_{\text{Ref}}) = (S_{\text{Ref}} - S_{\text{Abs}}) \frac{I_0 Z_{\text{Ant}}}{2R_{\text{Ant}}} \frac{\vec{E}_{\text{Ant}}}{I_{\text{Ant}}} \tag{6.6}$$

which is proportional to the difference between the reflection coefficients in the absorbing and reflecting modes. Inserting (6.6) in (6.4) completes the tag signal model:

$$h_{\text{RF}} = \vec{\ell}_{\text{e}} (S_{\text{Ref}} - S_{\text{Abs}}) \frac{I_0 Z_{\text{Ant}}}{2R_{\text{Ant}}} \frac{\vec{E}_{\text{Ant}}}{I_{\text{Ant}}} \frac{Z_{\text{L}}}{Z_{\text{Ant,R}} + Z_{\text{L}}} \tag{6.7}$$

This tag signal model allows designing of signal-pattern-based sensor tags as further discussed in Section 6.2.4.

6.2.3 RFID Tags

As indicated in (6.1), the backscattered tag signal at the chip input can be related to the reflection coefficients S_{Abs} and S_{Ref} defined by the respective antenna and chip impedances. Based on the reflection coefficients, the tag performance parameters τ and η_{Mod} can be calculated, that is, the power transmission coefficient and the modulation efficiency. The power transmission coefficient τ is a quality measure of the wireless power transfer toward the passive tag chip [19],

$$\tau = 1 - |S_{Abs}|^2 \text{ with } \tau \in [0,1] \tag{6.8}$$

while the modulation efficiency η_{Mod} is a measure of the modulation quality of the reflected tag signal [22],

$$\eta_{Mod} = \alpha |S_{Abs} - S_{Ref}|^2 \text{ with } \eta_{Mod} \in [0,1] \text{ and } \alpha = 0.25 \tag{6.9}$$

assuming a duty cycle of 50% [22]. Optimized antenna and chip impedances of UHF RFID tags maximize τ and η_{Mod} ensuring a good wireless power transfer from reader to tag and a sound data transmission toward the RFID reader [1].

6.2.3.1 Tag Chip

As shown in Figure 6.3, the chip can be characterized by the chip input impedances in the absorbing and reflecting modes, $Z_{Abs} = R_{Abs} + jX_{Abs}$ and $Z_{Ref} = R_{Ref} + jX_{Ref}$. These input impedances are generally nonlinear and vary with chip input power and operating frequency. The chip's RF frontend and the power consumption determine the power dependence. The frequency dependence is primarily due to the parasitic and packaging effects of the chip [14].

Figure 6.5 illustrates the block diagram of an example differential RF frontend of a custom-built tag chip in a 40 nm CMOS technology [23]. The RF terminals RF_1 and RF_2 of the frontend are directly connected to the tag antenna. Two pairs of antiparallel diodes form an electrostatic discharge (ESD) protection, which also protects the chip frontend from overvoltages at high levels of chip input power. An RF charge pump, also known as RF to direct current (DC) power converter, rectifies the RF signal received at the antenna and converts it into a DC voltage for chip power supply, that is, V_{DC}. A DC limiter regulates V_{DC} to prevent the generated V_{DC} voltage from reaching levels above the safe operating area of the on-chip circuits. The modulator of the tag chip is comprised of typical shunt transistors to switch the RF input to a low impedance state, that is, Z_{Ref}, modulating the reflected tag signal for backscatter communication. In the 40 nm CMOS technology, a dedicated RF limiter is used to regulate the RF input voltage at the chip [6, 24]. The chip uses an envelope detector to demodulate the amplitude-modulated signal transmitted by the reader. The modulator control signal $Data_{TX}$ and the demodulator output signal $Data_{RX}$ are connected to a digital control circuit, which handles the communication protocol

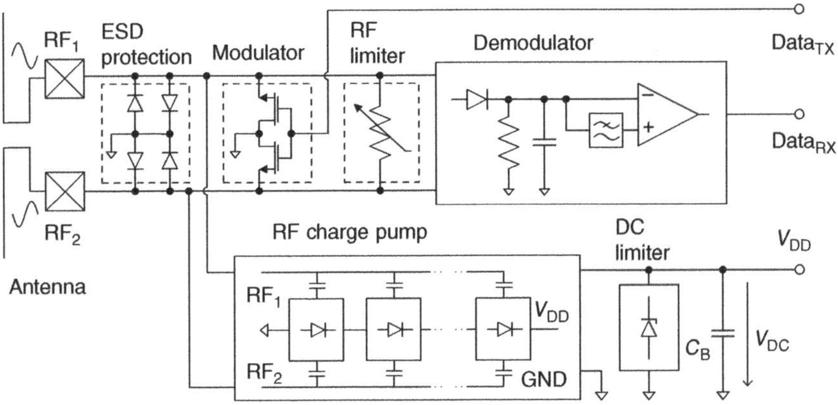

Figure 6.5 Block diagram of an exemplary differential frontend of a UHF RFID tag chip [1]: the RF charge pump rectifies the signal impinging at the antenna and provides a DC voltage at its output used as a voltage supply for the connected internal chip circuitry at node V_{DD}. The capacitor C_B decouples the RF part of the chip from the internal chip circuitry. Moreover, C_B presents one part of the buffer capacitance that buffers energy to supply the chip circuitry during modulation. The RF limiter and the ESD protection are typically implemented as a combined circuit to protect the circuit from overvoltages at the chip input [23].

and the access to nonvolatile memory. The tag ID is typically contained in only a few hundred bits stored in a nonvolatile on-chip electrically erasable programmable read-only memory (EEPROM), characterized by excellent retention of the stored data for a long time and without the necessity of a power supply.

Figure 6.6 shows the measured input impedance Z_{Abs} of the discussed exemplary tag chip [23]. The figure illustrates the typical characteristics of a chip in the absorbing mode $Z_{\mathrm{Abs}} = R_{\mathrm{Abs}} - jX_{\mathrm{Abs}}$ versus chip input power P_{Chip}. The chip shows a predominately capacitive input impedance of $Z_{\mathrm{Abs}} = (15 - j239)\ \Omega$ at an input power equal to the chip sensitivity $P_{\mathrm{min}} = -20$ dBm. The chip sensitivity P_{min} is determined by the RF frontend and the power consumption of the chip. Figure 6.6 shows that Z_{Abs} changes drastically as the input power increases. At high input power levels, the absorbing impedance converges to the reflecting impedance [25]. This behavior is due to the internal overvoltage protection circuits [6, 24].

6.2.3.2 Tag Antenna

As mentioned previously, antenna and chip impedances are optimized to maximize the power transmission coefficient τ as shown in (6.8). Inserting the reflection coefficient in the absorbing mode provided in (6.1) into (6.8) leads to [19]:

$$\tau = \frac{4R_{\mathrm{Abs}}R_{\mathrm{Ant}}}{\left|Z_{\mathrm{Abs}} + Z_{\mathrm{Ant}}\right|^2} \tag{6.10}$$

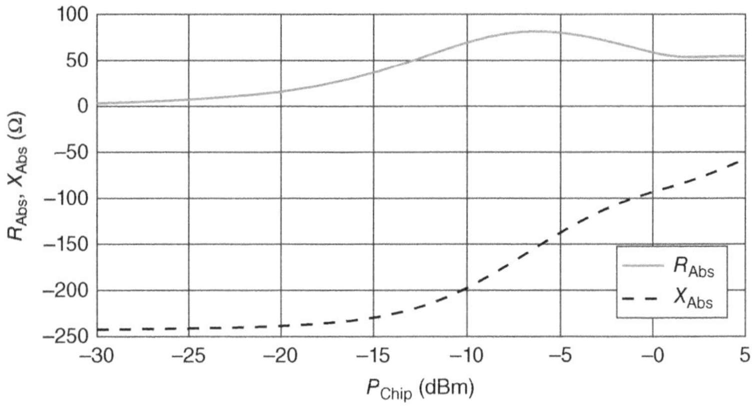

Figure 6.6 Real and imaginary parts of absorbing impedance $Z_{Abs} = R_{Abs} + jX_{Abs}$ of the passive RFID chip [23]: Z_{Abs} is measured versus chip input power P_{Chip} at 900 MHz. The chip sensitivity P_{min} is −20 dBm.

As the RF frontend design determines the input impedance of the chip, the antenna impedance is typically optimized to maximize τ and, thus, to maximize the wireless power transfer from the tag antenna toward the chip. Equation (6.10) shows that τ reaches its maximum if the antenna impedance is the complex conjugate of the chip absorbing impedance, that is, $Z_{Ant} = Z^*_{Abs}$ (at the specific operating frequency and chip sensitivity P_{min}).

Tag antennas in UHF RFID systems are typically based on half-wavelength dipole antennas, which include impedance-matching structures, such as T-match structures, to create a match between the tag chip and the antenna impedance [8, 25, 26]. During the antenna design, the tag environment has to be considered, as dielectric and conductive materials in the near field of an antenna affect the antenna properties considerably [19, 25]. This effect is exploited to create antenna-based sensing in UHF RFID tags (see Section 6.2.4).

6.2.3.3 Radar Cross Section

In the following, the differential radar cross section (RCS) of the tag antenna is calculated, which is another important performance parameter for UHF RFID tags that determines the power of the modulated backscattered tag signal [27]. The differential RCS can be calculated based on the differential scattered electric field $\Delta\vec{E}_{Scat}$ derived in (6.6).

The short-circuit current I_0 of the tag antenna in (6.6) can be further calculated by making use of the tag antenna equivalent circuit shown in Figure 6.3, in which the tag antenna is modeled by an open-circuit voltage $V_{oc,\,Tag}$ and its antenna impedance Z_{Ant} [27]:

$$I_0 = \frac{V_{oc,Tag}}{Z_{Ant}} = \frac{\vec{\ell}_e\,\vec{E}_{inc,Tag}}{Z_{Ant}} \tag{6.11}$$

where $\vec{E}_{\text{inc,Tag}}$ is the incident electric field at the tag and $\vec{\ell}_{\text{e}}$ is the vector effective antenna length of the tag antenna. In addition, the transmitted electric field of the tag antenna in (6.6) can be written as [19]

$$\vec{E}_{\text{Ant}} = -j\eta_0 \frac{I_{\text{Ant}}}{2\lambda r} \vec{\ell}_{\text{e}} \text{e}^{-j\beta r} \tag{6.12}$$

where η_0 is the free space impedance, $\beta = 2\pi/\lambda$ is the wave number, and r is the distance between the tag and the observation point. By using (6.11) and (6.12), the differential scattered field in (6.6) can be rewritten as:

$$\Delta\vec{E}_{\text{Scat}} = -j\left(S_{\text{Ref}} - S_{\text{Abs}}\right) \frac{\left|\vec{\ell}_{\text{e}}\right|^2 \vec{E}_{\text{inc,Tag}}}{2R_{\text{Ant}}} \frac{\eta_0}{2\lambda r} \text{e}^{-j\beta r} \tag{6.13}$$

In turn, $\left|\vec{\ell}_{\text{e}}\right|^2$ can be directly related to the tag antenna gain G_{Tag} in the reader direction by [21]

$$\left|\vec{\ell}_{\text{e}}\right|^2 = \frac{G_{\text{Tag}} \lambda^2 R_{\text{Ant}}}{\pi \eta_0} \tag{6.14}$$

which then leads to the following representation of $\Delta\vec{E}_{\text{Scat}}$:

$$\Delta\vec{E}_{\text{Scat}} = -j\left(S_{\text{Ref}} - S_{\text{Abs}}\right) \frac{G_{\text{Tag}} \lambda}{4\pi r} \vec{E}_{\text{inc,Tag}} \text{e}^{-j\beta r} \tag{6.15}$$

Using (6.15), the differential RCS of the tag antenna can be calculated by [20, 28]:

$$\sigma = \lim_{r\to\infty} 4\pi r^2 \frac{\left|\Delta\vec{E}_{\text{Scat}}\right|^2}{\left|\vec{E}_{\text{inc,Tag}}\right|^2} = \frac{\lambda^2 G_{\text{Tag}}^2}{4\pi} \left|S_{\text{Ref}} - S_{\text{Abs}}\right|^2 \tag{6.16}$$

This result is equal to the differential RCS as calculated by Nikitin et al. [27], implicitly validating the tag signal model derived in (6.7) in Section 6.2.2.

6.2.4 Antenna-based RFID Sensors

As mentioned before, one type of RFID sensor tags integrate passive transducers that change their RF parameters when the tag environment changes (e.g. temperature, curvature, liquid level) [1]. These parameter changes additionally modulate the digitally modulated tag signal in the analog domain detectable at the reader, see (6.7). A cost-effective and ultra-low-power approach to integrating a passive transducer into an RFID tag uses the tag antenna as the sensing device. One significant benefit of antenna-based sensing is that the sensor tag can still rely on off-the-shelf RFID tag chips and does not need an additional sensor device as, for example, a reed or bimetal switch as shown in Figure 6.2.

A literature review in [1] presents antenna-based sensor tag prototypes of recent years. The prototypes use analog modulation concepts to enable antenna-based sensing and high sensor system performance. The UHF RFID sensor system performance is affected by two significant factors [8]. The system performance degrades due to a non-sufficiently high wireless power transfer toward the batteryless tag chip, that is, τ gets smaller than one, see (6.10). Additionally, the sensor system performance degrades due to a dependence of the detected sensor data on the distance between the reader and tag, see (6.12), and thus on the wireless channel.

The initial prototypes from 2009 use an additional analog amplitude modulation to communicate the sensor states. However, this modulation type often leads to variations in the wireless power transfer toward the passive tag chip and, thus, to severe impairments in system performance. Therefore, in 2013, researchers started to control these variations in wireless power transfer by introducing constraints into the antenna design of the sensor tag [29], ensuring a sufficiently high τ. However, further research presented in 2014 and 2015 shows that pure phase modulation [30] or a combination of amplitude and phase modulation [31] achieves a considerably better system performance than pure amplitude modulation, as both modulation concepts inherently guarantee a high wireless power transfer. In 2016, researchers started exploiting off-the-shelf RFID tag chips that offer self-tuning capabilities [32, 33], ultimately ensuring a constant wireless power transfer in the case of pure amplitude modulation.

In addition to the need to provide a sufficiently high wireless power transfer, the other issue associated with most of the system prototypes presented so far is that they do not offer setup independence (i.e. independence of the reader-to-tag distance). Each time the reader-to-tag distance or the wireless channel changes, the systems need to be recalibrated; this is inconvenient in practical applications. In 2009, an analog identifier [34] solved the issue early on for amplitude-modulated systems. However, systems that exploit phase modulation and self-tuning capabilities of tag chips (in the nonlinear regime [32]) still suffer from this issue. Using multi-port tags solves the problem regarding phase-modulated sensor tags [35], however, with an increased cost and size of the sensor tags.

This tag signal model presented in (6.7) has been used to design sophisticated sensor tag antennas using signal patterns [1, 8, 9].

Finally, in 2018, a novel UHF RFID sensor system prototype was presented based on a combination of amplitude and phase-modulated sensor tag signals [8, 36]. The system relies on tag signal patterns to communicate the sensor state, that is, a specific tag signal pattern relates to a set of discrete sensor states, enabling, for example, binary sensing [1]. This system differs from the previously mentioned systems in which one specific tag signal is related to one discrete sensor state. As a result, the system constantly offers by a sophisticated antenna design a high wireless power transfer and setup independence [8, 9], that is, by and optimized antenna impedance, see (6.10), and optimized differential scattered electric field, see (6.13). Further analysis of signal pattern-based sensor systems shows, to a certain extent, the system's ability to operate robustly in harsh application environments, which are often subject to multipath propagation [1].

6.2.4.1 Harsh Application Environments

In harsh application environments, passive wireless UHF RFID sensor tags face a major reliability issue, that is, tight power constraints. This is due to the fact that the sensors operate without a battery. The same issue is present in terrestrial environments as well as in space environments [1].

As mentioned before, today, a backscatter RF technology is used for inventory procedures at the ISS [2]. In particular, surface acoustic wave (SAW) tags have been selected for the inventory application at ISS because they can be reliably detected even when placed on, behind, or between items with high liquid and metal content of randomly packed crew bags [2]. In addition, SAW backscatter RF tags are inherently robust against ionizing radiation since they have no electronics on board. However, SAW tags have significant limitations in contrast to modern backscatter RF tags equipped with state-of-the-art microchips based on low-cost CMOS technologies [1]. Due to these limitations, there have been investigations into using chip-based backscatter RF tags for identifying goods in space habitats in 2006. However, the reading reliability results in space habitats' harsh RF propagation environment did not warrant proceeding to an on-orbit test [2].

However, the recent development of modern backscatter RF tags might warrant proceeding to revisit the use of these devices in space habitats. Modern backscatter RF tags, that is, UHF RFID tags, provide a fully standardized communication protocol and anti-collision techniques. In addition, the performance of tag chip frontends has been improved constantly in terms of tag sensitivity from −10 to −20 dBm [1], relieving the system's power budget considerably. Last but not least, the addition of passive sensing capabilities to backscatter RF tags brings additional functionality without impacting the power budget at all.

6.3 Radiation Environments and Radiation Hardness

In the context of RFID applicability to ionizing environments for sensing and communication purposes, there are two key facts to observe. The first one is the multitude of ionizing radiation environments. These are briefly presented with an overview of the variety of radiation fields, where the most diversity is observed in natural space radiation. However, also characteristic radiation types of manmade sources are listed. A selection of the most common applications has been presented before in Section 6.1. The second observation is the fact that ionizing radiation causes damage to electronic components. Starting with the topic of interaction with matter through component and circuit damage mechanisms, an overview of radiation hardening techniques is given in this section.

6.3.1 Radiation Environments

6.3.1.1 Space Radiation

The sources of radiation can be either natural or manmade. The natural space radiation environment [37] provides the whole spectrum of ionizing radiation in terms of particle types and their energies. There are three distinct kinds of space radiation seen from our perspective: X-rays and particles produced during solar flares, cosmic rays originating from outside of the solar system, and charged particles trapped in the Earth's magnetosphere. Solar activity is driven by a 7-year-long cycle of flares with varying frequency and intensity, hence the term "space weather." The flares produce mainly protons but also electrons, alphas, and X-rays. The cosmic radiation from the sources outside of the solar system includes a similar composition, reaching higher energies though of lower intensity. Further, it includes heavy ions and other particles. As a result of the solar and cosmic radiation, the Earth's magnetosphere traps high-energy electrons and protons in the Van Allen Radiation Belts. These are zones around our planet where the radiation levels are particularly intense—the inner and the outer donut-shaped belts.

Particles or photons that possess sufficient energy to ionize atoms in the material they interact with are classified as ionizing radiation. This term encompasses, therefore, already the ultraviolet (UV) radiation. The shorter wavelength of electromagnetic radiation carries even higher energy. At a few hundred of eV, it is accepted to speak about the soft X-rays that easily interact with matter. Electromagnetic radiation above 5 keV is known as hard X-rays. The higher energy photons exist in a small intensity at ultra-high energies of hundreds of TeV. They are more penetrating, more difficult to detect, and therefore also less damaging. Dictated by different matter interaction phenomena, the heavily charged particles of extremely high energies will have a lower stopping power [38] and thus will travel through matter almost unnoticed. The primary concern for radiation effects on space and air-borne instruments is the natural space radiation in mid-energies. This radiation includes X-rays ranging from several keV to MeV and charged particles ranging from several to hundreds of MeV.

The Earth's magnetosphere and the atmosphere protect life on Earth from ionizing space radiation and its damaging effects on biological tissues. Nevertheless, charged particles of extremely low intensity reach the ground level. They can also be a secondary product of the interaction of other space-origin particles with the

Earth's atmosphere. Their effect on electronics will be discussed under the term single-event effects (SEEs) (see Section 6.3.3). It has to be considered, especially in applications of high functional safety grade [39]. The level of significance is similar to the background radiation source, the naturally present radioisotopes, with charged particles due to alpha and beta decay, which also can result in SEEs.

6.3.1.2 Manmade Radiation Sources

The manmade radiation includes the same types of radiation as those of the natural origins, observed in space. However, in their variety, the radiation sources produce only certain types of radiation and energies, with the resulting radiation environments much simpler than in space. Meanwhile, the photon or particle flux can be weaker or much stronger than the space radiation intensity. The sources to distinguish are X-ray generators, particle accelerators, and radionuclides. Radionuclides, besides the primordial ones that are residues from the Big Bang, are produced in nuclear fission and synthesized in accelerators or in nuclear reactors. When electrons are accelerated to high energy and hit a target, X-rays can be produced. Electrons slow down in the high atomic number target material, thus producing bremsstrahlung [40], X-rays in a continuum energy spectrum. The maximum energy cannot exceed the high voltage applied for electron acceleration. The commercial X-ray tubes operate at voltages up to 300 kV. X-rays can also be produced in accelerators, where a particle accelerated to high energy is deflected, again resulting in bremsstrahlung photons. However, there are different types of accelerators for different purposes and producing different radiation compositions, including heavy ions [41].

6.3.2 Interactions with Materials

Ionizing radiation will penetrate the environment and materials and it will interact or not. What happens is dependent on the type of radiation, the material of concern—its density, atomic number, and the energy of incident radiation, as conceptually illustrated in Figure 6.7. This together gives certain interaction probability. The next question is, what is the product of the interaction? This can be simply ionization of the material with the initial photon or particle energy completely absorbed locally in the form of free charge carriers and thermal energy. However, many interactions, like the Compton effect, pair production, or nuclear scattering, leave a product of secondary particles or photons. These may further interact locally or leave the material without further energy deposition. From the ionizing dose point of view, if the energy is not deposited, the radiation has penetrating properties. For example, 10 keV photons will deposit more energy in a water sample than the same flux, measured in photons per unit area per second, of 100 keV X-rays. In semiconductor devices,

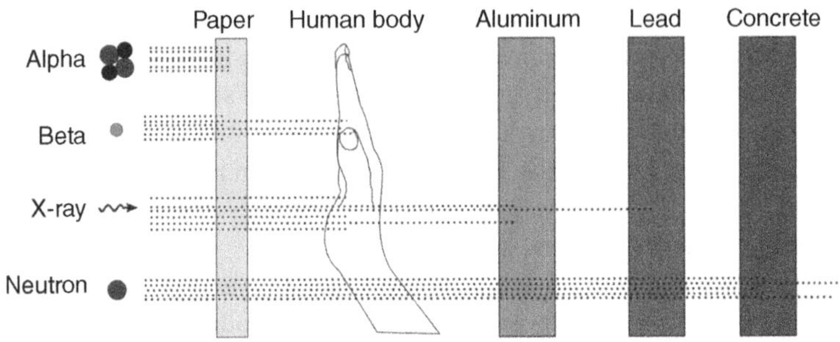

Figure 6.7 Absorption of ionizing radiation depends on the radiation type, its energy, as well as the material composition, its density, and atomic number. These parameters will also determine the probability of a given interaction, for example, the photoelectric effect, Compton effect, pair production, or elastic scattering. As a result of interaction, there will be emission of electron–hole pairs, micro-displacement of atoms in the material, increased thermal energy, or emission of secondary interaction products, photons, or particles.

the deposited ionizing dose results in the generation of charge carriers. This charge can have damaging effects, for example, by getting trapped in insulating layers and inducing permanent electric fields that modify device characteristics. This type of damage is related to total ionizing dose (TID). In a biased device, it is possible that the carriers are discharged in an uncontrolled rapid local current pulse that can cause damage. Typically this is a result of charged particles, and the phenomenon is known as a single-event effect (SEE). Another result of the interaction is nonionizing energy deposition. This is a possible result for incident particles with mass, for example, protons, neutrons, or heavy ions. In the case of the crystal structure of semiconductors used as the base material for all electronic devices, this leads to the displacement of atoms in the crystal structure, changing the electrical parameters: effective doping, resistivity, and charge transport properties. This type of damage is known as displacement damage (DD), and the severity is measured by a total nonionizing dose (TNID), also referred to as nonionizing energy loss (NIEL).

6.3.3 Radiation Effects in Electronic Devices

The radiation effects are generally grouped into cumulative effects and SEEs. The first group is similar to the aging process of electronic components. However, in addition to accelerating factors of temperature, electric field, and current flow, there is also the accelerating factor of certain dose rates: TID or TNID, due to ionizing radiation. The second group is comprised of SEEs, where at any point in the lifetime, a single charged particle depositing a track of ionized charge leads to malfunction or damage of a semiconductor device or circuit. It has been shown that in some cases, the probability of SEE or the severity of induced damage increases if the cumulative damage happened in the first place. This is the case, for example, for single event latch-up (SEL) after applying NIEL [42].

6.3.3.1 Total Ionizing Dose

Starting with TID, typically, the MOSFET transistor is known to suffer a lot from this type of damage. TID effects occur mainly in insulators, with the effects being more severe under an applied bias voltage. Incident ionizing radiation leads to the generation of electron–hole pairs. In the conductive layers, this leads to photocurrents which are usually negligible. The more alarming situation is related to the insulating parts in the device structure, and in particular, the MOSFET gate. As in silicon dioxide, the electrons are much more mobile than holes. They are easily swept out from the oxide by tunneling. This leaves behind a positive charge, which gets trapped or under applied electric field moves by a hopping transport to the semiconductor-insulator interface. At the interface, it may lead to the release of hydrogen ions. These are typically introduced in CMOS circuits to improve the silicon surface quality. Without hydrogen passivation, there is more surface charge trapping in the MOSFET oxide [43], leading to higher noise and off-state leakage current. The parasitic electric field induced by oxide-trapped charge also results in shifts in transistor parameters, as conceptually illustrated in Figure 6.8. In consequence, the following is observed: (i) shift in threshold voltage, with the absolute value higher in p-channel and lower in n-channel MOSFET, (ii) degradation of mobility and consequently (iii) lower transconductance, (iv) lower on-state current, and (v) worse profile of the subthreshold slope [44]. The effects occurring in insulators have an influence also on the operation of diodes and bipolar transistors in planar processes, silicon-on-insulator-based MOSFET, and potentially also other semiconductor sensors and devices, mainly because of the phenomena at the silicon–insulator interface, whether it is silicon dioxide or silicon nitride [45].

6.3.3.2 Total Non-ionizing Dose

The second listed radiation damage effect, displacement damage, is also a cumulative effect due to TNID caused by high-energy particles with mass penetrating the material. The incident radiation acts as pressure on the atoms of the semiconductor crystal, leading to their displacement, single damage, or even production in cascades. Newly produced crystal defects, in the form of a Frenkel pair, form charge traps that facilitate generation and recombination, resulting in increased noise in MOSFET transistors, higher dark current in off-states field effect and bipolar devices, as well as gain degradation in bipolar transistors [46]. A Frenkel pair is a missing atom and an interstitial atom in a nearby location. For it to be created, the energy of the incident radiation has to be above the displacement threshold energy.

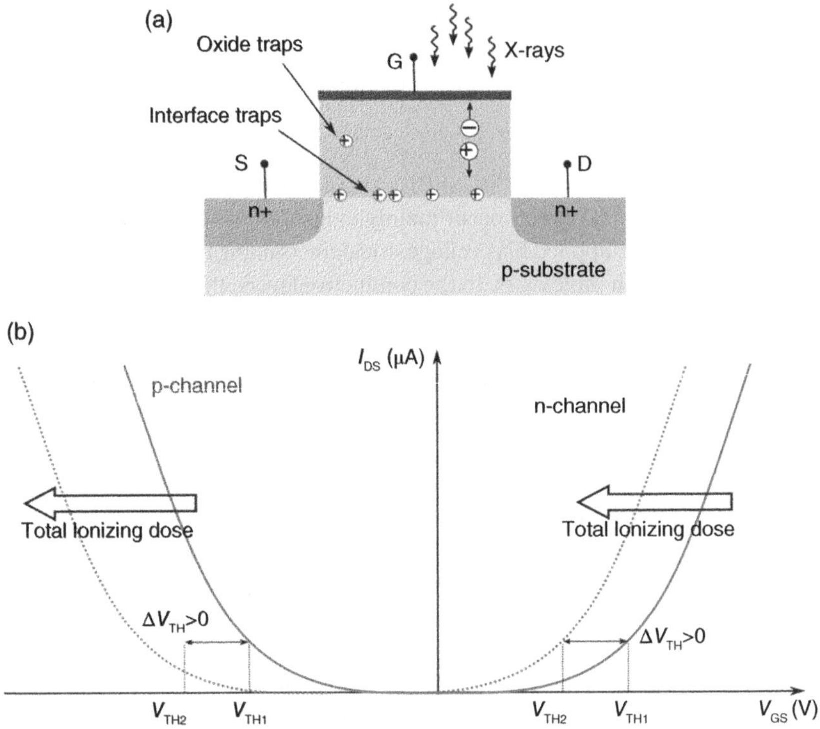

Figure 6.8 Ionization leads to the generation of free electron–hole pairs. In oxide, the charge carriers are less mobile, especially holes will not tunnel as easily as electrons. This leaves behind an uncompensated charge in the oxide. Also, interface traps are created in the dynamic process. In consequence, the MOSFET parameters degrade. One of the effects is the MOSFET threshold voltage shift V_{TH}, observed on the drain current I_{DS} versus gate–source voltage V_{GS} characteristics. It is typically greater in p-channel MOSFET.

6.3.3.3 Single-event Effects

One more group of radiation effects to cover is the SEEs. In the case of charged particles moving at relativistic speed, their energy is partially deposited in the material through Coulomb elastic scattering, where atoms are ionized along the particle track while the particle loses its energy. The energy loss depends on the particle, its energy, and the material that it interacts with. It is measurable and known as linear energy transfer, which is deposited energy per unit length. Knowing the energy and semiconductor electron–hole creation energy, the mean number of generated charge carriers can be calculated. For particles in the MeV energy range, there may be enough energy available locally that disturbs the chip functionality. The deposited charge will appear in the circuit as an instantaneous current pulse. Depending on the discharge path, various consequences can follow. In the most severe case, a single event may lead to latch-up, damaging the chip. The root cause for that is the parasitic thyristor structure in the CMOS circuit that triggers the opening of the path between the upper and lower power supply. This effect is shown in Figure 6.9. Another possibility is that the pulse is read out as a false signal. This effect is known as single-event

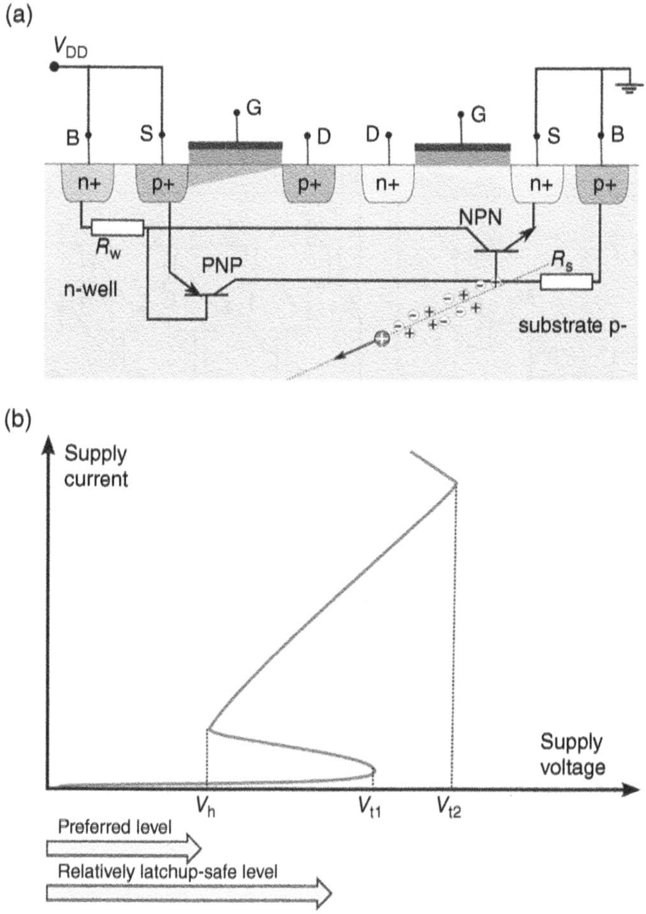

Figure 6.9 A track of ionized atoms with electron–hole pairs of free charge carriers along the passage trajectory of a single charged particle (e.g., a heavy ion) can cause various problems in the circuit. These all fall into the category of single-event effects. One of them is a single-event latch-up, where a low-ohmic current path through a parasitic CMOS thyristor is triggered (V_{t1}). A high supply current can eventually lead to a catastrophic thermal runaway that damages the circuit (V_{t2}). This condition can be stopped by reducing the power supply level below the holding voltage (V_h).

transient (SET). If powerful enough, a transient event can change the state in the static random access memory (SRAM) cell. A situation like that is referred to as a single-event upset (SEU). Both SET and SEU are so-called soft errors that only disturb a circuit's functionality temporarily. The examples are illustrated in Figure 6.10. A comprehensive review of SEEs on electronic components is covered in [47] and [48].

6.3.4 Radiation Hardening Techniques

Depending on the damage mechanism, it is possible to mitigate the effects of ionizing radiation on electronic devices. This can be approached at several levels of

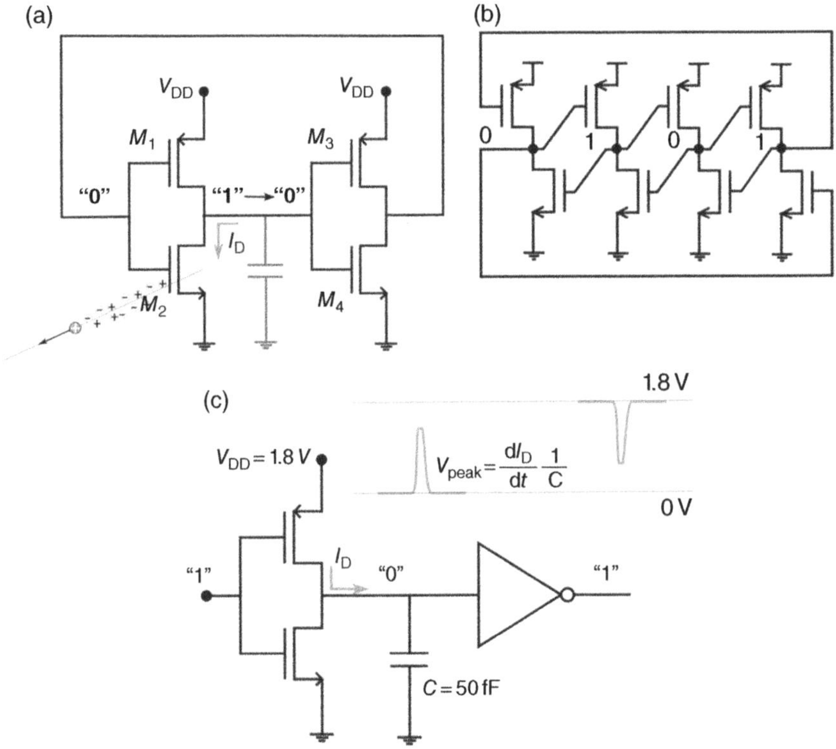

Figure 6.10 A single event can lead to changing the state of an active memory cell (a). This effect is called single-event upset. A dual-interlocked storage cell is a solution to increased radiation hardness of SRAM (b). In other parts of the circuit, the collection of SEE-induced charge can lead to transient noise, the effect called single-event transient, which can be wrongly interpreted by the system as a signal (c). Both SEU and SET are so-called soft failures since no permanent damage happens to the system, though both might lead to serious high-level consequences.

the product concept: the technology process level, the physical layout level, and the architecture level. The latter two, however, are most often associated with increased layout or circuit complexity.

6.3.4.1 Technology Level

Consider two IC components with the same feature size, that is—the same technology node and the same transistor dimensions. However, they are fabricated in two different CMOS foundries. Even if the fabrication processes are well adjusted so that the component electrical performance parameters are very similar, they will likely have different radiation tolerances. This is because the recipes for fabricating integrated devices are uniquely tuned, with differences in gate oxide quality, the level of implantation-induced damage, or local resistivity. Usually, a modification of a fabrication process to make it more radiation hard is not an available option. Therefore, the technology-level mitigation considers mainly a careful selection of the fabrication process so that it is intrinsically possibly radiation hard. An experimental comparison of the available options is usually necessary. Further additional process options or features can also be considered. For example, in the case where a single-event induced latch-up is a potential issue in the product application, a low-resistivity wafer with a top epitaxial layer can be selected, reducing the parasitic voltage drops in the substrate, which can lead to latch-up triggering.

The aspects of process-level mitigation also include the considerations of the technology node. Generally, smaller CMOS technology nodes characterize a thinner layer of gate silicon dioxide. This is beneficial since, for the radiation-induced charge, it is easier to escape the insulating layers via tunneling mechanisms. Consequently, the scaled-down processes tend to be more robust to the effects of the TID [49]. Lower supply voltages in scaled technology nodes mean better robustness against SEL [50]. However, the same low power supply level in combination with reduced node capacitances results in increased sensitivity to SET and SEU in flip-flops or in SRAM memory cells [49]. Also, with the reduced physical size, the circuit nodes are closer to each other, and a charge cloud originating from a single event can easily be collected in a shared manner between several nodes, potentially resulting in multiple-bit SEU. An intended spatial separation in the circuit layout can significantly reduce the probability of multiple-bit upsets.

6.3.4.2 Physical Layout Level

The mitigation techniques related to the physical layout of CMOS structures rely on differences in response to radiation effects depending on the components' placement and the geometrical shapes. Similarly, as in the case of the technology-level mitigation techniques, both cumulative and SEEs should be considered.

The most efficient method for reducing the influence of TID on MOSFETs is the enclosed layout transistor (ELT) [51]. This layout approach eliminates the shallow trench isolation (STI) edge that connects the drain and source on both sides of the gate in a rectangular transistor, as illustrated with dashed lines on both sides along the gate in Figure 6.11. STI is a thick, thermally grown oxide, typically of poor quality. Therefore radiation-induced traps will form easily on the interface between this material and silicon, leading to increased leakage current and to variations in threshold voltage, in particular, in n-channel transistors [52]. Transistors drawn with ELT geometry without the STI edge along the gate show significantly lower radiation degradation. Importantly, in the case of such non-standard gate geometries, one has to take good care of modeling the transistor for simulation purposes. Usually, it is done by calculating the equivalent dimensions of a standard MOSFET gate geometry [53]. CMOS process nodes below 65 nm are known to be very conservative in the layout design rules due to complex photolithography, and therefore in scaled technologies, ELT is not possible because of physical design rules violations.

Regarding SEE radiation-tolerant layout design techniques, the first one to discuss is SEL (see Figure 6.9). One can do a lot in the physical placement to reduce the gain of the parasitic bipolar transistors and to reduce their base–emitter voltage drops. This is done, for example, by applying safe spacing between n-channel and p-channel MOSFETs and placement of guard rings around localized transistors sharing the same MOSFET-body bias region (e.g. NWELL or substrate), with possibly low resistance between the body bias terminal and the source/drain terminals of the MOSFETs inside the guard ring [54]. In terms of SETs and SEUs,

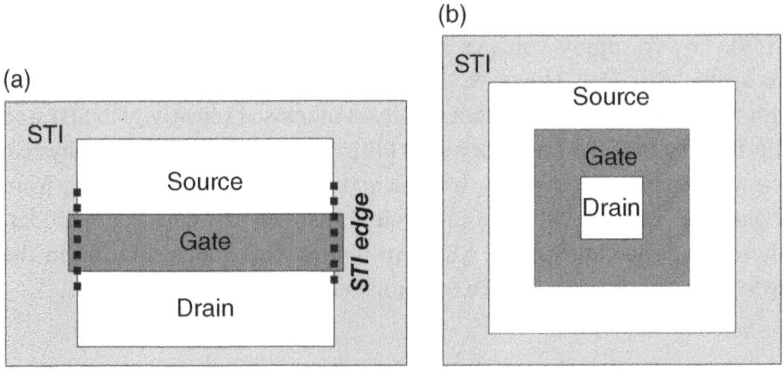

Figure 6.11 In a traditional rectangular layout transistor (a), the channel under the gate sees the edge of the shallow trench isolation region, which fills the space between integrated devices. This region is effective in collecting trapped charge when exposed to cumulative TID, resulting in further degradation of transistor parameters. Enclosed layout transistor (b) eliminates the edge, resulting in a more radiation-tolerant device.

what is to consider is the situation where the charge particle deposits charge in a larger area, where several nodes of the circuit share the charge collection. In digital circuits with storage cells, this might lead to a simultaneous upset of several memory cells. Therefore, a planned spatial spacing would be beneficial. On the other hand, in differential architectures, symmetrical charge collection in parallel differential paths would result in partial cancellation of the single event transient. Carefully planned proximity of differential transistors and signal paths can increase the robustness against SET.

6.3.4.3 Architecture Level

Regarding the third dimension of hardening an IC against ionizing radiation, the architecture level, one can do a lot to improve the robustness. For hardening against TID, the recipes are intuitive, following good design practices. These include adaptive compensation to correct for the parameter shift [55]. Also, chopper stabilization and autozero techniques are helpful to mitigate radiation-induced mismatch [56]. In the case of SEEs, there are already dedicated to that purpose design techniques, like, for example, the well-known dual-interlocked storage cell [57], shown in Figure 6.9, as SEU-robust SRAM storage cell. Meanwhile, to reduce the probability of errors due to SET, the most common method is triple modular redundancy [58], applicable to digital circuits. The voter circuit that decides which output is valid can rely on the majority voting or may be enriched by the temporal filtering technique to reduce the probability of soft error in the case of a distributed event [59]. A delay combined with a guard gate is another temporal analysis method for local SET mitigation. This approach, without the need for a large redundancy block, has been proposed by [60]. An analog counterpart of the redundancy technique can also be considered, as proposed by [61] with the differential charge cancellation circuit. In both cases, the penalty is an increased circuit area and complexity.

For radiation-hardening-by-design, the circuit concept has to be custom, that is, considered individually block-by-block. Also, the level of functional safety required by the application has to be considered. There are considerable trade-offs that have to be balanced. These include power consumption, circuit die area, complexity influencing also the system level, and a clocked versus continuous mode approach. Once the design choices enter the system level, there are certain considerations beyond the IC architecture, that is—the operating conditions. Defining the periods of operation and power-down phases can have a significant impact on the degradation rate or the risk of failure. As an example, every time the Hubble Telescope flies over the South Atlantic anomaly [37, 62], a region with unusually high radiation levels in Van Allen radiation belts, it interrupts the data acquisition to reduce the risk of single event-related damage or data loss [63]. This is the case for all electronic devices, that at power down, there is no risk of SEL. Furthermore, the influence of TID on device characteristics is smaller since radiation damage mechanisms in MOSFET are known to bias voltage dependent [64, 65, 66]. With a positive electric field at the gate with respect to the source, one observes the worst-case damage in n-channel MOSFETs.

6.3.5 Radiation Assurance Testing

What makes radiation hardening a very challenging task is the difficulty of quantifying the improvement without the actual tests. Proper radiation hardness qualifications are required to show whether the circuit or system is suitable for a given application and environment. For TID and SEE testing, there are standards describing the test environment, sample selection and preparation, dosimetry, as well as the complete procedure of stress and testing. All the procedures ensuring radiation hardness are encompassed by the term radiation assurance testing (RHA). The TID stress test is described in the European Space Components Coordination (ESCC) 22900 standard and the United States Military Standards MIL-STD-883, as well as MIL-STD-750-1. The commonly used standards for SEE testing are ESCC 25100 and the JEDEC Global Standards for the Microelectronics Industry JESD57 and JESD234, as well as the already mentioned MIL-STD-883 and MIL-STD-750-1 standards.

Standards have their limitations and challenges. In the case of TID, the performance tests have to be performed within two hours from the halt of the irradiation. This requirement stems from the fact that radiation damage is a complex process in solid-state materials, where the recovery or degradation continues even when the irradiation stops. In case of a complex system or difficulty with in situ tests of post-TID performance, this relatively short time interval of two hours might be very difficult to meet. One of the main challenges in SEE testing with heavy ions is the fact that the IC sample package has to be open for tests while maintaining its functional performance. In a ceramic package, lid removal is easier compared to plastic while maintaining full testing functionality. The removal is necessary in order to expose the IC die directly to the linear energy transfer (LET) known for a given ion type of a given energy passing through silicon. The presence of a package of possibly unknown thickness and composition would modify LET. Meanwhile, the knowledge of LET is essential in SEE to calculate the SEU or SEL event cross section. Another concern is that it may take many years from the moment a new process or technology is introduced until the testing standards are adopted.

Finally, for some specific applications, the test or stress parameters are beyond the norms defined by the standards. For example, in an application where the electronics absorb with a very large TID, above the 1 *Mrad* range, the testing time would get unreasonably long with respect to the overall product development

(a) (b)

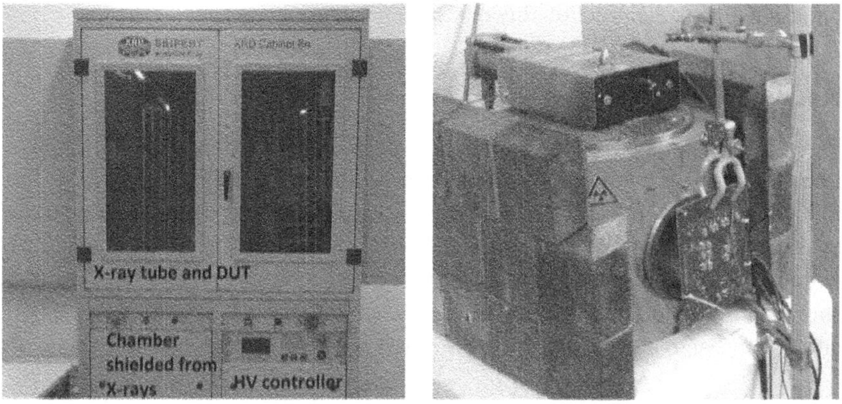

Figure 6.12 (a) X-ray irradiator. X-rays enable efficient and low-cost testing against total ionizing dose, which is especially valuable at the product development stage, to compare different test structures or to reach very high levels of TID above 1 Mrad. However, the X-ray beam is partially attenuated in the package. Therefore for an accurate estimation of the absorbed dose, the bare die of the chip should be exposed. (b) Cobalt-60 source. It is a standardized TID testing method used for space applications. The dose rate adjustment is done by changing the distance from the source. The area irradiated is very large. Therefore, larger electronic systems can be tested.

time. The reason is the fact that the TID accelerated stress test is done with a Cobalt-60 radioisotope. The intensity of this source is usually very limited because of the difficulty in maintaining its safety. TID tests with X-ray generators are a very good solution to overcome this problem since much higher doses are achievable. The disadvantage is that for correct dosimetry, the IC package has to be open, similarly as described above in SEE assurance testing. The important benefit of X-ray testing is the uncomplicated access, safety, and costs. All in all, it is a very good tool for tracking the radiation hardness of a product under development or to compare several products or prototypes. The irradiation environment for both TID test methodologies is illustrated in Figure 6.12.

6.4 RFID Sensors in Radiation Environments

As mentioned before, one reliability issue of passive wireless RFID sensors in harsh environments is radiation hardness of ionizing radiation. Thus, this section discusses modern backscatter UHF RFID sensor tags based on CMOS chips, as presented in Section 6.2, and evaluates perspectives for their radiation hardness, as discussed in Section 6.3. In particular, the effects of ionizing radiation have to be investigated for the UHF RFID chip and its integrated circuits, as the ionizing radiation affects the functionality of semiconductor devices.

6.4.1 RFID Tag Chip

Figure 6.5 illustrates the block diagram of an exemplary differential RF frontend of a custom-built UHF RFID tag chip in a 40 nm CMOS technology [1, 23]. The following discussion focuses on the RF charge pump, as only its careful design can guarantee excellent system performance [67]. A full-wave rectifier realizes the energy conversion function in a single stage of the RF charge pump. This rectifier shown in Figure 6.13a is also known as a Schenkel rectifier. An alternative differential approach is a diode bridge, known as a Graetz bridge, illustrated in

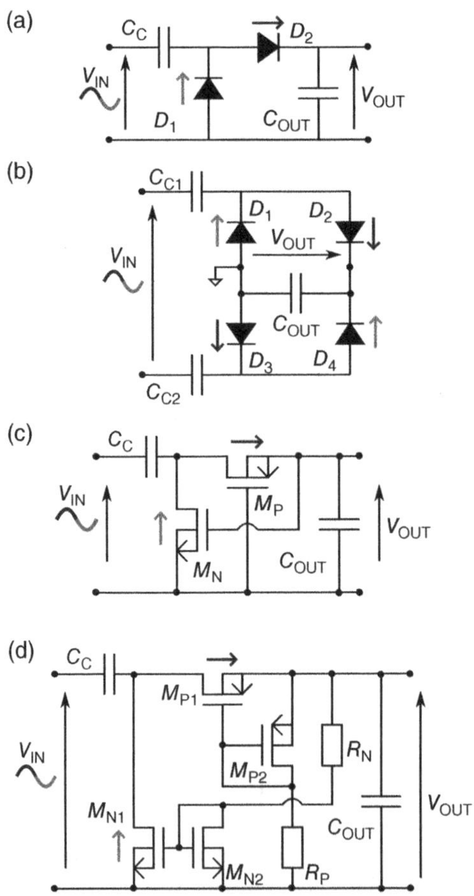

Figure 6.13 (a) The schematic of a full-wave rectifier as a single stage of the RF charge pump; (b) the schematic of a differential Graetz bridge rectifier; (c) the schematic of a MOSFET-based rectifier with pre-charge gate voltage; and (d) the schematic of a MOSFET-based rectifier with stabilized pre-charge gate voltage. *C* denotes capacitances, *D* denotes diodes, *M* denotes transistors, and *R* denotes resistances. Figure from [1]. *Source:* (c) Adapted from [72]; (d) Adapted from [73].

Figure 6.13b. Several rectifier stages are stuck in series to achieve a voltage multiplication with the RF charge pump. For this purpose, the multi-stage Dickson topology is widely used in CMOS applications [68]. The number of stages has to be optimized, as too many stages can reduce the conversion efficiency, linked with the parasitics added by each stage [69]. The number of available stages also determines the chip sensitivity $P_{\min}$.

One critical charge pump design aspect is the diode component (see D_1 and D_2 in Figure 6.13a). A standard p–n junction diode has a significant voltage drop, resulting in considerable power losses. A Schottky diode [70] with a much lower built-in voltage is better suited from this perspective; however, it is not always an available option in CMOS processes. Further, Schottky diodes often suffer from a higher reverse leakage current. Metal-oxide-semiconductor field-effect transistor (MOSFET) adaptation might be beneficial; the difficulty is that a diode-connected MOSFET needs a specific threshold voltage to conduct the rectifier current. Thus, various threshold voltage compensation techniques have been proposed [69, 71]. Figure 6.13c shows one of the most straightforward solutions where gate terminals of the n-channel and the p-channel MOSFETs are directly connected to the output terminals [72]. This approach has a disadvantage if the single-stage charge pump output voltage reaches a specific high-enough DC value and the diode transistors are over-biased, resulting in significant reverse currents that degrade the power conversion efficiency. Introducing a pre-charge voltage based on a simple regulation stabilizes the gate bias voltage of the rectifying transistors, as illustrated in Figure 6.13d [73]. Since the power efficiency of the RFID system highly depends on the charge pump design, it will be essential to consider design challenges related to radiation effects.

6.4.2 Radiation Effects in RFID Tags

The effects caused by ionizing radiation can have severe consequences for the operation of passive RFID circuits. Therefore, the perspectives of radiation-hard RFID ICs are discussed based on an analysis of existing passive RFID building blocks and technology requirements for such applications. As mentioned above, the RF charge pump is a critical block in the RFID frontend, and a change in its characteristics might result in a degradation of power efficiency or, ultimately, in lower available output power for a given RF input. Specifically, in the case of a diode rectifier, this situation can be related to radiation-induced degradation of the ideality factor and increased saturation current [74]. Similarly, the performance of a MOSFET-based rectifier, as shown in Figure 6.13c, is expected to change because of radiation-induced threshold voltage shifts (see Figure 6.8) and an increased off-state current [75, 76].

Another concern at the frontend level is the input impedance of the tag chip. The chip absorbing impedance and the tag antenna impedance must be power matched for maximizing wireless power transfer (see Section 6.2). A shift of the threshold voltages upon irradiation changes the MOS capacitance in its *C–V* characteristics [77], possibly leading to an unintended impedance mismatch. Consequently, the frontend input impedance must be examined after ionizing radiation stress in a given application.

So far, most of the attention has been devoted to the analog frontend. However, the digital part and the memory block included in each RFID tag chip must also be considered. One of the SEEs that are a big concern in radiation environments is latch-up (see Figure 6.9). RFID chips, because of the limited energy stored through energy harvesting, have the potential to be robust against the destructive effects of latch-up. However, such an event, being possible, might lead to soft errors or disrupted communication at the system level.

Another relevant SEE is the SEU (see Figure 6.10). In the processing memory, like flip-flops, such an event would be considered a soft failure. A far more critical location is the EEPROM, where the tag ID is stored next to other configuration parameters. EEPROM in many passive RFID tags is based on a floating-gate nonvolatile memory [78, 79]. In the case of a single event passing through the floating-gate adjacent oxide, the radiation-induced highly concentrated electron–hole pairs might create a leaky pipe, which is a leakage path for the stored charge [80]. Other types of EEPROM nonvolatile memories eliminate that risk, that is, memories based on charge trapping mechanisms, where the stored charge is distributed in an insulating layer, like silicon-oxide-nitride-silicon (SONOS) [81], NROM [82], or sidewall spacer [83]. Whether floating gate or charge trapping-based, also the TID effects need to be addressed in any nonvolatile memory. In the case of the sidewall spacer, promising results are obtained [83], in addition to demonstrating an approach to monitor the device degradation over a lifetime by using memory sections with different values of the bit-cell programming time. Naturally, the selection of the EEPROM type is typically limited to a single component that a specific CMOS process offers. Despite these design constraints, it is crucial to recognize the differences between different EEPROM schemes and, above all, the risks in the context of radiation reliability.

6.4.3 Radiation Hardening of RFID Tags

Engineers can exploit device layout techniques for radiation hardening where applicable (see Section 6.3.4). In passive RFID tag ICs, the forward diode of the rectifier can be improved by realizing it as a dynamic-threshold MOS (DTMOS) diode, initially proposed as a solution for ultra-low-power circuits thanks to its low threshold [84]. The low-threshold feature is achieved using a conventional MOSFET and tying bulk and gate together, resulting in a lowered surface potential under the gate. As a result, the built-in voltage of the drain-bulk diode will be much lower than in a conventional p–n diode. This technique has already been implemented in RFID tag ICs [85]. However, the prototype includes only a standard rectangular transistor, where the radiation-susceptible STI edge is present between the drain and the source (see Figure 6.8). A radiation-hard approach can be adopted for ionizing environments, that is, the same technique used in a forward diode for bandgap reference circuits. The DTMOS is then placed in an enclosed layout [86].

As mentioned before, the ultra-low-power capability of state-of-the-art passive RFID tags is enabled thanks to voltage-threshold compensation techniques in the rectifiers used in the RF charge pump. For the same reason, this part of the circuit is susceptible to shifts in the threshold voltage. In terms of radiation effects, if the p-channel transistor pair and the n-channel transistor pair in the rectifier from Figure 6.13d are matched and degrade simultaneously, this block architecture can be considered an adaptive compensation mitigation approach. If the threshold voltage of p-channel MOSFET M_{P1} increases with irradiation, it needs a larger gate bias to meet the same operating point as when unstressed. Simultaneously the threshold voltage of the p-channel MOSFET generating the pre-charge-voltage M_{P2} also increases with irradiation, providing the required higher gate-to-source voltage of transistor M_{P1}.

There is one more critical aspect besides the transistor-level design and layout: technology choice and the mode of operation. The observation that scaled processes tend to be more robust against TID is beneficial for passive RFID chips since they follow the scaling trend because of cost benefits due to a decreased chip size [87]. Typically, radiation-induced parameter drifts in CMOS processes are characterized based on DC measurements of MOSFET drain current–gate voltage I_D–V_G characteristics, which include voltage threshold, off-state current, on-state current, and transconductance. In the case of RFID tags, the RF parameters of the circuit components are also critical. A review of the RF performance under TID stress in CMOS technologies is covered in [1]. These are mostly silicon-on-insulator (SOI) technologies, which are more expensive than bulk CMOS. Indeed, more knowledge is still needed in this area.

6.5 Conclusions

The technological advancement in wireless sensor technology to batteryless systems enables new applications, like proximity communication, identification, or environmental monitoring. Consequently, wireless communication and sensing devices have become more widespread and easily accessible. These functional capabilities of wireless sensor technology would have a lot to offer, particularly in harsh environments, with restricted human access. Interesting settings are the broad-sense industrial, medical, and space environments, where monitoring instruments are exposed to extreme temperatures, vibrations, humidity, and ionizing radiation. This chapter reviews existing work on batteryless wireless sensors based on the UHF RFID technology for their reliable ultra-low-power operation and evaluates the perspectives for their radiation hardness. This discussion links the fundamentals of radiation effects in general and in CMOS circuits in particular with the architectural characteristics and operation features of wireless frontends.

Currently, no modern backscatter RFID sensor tags are used in harsh environments, as the main issues for reliable operation have not yet been solved. However, the review of existing sensor tag prototypes shows good perspectives that these issues, that is, tight power constraints due to the batteryless operation and harsh radiation environments due to ionizing radiation, can be overcome in the future. In particular, signal-pattern-based sensor tag systems show promising results in wireless power transfer, setup independence, and robustness in moderate static multipath environments. In addition, the trend of using scaled CMOS technologies for RFID tag chips and their passive operation promises radiation hardness to a certain extent. Ultimately, radiation-hard batteryless tags will be an essential technological advancement to applications in harsh industrial, medical, and space environments, where wireless sensors are exposed to ionizing radiation.

6.6 Biographies

Jasmin Grosinger is an Associate Professor at the Graz University of Technology in Austria and a Visiting Associate Professor at the Tohoku University in Sendai, Japan. She earned her MSc from the Vienna University of Technology and worked as a Project Assistant with the Institute of Telecommunications. She later became a Laboratory Associate with Disney Research in Pittsburgh, USA. In 2012, she received her PhD from the Vienna University of Technology. Since 2013, she has researched (ultra-)low-power microwave components and systems at Graz University of Technology's Institute of Microwave and Photonic Engineering. Jasmin has also been a Guest Professor at Friedrich-Alexander-University Erlangen-Nuremberg's Institute of Electronics Engineering. She is an IEEE Senior Member, and her research has resulted in more than 80 peer-reviewed publications and one US patent. Jasmin is actively involved in the Technical Program and Steering Committees of various microwave-related conferences and is an Associate Editor of the IEEE Microwave and Wireless Technology Letters. She is also a member of the IEEE Microwave Theory and Techniques Society (MTT-S) and was selected as a Distinguished Microwave Lecturer. In 2022, Jasmin served as the

MTT-S Administrative Committee (AdCom) Secretary. Since 2023, she has served as an Elected Voting Member of the IEEE MTT-S AdCom, chairing the Meetings and Symposia Committee in 2024.

Alicja Michalowska-Forsyth received a Master of Science degree in electronics and telecommunication from the Technical University of Lodz in Poland and a PhD in physics from the University Paris-Sud XI in France in 2006 and 2013. She has several years of industrial experience at austriamicrosystems (currently ams-osram) as an R&D analog designer of application-specific integrated circuits (ASICs). From 2009, she worked with the research group at CEA-Saclay in France on readout frontends for solid-state detectors used for spectro-imaging of X- and gamma-rays in space applications. Next to functionality and performance, reliability in the ionizing environment was an essential part of these studies. Since 2014, Dr. Michalowska-Forsyth has been a scientific employee at the Institute of Electronics at Graz University of Technology in Austria, where she is carrying on research in the area of reliability and, in particular, radiation effects and hardening of integrated circuits and devices. Currently, she holds the position of Assistant Professor, leading a research group and being active in teaching.

References

1 Grosinger, J. and Michalowska-Forsyth, A. (2022). Space tags: ultra-low-power operation and radiation hardness for passive wireless sensor tags. *IEEE Microwave Magazine* 23 (3): 55–71. https://doi.org/10.1109/MMM.2021.3130686.

2 Brown, P., Hartmann, P., Schellhase, A. et al. (2007). Asset tracking on the international space station using global SAW tag RFID technology. *2007 IEEE Ultrasonics Symposium Proceedings*, 72–75. https://doi.org/10.1109/ULTSYM.2007.31.

3 Finkenzeller, K. (2003). *RFID Handbook: Fundamentals and Applications in Contactless Smart Cards and Identification*. Chichester: Wiley.

4 Bhattacharyya, R., Floerkemeier, C., and Sarma, S. (2010). Low-cost, ubiquitous RFID-tag-antenna-based sensing. *Proceedings of the IEEE* 98 (9): 1593–1600. https://doi.org/10.1109/JPROC.2010.2051790.

5 Grosinger, J. (2023). Robustly operating: passive near-field communication systems in metal environments. *IEEE Microwave Magazine* 24 (4): 30–39. https://doi.org/10.1109/MMM.2022.3233508.

6 Grosinger, J., Deutschmann, B.J.B., Zöscher, L. et al. (2022). HF RFID tag chip impedance measurements. *IEEE Transactions on Instrumentation and Measurement* 71: 1–11. https://doi.org/10.1109/TIM.2021.3130664.

7 Fischbacher, R., Görtschacher, L., Amtmann, F. et al. (2020). Localization of UHF RFID magnetic field sensor tags. *2020 IEEE Radio Wireless Conference*, 169–172, January 2020.

8 Görtschacher, L. and Grosinger, J. (2019). UHF RFID sensor system using tag signal patterns: prototype system. *IEEE Antennas and Wireless Propagation Letters* 18 (10): 2209–2213. https://doi.org/10.1109/LAWP.2019.2940336.

9 Görtschacher, L. and Grosinger, J. (2019). Localization of signal pattern based UHF RFID sensor tags. *IEEE Microwave and Wireless Components Letters* 29 (11): 753–756. https://doi.org/10.1109/LMWC.2019.2940082.

10 Curty, J.-P., Declercq, M., Dehollain, C., and Joehl, N. (2007). *Design and Optimization of Passive UHF RFID Systems*. New York: Springer Science+Business Media.

11 Molisch, A. (2005). *Wireless Communications*. Wiley.

12 Nikitin, P. and Rao, K. (2006). Performance limitations of passive UHF RFID systems. *2006 IEEE Antennas and Propagation Society International Symposium*, 1011–1014, July 2006. https://doi.org/10.1109/APS.2006.1710704.

13 Görtschacher, L., Grosinger, J., Khan, H. et al. (2016). SDR based RFID reader for passive tag localization using phase difference of arrival techniques. *2016 IEEE MTT-S International Microwave Symposium (IMS)*, 1–4, May 2016. https://doi.org/10.1109/MWSYM.2016.7538227.

14 Nikitin, P. and Rao, K. (2006). Theory and measurement of backscattering from RFID tags. *IEEE Antennas and Propagation Magazine* 48 (6): 212–218. https://doi.org/10.1109/MAP.2006.323323.

15 Kurokawa, K. (1965). Power waves and the scattering matrix. *IEEE Transactions on Microwave Theory and Techniques* 13 (2): 194–202. https://doi.org/10.1109/TMTT.1965.1125964.

16 Andersen, J. and Vaughan, R. (2003). Transmitting, receiving, and scattering properties of antennas. *IEEE Antennas and Propagation Magazine* 45 (4): 93–98. https://doi.org/10.1109/MAP.2003.1241316.

17 Saxl, G., Görtschacher, L., Ussmueller, T., and Grosinger, J. (2021). Software defined RFID readers: wireless reader testbeds exploiting software-defined radios for enhancements in UHF RFID systems. *IEEE Microwave Magazine* 22 (3): 46–56. https://doi.org/10.1109/MMM.2020.3042408.

18 Angerer, C., Langwieser, R., and Rupp, M. (2010). RFID reader receivers for physical layer collision recovery. *IEEE Transactions on Communications* 58 (12): 3526–3537. https://doi.org/10.1109/TCOMM.2010.101910.100004.

19 Balanis, C. (2016). *Antenna Theory: Analysis and Design*. Wiley.

20 Green, R. (1963). The general theory of antenna scattering. PhD thesis. The Ohio State University.

21 Hansen, R. (1989). Relationships between antennas as scatterers and as radiators. *Proceedings of the IEEE* 77 (5): 659 662. https://doi.org/10.1109/5.32056.

22 Nikitin, P. and Rao, K. (2008). Antennas and propagation in UHF RFID systems. *2008 IEEE International Conference on RFID*, 277–288, April 2008. https://doi.org/10.1109/RFID.2008.4519368.

23 Zöscher, L., Herkess, P., Grosinger, J. et al. (2017). Passive differential UHF RFID front-ends in a 40 nm CMOS technology. *2017 47th European Microwave Conference (EuMC)*, 105–108, October 2017. https://doi.org/10.23919/EuMC.2017.8230810.

24 Zöscher, L., Grosinger, J., Muehlmann, U. et al. (2015). RF voltage limiters for passive differential UHF RFID front-ends in a 40 nm CMOS technology. *2015 IEEE MTT-S International Microwave Symposium (IMS)*, 1–4, May 2015. https://doi.org/10.1109/MWSYM.2015.7166839.

25 Grosinger, J., Mecklenbräuker, C., and Scholtz, A. (2010). UHF RFID transponder chip and antenna impedance measurements. *2010 International EURASIP Workshop on RFID Technology*, 1–4, September 2010.

26 Marrocco, G. (2008). The art of UHF RFID antenna design: impedance-matching and size-reduction techniques. *IEEE Antennas and Propagation Magazine* 50 (1): 66–79. https://doi.org/10.1109/MAP.2008.4494504.

27 Nikitin, P., Rao, K., and Martinez, R. (2007). Differential RCS of RFID tag. *Electronics Letters* 43 (8): 431–432. https://doi.org/10.1049/el:20070253.

28 Görtschacher, L. (2019). Novel UHF RFID tracking and sensing systems. PhD thesis. Graz University of Technology.

29 Occhiuzzi, C. and Marrocco, G. (2013). Constrained-design of passive UHF RFID sensor antennas. *IEEE Transactions on Antennas and Propagation* 61 (6): 2972–2980. https://doi.org/10.1109/TAP.2013.2250473.

30 Caccami, M., Manzari, S., and Marrocco, G. (2015). Phase-oriented sensing by means of loaded UHF RFID tags. *IEEE Transactions on Antennas and Propagation* 63 (10): 4512–4520. https://doi.org/10.1109/TAP.2015.2465891.

31 Grosinger, J., Görtschacher, L., and Bösch, W. (2016). Passive RFID sensor tag concept and prototype exploiting a full control of amplitude and phase of the tag signal. *IEEE Transactions on Microwave Theory and Techniques* 64 (12): 4752–4762. https://doi.org/10.1109/TMTT.2016.2623610.

32 Caccami, M. and Marrocco, G. (2018). Electromagnetic modeling of self-tuning RFID sensor antennas in linear and nonlinear regimes. *IEEE Transactions on Antennas and Propagation* 66 (6): 2779–2787. https://doi.org/10.1109/TAP.2018.2820322.

33 Zannas, K., El Matbouly, H., Duroc, Y., and Tedjini, S. (2018). Self-tuning RFID tag: a new approach for temperature sensing. *IEEE Transactions on Microwave Theory and Techniques* 66 (12): 5885–5893. https://doi.org/10.1109/TMTT.2018.2878568.

34 Marrocco, G. (2010). Pervasive electromagnetics: sensing paradigms by passive RFID technology. *IEEE Wireless Communications* 17 (6): 10–17. https://doi.org/10.1109/MWC.2010.5675773.

35 Caizzone, S., Giampaolo, E., and Marrocco, G. (2017). Setup-independent phase-based sensing by UHF RFID. *IEEE Antennas and Wireless Propagation Letters* 16: 2408–2411. https://doi.org/10.1109/LAWP.2017.2721432.

36 Görtschacher, L., Bösch, W., and Grosinger, J. (2018). UHF RFID sensor tag antenna concept for stable and distance independent remote monitoring. *2018 IEEE International Microwave Biomedical Conference (IMBioC)*, 16–18, June 2018. https://doi.org/10.1109/IMBIOC.2018.8428853.

37 Velazco, R., McMorrow, D., and Estela, J. (2019). *Radiation Effects on Integrated Circuits and Systems for Space Applications*. Springer.

38 Ziegler, J.F., Ziegler, M.D., and Biersack, J.P. (2010). SRIM The stopping and range of ions in matter (2010). *Nuclear Instruments and Methods in Physics Research Section B: Beam Interactions with Materials and Atoms* 268 (11–12): 1818–1823.

39 Vankeirsbilck, J., Hallez, H., and Boydens, J. (2015). Integration of soft errors in functional safety: a conceptual study. *Annual Journal of Electronics* 8: 108–111.

40 Heitler, W. (1984). *The Quantum Theory of Radiation*. Courier Corporation.

41 Wiedemann, H. (2015). *Particle Accelerator Physics*. Springer Nature.

42 Linten, D., Vassilev, V., Thijs, S. (2014). Anti-series GGNMOS ESD clamp for space application IC's. *Electrical Overstress/Electrostatic Discharge Symposium Proceedings 2014*, 1–7. IEEE.

43 Okuyama, R., Kadono, T., Onaka-Masada, A. et al. (2020). Hydrogen passivation for reduction of SiO_2/Si interface state density using hydrocarbon-molecular-ion-implanted silicon wafers. *Japanese Journal of Applied Physics* 59 (12): 125502.

44 Barnaby, H.J. (2006). Total-ionizing-dose effects in modern CMOS technologies. *IEEE Transactions on Nuclear Science* 53 (6): 3103–3121.

45 Pease, R.L. (2003). Total ionizing dose effects in bipolar devices and circuits. *IEEE Transactions on Nuclear Science* 50 (3): 539–551.

46 Srour, J.R., Marshall, C.J., and Marshall, P.W. (2003). Review of displacement damage effects in silicon devices. *IEEE Transactions on Nuclear Science* 50 (3): 653–670.

47 Messenger, G.C. and Ash, M.S. (1997). Single event phenomena I. In: *Single Event Phenomena*, 179–231. Springer.

48 Nicolaidis, M. (2010). *Soft Errors in Modern Electronic Systems*, vol. 41. Springer Science+Business Media.

49 Dodd, P.E., Shaneyfelt, M.R., Schwank, J.R., and Felix, J.A. (2010). Current and future challenges in radiation effects on CMOS electronics. *IEEE Transactions on Nuclear Science* 57 (4): 1747–1763.

50 Boselli, G., Reddy, V., and Duvvury, C. (2005). Latch-up in 65 nm CMOS technology: a scaling perspective. *2005 IEEE International Reliability Physics Symposium, 2005. Proceedings. 43rd Annual*, 137–144. IEEE.

51 Snoeys, W., Faccio, F., Burns, M. et al. (2000). Layout techniques to enhance the radiation tolerance of standard CMOS technologies demonstrated on a pixel detector readout chip. *Nuclear Instruments and Methods in Physics Research Section A: Accelerators, Spectrometers, Detectors and Associated Equipment* 439 (2): 349–360. https://doi.org/10.1016/S0168-9002(99)00899-2.

52 Faccio, F., Barnaby, H., Chen, X. et al. (2008). Total ionizing dose effects in shallow trench isolation oxides. *Microelectronics Reliability* 48 (7): 1000–1007. https://doi.org/10.1016/j.microrel.2008.04.004.

53 Bezhenova, V. and Michalowska-Forsyth, A. (2018). Modeling of annular gate MOS transistors. In *2018 18th European Conference on Radiation and Its Effects on Components and Systems (RADECS)*, 1–4. https://doi.org/10.1109/RADECS45761.2018.9328697.

54 Dodds, N.A., Hooten, N.C., Reed, R.A. et al. (2012). Effectiveness of SEL hardening strategies and the latchup domino effect. *IEEE Transactions on Nuclear Science* 59 (6): 2642–2650. https://doi.org/10.1109/TNS.2012.2224374.

55 De Vos, J., Kilchytska, V., Flandre, D., and Bol, D. (2014). Compensation of total ionizing dose effects in ULV SoCs through adaptive voltage scaling. *2014 SOI-3D-Subthreshold Microelectronics Technology Unified Conference (S3S)*, 1–2. IEEE.

56 Enz, C.C. and Temes, G.C. (1996). Circuit techniques for reducing the effects of op-amp imperfections: autozeroing, correlated double sampling, and chopper stabilization. *Proceedings of the IEEE* 84 (11): 1584–1614.

57 Calin, T., Nicolaidis, M., and Velazco, R. (1996). Upset hardened memory design for submicron CMOS technology. *IEEE Transactions on Nuclear Science* 43 (6): 2874–2878.

58 Morgan, K.S., McMurtrey, D.L., Pratt, B.H., and Wirthlin, M.J. (2007). A comparison of TMR with alternative fault-tolerant design techniques for FPGAs. *IEEE Transactions on Nuclear Science* 54 (6): 2065–2072.

59 Mavis, D.G. and Eaton, P.H. (2002). Soft error rate mitigation techniques for modern microcircuits. *2002 IEEE International Reliability Physics Symposium. Proceedings. 40th Annual (Cat. No. 02CH37320)*, 216–225. IEEE.

60 Baze, M.P., Wert, J., Clement, J.W. et al. (2006). Propagating set characterization technique for digital CMOS libraries. *IEEE Transactions on Nuclear Science* 53 (6): 3472–3478.

61 Olson, B.D., Holman, W.T., Massengill, L.W. et al. (2008). Single-event effect mitigation in switched-capacitor comparator designs. *IEEE Transactions on Nuclear Science* 55 (6): 3440–3446.

62 Heirtzler, J.R. (2002). The future of the South Atlantic anomaly and implications for radiation damage in space. *Journal of Atmospheric and Solar-Terrestrial Physics* 64 (16): 1701–1708.

63 Adler, D.S. (2012). Keeping the Hubble Space Telescope operating efficiently in its third decade. In: *Observatory Operations: Strategies, Processes, and Systems IV*, vol. 8448 (ed. A.B. Peck, R.L. Seaman, and F. Comeron), 309–316. International Society for Optics and Photonics, SPIE. https://doi.org/10.1117/12.924363.

64 Gonella, L., Faccio, F., Silvestri, M. et al. (2007). Total ionizing dose effects in 130-nm commercial CMOS technologies for HEP experiments. *Nuclear Instruments and Methods in Physics Research Section A: Accelerators, Spectrometers, Detectors and Associated Equipment* 582 (3): 750–754.

65 Liu, Z., Hu, Z., Zhang, Z. et al. (2011). Analysis of bias effects on the total ionizing dose response in a 180 nm technology. *Nuclear Instruments and Methods in Physics Research Section A: Accelerators, Spectrometers, Detectors and Associated Equipment* 644 (1): 48–54.

66 Borghello, G., Faccio, F., Termo, G. et al. (2021). Effects of bias and temperature on the dose-rate sensitivity of 65-nm CMOS transistors. *IEEE Transactions on Nuclear Science* 68 (5): 573–580.

67 Soyata, T., Copeland, L., and Heinzelman, W. (2016). RF energy harvesting for embedded systems: a survey of tradeoffs and methodology. *IEEE Circuits and Systems Magazine* 16 (1): 22–57. https://doi.org/10.1109/MCAS.2015.2510198.

68 Dickson, J. (1976). On-chip high-voltage generation in MNOS integrated circuits using an improved voltage multiplier technique. *IEEE Journal of Solid-State Circuits* 11 (3): 374–378. https://doi.org/10.1109/JSSC.1976.1050739.

69 Shetty, D., Steffan, C., Holweg, G. et al. (2021). Submicrowatt CMOS rectifier for a fully passive wake-up receiver. *IEEE Transactions on Microwave Theory and Techniques* 69 (11): 4803–4812. https://doi.org/10.1109/TMTT.2021.3098694.

70 Sze, S.M., Li, Y., and Ng, K.K. (2021). *Physics of Semiconductor Devices*. Wiley.

71 Zöscher, L., Herkess, P., Grosinger, J. et al. (2017). A differential threshold voltage compensated RF-DC power converter for RFID tag ICs. *2017 Integrated Nonlinear Microwave and Millimetre-wave Circuits Workshop (INMMiC)*, 1–3, April 2017. https://doi.org/10.1109/INMMIC.2017.7927297.

72 Kotani, K. and Ito, T. (2007). High efficiency CMOS rectifier circuit with self-Vth-cancellation and power regulation functions for UHF RFIDs. *2007 IEEE Asian Solid-State Circuits Conference*, 119–122. https://doi.org/10.1109/ASSCC.2007.4425746.

73 Nakamoto, H., Yamazaki, D., Yamamoto, T. et al. (2007). A passive UHF RF identification CMOS tag IC using ferroelectric RAM in 0.35-μm technology. *IEEE Journal of Solid-State Circuits* 42 (1): 101–110. https://doi.org/10.1109/JSSC.2006.886523.

74 Nelson, D.L. and Sweet, R.J. (1966). Mechanisms of ionizing radiation 60 surface effects on transistors. *IEEE Transactions on Nuclear Science* 13 (6): 197–206. https://doi.org/10.1109/TNS.1996.4324362.

75 Schwank, J., Shaneyfelt, M., Fleetwood, D. et al. (2008). Radiation effects in MOS oxides. *IEEE Transactions on Nuclear Science* 55 (4): 1833–1853. https://doi.org/10.1109/TNS.2008.2001040.

76 Faccio, F., Michelis, S., Cornale, D. et al. (2015). Radiation-induced short channel (RISCE) and narrow channel (RINCE) effects in 65 and 130 nm MOSFETs. *IEEE Transactions on Nuclear Science* 62 (6): 2933–2940. https://doi.org/10.1109/TNS.2015.2492778.

77 Ferńandez-Martínez, P., Cortés, I., Hidalgo, S. et al. (2011). Simulation of total ionising dose in MOS capacitors. In *Proceedings of the 8th Spanish Conference on Electron Devices, CDE'2011*, 1–4. IEEE.

78 Lu, C., Li, Y., Ou, J. et al. (2016). A low power EEPROM for passive RFID tag IC. In *2016 IEEE International Conference on RFID Technology and Applications (RFID-TA)*, 49 54. https://doi.org/10.1109/RFID-TA.2016.7750732.

79 Lee, K., Chun, J., and Kwon, K. (2010). A low power CMOS compatible embedded EEPROM for passive RFID tag. *Microelectronics Journal* 41 (10): 662–668. https://doi.org/10.1016/j.mejo.2010.06.006.

80 Cellerea, G. and Paccagnella, A. (2006). Subpicosecond conduction through thin SiO_2 layers triggered by heavy ions. *Journal of Applied Physics* 99 (7): https://doi.org/10.1063/1.2181313.

81 White, M., Adams, D., and Bu, J. (2000). On the go with SONOS. *IEEE Circuits and Devices Magazine* 16 (4): 22–31. https://doi.org/10.1109/101.857747.

82 Eitan, B., Pavan, P., Bloom, I. et al. (2000). NROM: A novel localized trapping, 2-bit nonvolatile memory cell. *IEEE Electron Device Letters* 21 (11): 543–545. https://doi.org/10.1109/55.877205.

83 Vincenzi, T., Schatzberger, G., and Michalowska-Forsyth, A. (2019). Program time effects on total ionizing dose tolerance of sidewall spacer memory bit cell. In *2019 Austrochip Workshop on Microelectronics (Austrochip)*, 55–58. https://doi.org/10.1109/Austrochip.2019.00021.

84 Assaderaghi, F., Sinitsky, D., Parke, S.A. et al. (1997). Dynamic threshold-voltage MOSFET (DTMOS) for ultra-low voltage VLSI. *IEEE Transactions on Electron Devices* 44 (3): 414–422. https://doi.org/10.1109/16.556151.

85 Chouhan, S. S. and Halonen, K. (2014). The design and implementation of DTMOS biased all PMOS rectifier for RF energy harvesting. In *2014 IEEE 12th International New Circuits and Systems Conference (NEWCAS)*, 444–447. https://doi.org/10.1109/NEWCAS.2014.6934078.

86 Gromov, V., Annema, A., Kluit, R. et al. (2007). A radiation hard bandgap reference circuit in a standard 0.13 μm CMOS technology. *IEEE Transactions on Nuclear Science* 54 (6): 2727–2733. https://doi.org/10.1109/TNS.2007.910170.

87 Zöscher, L., Spreitzer, R., Gross, H. et al. (2016). HF/UHF dual band RFID transponders for an information-driven public transportation system. *Elektrotechnik und Informationstechnik* 133 (3): 163–175.

Section 2

Chipless

7

SAW Devices Combining RFID and Sensor Functionalities for Harsh Environments

Omar Elmazria, Cécile Floer, Thierry Aubert, and Sami Hage-Ali

Université de Lorraine, CNRS, IJL, Nancy, France

7.1 Introduction

Conventional surface acoustic wave (SAW) devices are widely used as filters, resonators, or delay lines in electronic systems for a wide range of applications: mobile communications, radars, and stable resonators for clock generation. Several billion SAW units are produced each year making their cost very competitive. Furthermore, the resonant frequency and the signal transmission delay of SAW devices can be very sensitive to the physical parameters of the environment, requiring packaging for protection against disturbances and external pollution and even the control of thermal conditions to ensure optimal operation. These defects and limitations have been cultivated to generate new fields of applications. SAW devices have therefore been intensively studied to produce miniature sensors allowing the measurement of a wide variety of parameters such as temperature, pressure, deformation, magnetic field, radiations, concentration of a species in solution, composition of an atmosphere, etc. In addition to being small, simple to produce (low cost) and robust, SAW devices can be batteryless, wirelessly interrogated [1], possibly packageless [2, 3], and offer multi-tagging (IDTAG) capabilities [4, 5]. SAW devices being widely used as standard components of RF communication, SAW sensors and their reader units can be inexpensive and several solutions are commercially available [6–8]. Because SAW sensors can be operated wirelessly, being used in backscattering mode and controlled by the RF electronic reader located in a safe area, they can operate in extreme conditions such as high temperatures up to 1 000 °C (glass or

Wireless Identification and Sensing Systems for Harsh and Severe Environments,
First Edition. Edited by Smail Tedjini and Valentina Palazzi.

Published 2025 by John Wiley & Sons, Inc.

steel industry) or radioactive environments (nuclear plant), in which no other wireless sensor can operate [9]. SAW sensors are also relevant for environments where the longevity of the sensor cannot be guaranteed by a local energy source. Moreover, they bring original solution to monitor torque, vibration, and temperature of rotating parts [10]. Thus, several applications could be addressed in the fields of aeronautic, railway, aeronaval, wind turbine, etc.

The achievement of the sensor is therefore completely linked to the application envisaged. Indeed, the choice of materials constituting it must allow the best performance while being compatible with the environment and the operating conditions. The measurement chain must also be adapted to the performance of the transducer to get the best out of it. The design of the sensor therefore requires a system approach that takes into account all the bricks of the measurement chain. In this chapter, we will describe the principle of the SAW sensor, and the two configurations allowing it to be interrogated remotely, namely the SAW resonator (RSAW) and the reflective delay line (R-DL). This last configuration allows in addition to the sensing functionality its use as a radio frequency identification tag (RFID) enabling the achievement of SAW sensor networks [4].

7.2 Saw Sensor Principle

The key part of the sensor is a small piece of a polished piezoelectric substrate (Figure 7.1) and an interdigital transducers (IDTs) formed by a standard photolithographic process with thin-film metallic material (generally aluminum) [11]. Due to the piezoelectric effect, an RF electric signal at the input of the IDT will launch an acoustic wave at the surface of the piezoelectric substrate. Vice versa, the surface acoustic wave causes an electrical charge distribution at the receiving IDT and hence an electric signal. The operating frequency of the device is fixed by the elastic velocity (v) of the substrate and spatial periodicity of the IDT (Figure 7.1) that define the wavelength (λ) [12]:

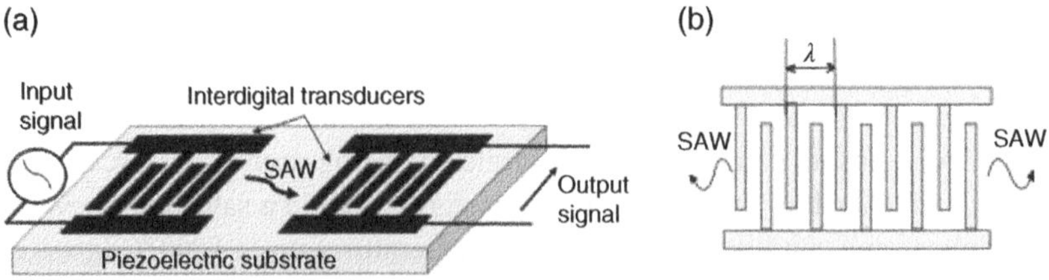

Figure 7.1 Basic structure of SAW device. (a) Two port delay line, and (b) single electrode type of IDT.

$$f = \frac{v}{\lambda} \tag{7.1}$$

Because the SAW velocity is five orders of magnitudes smaller than the velocity of light, the wavelength is in the range of μm for operating frequencies in GHz range. This leads to achieving small-size delay line of few mm when a hundred km is required with coaxial or optical fiber. Thus, the total area of SAW devices is of few mm^2 and the standard substrate thickness is of 125 μm. The development of new composites, piezoelectric on insulator (POI), and substrates allow the use of piezoelectric layers with a thickness of the order of mm. The silicon substrate can be thinned to adapt the thickness of the composite to the intended application by controlling, for example, the sensitivity of the sensor.

Knowing that the acoustic velocity strongly depends on the physical and chemical parameters of the environment, the measurand (temperature, pressure, strain, torque, etc.) affects the propagation of the SAW in the sensor and then the resonance frequency, attenuation and the delay of the signal. The sensitivity of the sensor depends on the used material, on its crystallographic cut and also, on the operating frequency. The partial derivation of frequency to a given perturbation defines the sensitivity of acoustic wave devices. It depends generally on the dispersion curves behavior under perturbation of the investigated mode. Mass effect is the most intuitive principle in various acoustic wave sensors and Quartz Crystal Microbalance (QCM) is the most popular.

Relative frequency variation could be expressed as a function of variation of mass (m), electrical field (E), mechanical stress (σ), Temperature (T), and any other perturbation of the resonator environment.

$$\frac{f}{f_0} \cong \frac{1}{f_0}\left(\frac{\partial f}{\partial m}\Delta m + \frac{\partial f}{\partial E}\Delta E + \frac{\partial f}{\partial \sigma}\Delta \sigma + \frac{\partial f}{\partial T}\Delta T + \ldots\right) \tag{7.2}$$

To achieve high-quality sensors with the best performances, the choice of the piezoelectric substrate and the crystalline orientation must therefore take into account a multitude of complex parameters to be optimized depending on the working environment and technological constraints:

- Working frequency and available bandwidth
- Sensitivity of the phase velocity of the elastic wave to each considered or interfering quantity
- Nature of the wave (polarization, guidance)

Table 7.1 Summary of the SAW parameters (K^2, TCF and elastic velocity) of the most popular piezoelectric materials.

Piezoelectric material	Euler angles	V_{SAW} (m/s)	K^2 (%)	TCF (ppm/°C)	References
Quartz	(0°, 160°, 80°)	3677	0.03	−56	Zhgoon et al. [13]
	(90°, 104°, −168°)	3185	0.22	33	
$LiNbO_3$	(0°, 90°, 90°)	3488	4.817	−93.4	Bergman et al. [14], Li et al. [15]
	(0°, 218°, 0°)	3995	5.34	−72.02	
LGS	(0°, 90°, 0°)	2340	0.4	40	Campbell [12], Hornsteiner et al. [16]
	(0°, 139°, 24°)	2729	0.39	−32.2	
	(0°, 142°, 21°)	2729	0.5	−29.36	
ZnO		2700	≥1	−40	Nicolay [17], Ieki and Kadota [18], Assouar [19]
GaN		3700	0.3	−30	Justice et al. [20]
AlN		5600	0.3	25	Caliendo [21], Caliendo [22]

Both crystals (Quartz, LGS and $LiNbO_3$) and thin films (ZnO, AlN, GaN) are mentioned.

- Electromechanical coupling (K^2) and temperature coefficient of frequency (TCF) of the material
- Quality factor and electrical impedance of the device

The key parameters of SAW devices are summarized in Table 7.1, for the most popular piezoelectric materials. Are considered crystals with several crystallographic cuts and thin film materials.

7.3 Principle of Wireless Sensors Including RFID Code

Wireless SAW sensors are suitable and become competitive wherever conventional sensor systems come to their limits. It is the case of applications in severe environments, fast movement of rotating parts, or when no power supply is possible. Two configurations can be considered for wireless SAW sensors: resonator (RSAW) and reflective delay lines (R-DL). The interrogation principle is based on radar technique and a variety of architectures such as time domain sampling (TDS), frequency domain sampling (FDS), or hybrid concepts for both R-SAW and R-DL sensors have been investigated [23].

7.4 Resonator

An IDT is placed between two acoustic Bragg mirrors, to form a high-quality factor resonant cavity (Figure 7.2). The resonator can be excited wirelessly by the interrogation system at a frequency close to the resonant frequency of the cavity. When the excitation is switched off, the resonator keeps oscillating freely for a period of time at its eigenfrequency, which depends on the environmental parameters. The detection of these free oscillations will be easier with high Q resonators, since the higher the quality factor Q is, the higher the frequency resolution will be. Also very important is the figure of merit (FOM) equal to Q^*K^2, where K^2 is the electromechanical coupling coefficient of the SAW piezoelectric substrate since it determines the power reradiation efficiency of the resonator as well as the optimal matching [24]. Hence, the quality factor and the figure of merit are essential parameters for the production of efficient wireless SAW resonators. However, it has been demonstrated that Q and K^2 are two antagonistic parameters [25, 26]. Trying to improve one will be at the expense of the second. Thus, there is a trade-off to obtain the highest FOM while choosing the materials and designs that best suit the intended application and the constraints imposed by the environment and legislation (operating frequency, bandwidth, ...).

7.5 Reflective Delay Line (R-DL)

7.5.1 Conventional Configuration of R-DL

A radio frequency (RF) interrogation signal is sent to the sensor, where it is picked up by an antenna and converted into surface acoustic waves (SAWs), by the IDT [27]. The SAWs then propagate on the surface of the chip. In a specific configuration, called "Reflective Delay Line" or R-DL (Figure 7.3), a fraction of the acoustic energy is reflected back toward the IDT, by thin metal gratings or grooves structured along the propagation path. The reflected signals are converted back into successive RF pulses by the IDT and form a series of echoes in the RF impulse response of the R-DL. The sequence of echoes can easily be made unique and is

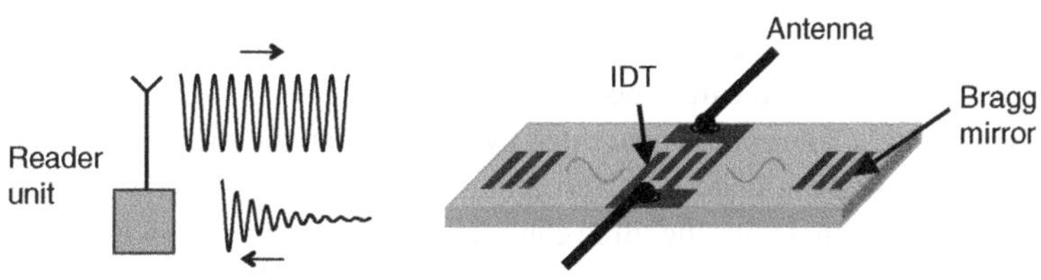

Figure 7.2 Principle of a wireless SAW sensor in resonator configuration.

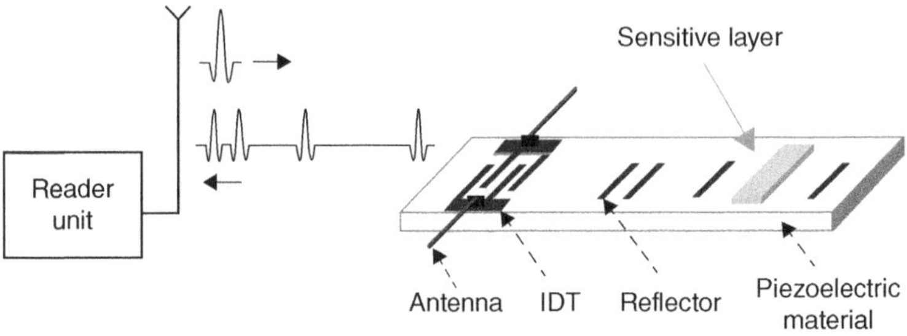

Figure 7.3 Principle of a wireless SAW sensor in reflective delay line configuration.

equivalent to a bar code. It is then possible to strictly identify the R-DL, which is now a "RFID R-DL." As the position of the peaks also depends on the temperature and strain state of the chip, it becomes eventually possible to use the devices as very robust RFID sensors, with no embedded electronics. When adding on the propagation path of the waves, a thin layer sensitive to another physical or chemical parameter (Figure 7.3), it becomes possible to measure this parameter, whose variations will also affect the propagation time of the acoustic waves (e.g. magnetic fields [28]).

7.5.2 R-DL with Connected IDTS

As mentioned in the previous section, the ID-tag configuration is particularly interesting for remote measurements in a sensor network. This configuration nevertheless requires a high-signal amplitude which can be achieved with a high coupling coefficient (K^2) and low propagation losses (α). Indeed, the higher the amplitude of the echoes, the longer the interrogation distance. However, harsh environments usually affect the devices' performances and therefore degrade the quality of the detected signal. Also, the well-known substrates for harsh environments (see Section 7.6 for more details) such as langasite (LGS), langatate (LGT), or AlN/Sapphire have limited performances in terms of K^2 and α [29].

In this context, an original configuration with connected IDTs has been investigated [5]. The main interest is to use the direct transmission between IDTs instead of the reflection from passive reflectors, thus leading to an increase in the detected amplitude. The configuration can hence be useful to mitigate the adverse effects of harsh environments. The code is generated by the number and the gap between the IDTs. As an example, a code with three main peaks (at 2d, 2D and 2d+2D distances) can be generated with the three connected IDTs configuration resulting from the direct transmissions, whereas three reflectors are needed to obtain three peaks with single reflections (Figure 7.4).

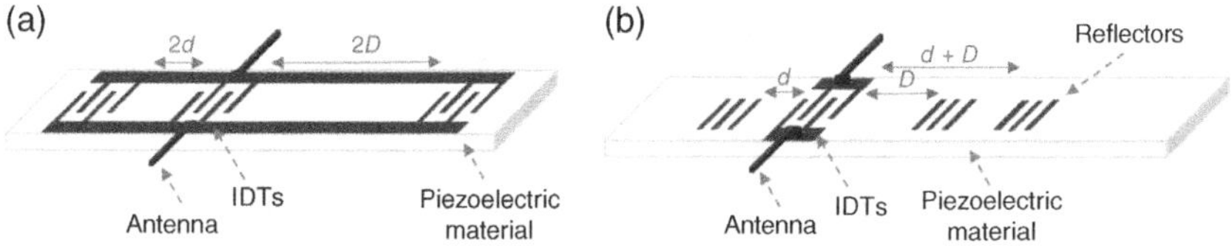

Figure 7.4 Illustration of a configuration (a) with three connected IDTs, and (b) its equivalent with reflectors.

Besides the characteristics of each sensor type, regarding the RFID communication chain, there are some more considerations that should be taken into account.

This configuration with connected IDTs is of great interest for the acoustic performances but some drawbacks should be known. Indeed, the total size of the device with connected IDTs is larger than the one from the conventional structure with reflectors. Also, there is a maximum number of IDTs that can be connected together as the RF power of the interrogation signal is split between the IDTs. This means that beyond a given number of connected IDTs, the signal amplitude of the direct transmission will be lower than the one reflected in the standard ID-tag configuration. The higher the number of connected IDTs, the more difficult it can be to properly identify the direct transmission peaks in the code. The electrical interaction between connected IDTs also needs to be considered to solve the antenna impedance matching issue which is particularly important for remote interrogation. Hence, it is important to find a trade-off between the amplitude level increase, the space requirements, the antenna matching and the ease of data processing. To overcome the aforementioned drawbacks, it is therefore recommended to aim for a limited number of connected IDTs.

This idea of improving the performances thanks to work on the IDTs design that has already been the subject of several (mainly numerical) studies and has led to structures similar to one of the connected IDTs. Figure 7.5 gives some examples [4, 30]. However, these structures have not been experimentally validated for a tag application, unlike the above-mentioned connected IDTs structure.

7.6 Saw Sensor for Harsh and Severe Environments

7.6.1 Piezoelectric Material for High Temperature

Very few piezoelectric materials can keep their properties under high-temperature conditions. Regarding conventional piezoelectric materials used in SAW applications taking place at room temperature, only lithium niobate ($LiNbO_3$) crystals can be operated above 400 °C. Indeed, quartz crystals undergo a phase transition

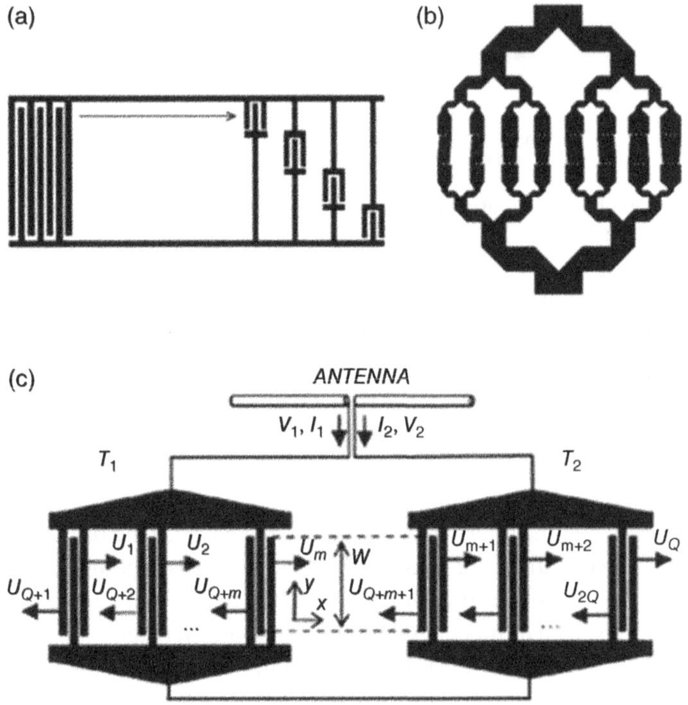

Figure 7.5 Examples of IDTs improvements. (a) 1-track design where each coding structure is passed only once, (b) design with an output summing of 8 IDTs and (c) tapped-input-tapped-output design (4,31). *Source:* Plessky and Reindl [4] and Sveshnikov and Shitvov [30].

at 573 °C, from the alpha-phase to the poorly piezoelectric beta-phase. However, the Q-factor of quartz resonators decreases significantly from 300 °C because of an increase in the structural disorder [31]. On the other side, ferroelectric lithium tantalate crystals have never been studied as SAW substrates for high-temperature conditions, likely because of their low Curie temperature (about 600 °C) and their strong pyroelectric properties [32].

Congruent lithium niobate (cLN) ferroelectric crystals are much more adapted for high-temperature applications, having a Curie temperature of about 1 200 °C and much smaller pyroelectric properties. However, this material, which shows exceptionally high electromechanical coupling coefficients K^2 of several percent, also knows serious limitations for applications at high temperatures. First of all, like any semiconductor material, its electrical resistivity decreases with temperature, reaching values as low as $10^5\ \Omega\cdot$cm at 600 °C [33]. Secondly, cLN crystals are only metastable at room temperature. When the temperature increases, the kinetics of the segregation process rise, so that cLN crystals evolve toward the thermodynamical equilibrium, that is, a mix between $LiNbO_3$ and lithium triniobate $LiNb_3O_8$ which is richer in niobium element [34]. The kinetics of the segregation process becomes significant at meaningful human timescale from 300 to 350°C. Thus explaining why Born et al. observed no degradation of cLN crystals after 22 days at 300 °C, but the apparition of $LiNb_3O_8$ micronic crystals in less than one day at 750 °C under air atmosphere [35]. This limitation can be avoided by the use of stoichiometric lithium niobate crystals (sLN) which are thermodynamically stable between the ambient and their melting temperature at 1 170 °C [36]. However, high-quality sLN crystals are much more difficult to obtain than cLN ones. Consequently, very few sLN orientations are currently available on the market.

On the other side, cLN crystals have great electromechanical properties: large K^2 and SAW velocities (e.g. 5.5% and 3 992 m/s, respectively for the Rayleigh wave propagating in the Y_{128}-X direction), significantly lower propagation losses at high frequencies compared to some other piezoelectric materials [37] and great temperature coefficient of frequency (TCF), suitable to make temperature sensors. Consequently, thanks to this unique combination of high-temperature stability and great electromechanical performances, cLN crystals have been intensively studied in the early twenties to make SAW R-DL sensors working in the 2.45 GHz ISM band, whose large bandwidth is suitable for this technology. 2.45 GHz SAW R-DL sensors based on cLN substrates and Al electrodes have been commercialized for more than a decade. They can operate for several days at 350 °C, but only some hours at 400 °C. This limited lifetime seems to be more related to the oxidation of the Al electrodes than to the cLN crystal segregation [38].

Thus, in the last paragraphs, it has been evidenced that no conventional piezoelectric substrate can be operated for SAW measurements upon 500 °C, whereas many industrial applications take place in the 500–1 000 °C range. For this reason, several new materials have been developed since the nineties [39]. One of them has gradually prevailed, namely langasite $La_3Ga_5SiO_{14}$ (LGS). This crystal was synthesized first in USSR in 1981 [40]. The growing method has been constantly improved until it became possible to obtain 4″ good-quality LGS wafers of different orientations.

LGS crystals show outstanding stability at high temperatures since they keep their piezoelectric properties up to their melting temperature, at 1470 °C [41]. Many studies have been dedicated to high-temperature LGS-based SAW devices and some impressive results have been obtained. In particular, Pereira da Cunha et al. operated a LGS-based SAW resonator for more than five months at 800 °C in air [42], while Aubert et al. measured a SAW signal on a LGS-based delay line up to 1 140°C (Figure 7.6) [43].

Despite this exceptional stability at high temperatures, LGS crystals show several limitations. The first one is related to the frequency. Indeed, LGS is not suitable for high-temperature applications in the GHz range because propagation losses in

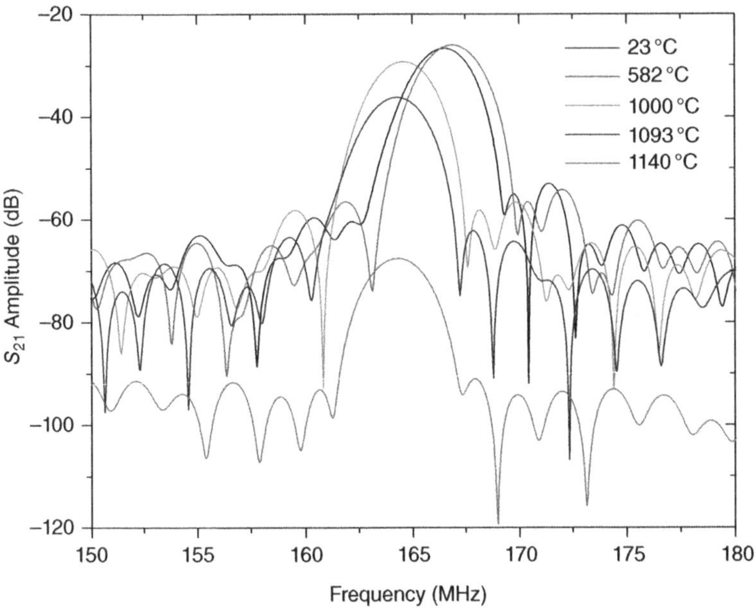

Figure 7.6 LGS-based delay-line response up to 1 140 °C. *Source:* Aubert et al. [43].

LGS increase very quickly with temperature and frequency [37]. Moreover, SAW velocity in LGS is in the range from 2 500 to 3 000 m/s, which makes it necessary to use wavelengths hardly above 1 μm to reach the 2.45 GHz ISM band, incompatible with standard UV photolithography process. Anyway, the low K^2 of LGS (<0.5%) makes this substrate not suitable to make 2.45 GHz R-DL sensors. Thus, the best use of LGS for high-temperature applications is made with 434 MHz resonators, or at a push, 915 MHz resonators. One other limitation of LGS is related to its oxide nature which makes it unstable under low-O environments at high temperature, just like $LiNbO_3$ by the way (for that matter $LiNbO_3$ crystals are annealed under N_2 or H_2 atmosphere to obtain "black" cLN. The latter have O vacancies that make the crystal less electrically resistive which prevents the formation of sparks related to pyroelectricity, during the photolithography process [44]). In such conditions, LGS crystals undergo Ga and O losses from 650 °C [45]. This phenomenon also happens under air atmosphere but at much higher temperatures, that is, around 1000 °C [46]. And finally, the resistivity of LGS crystals decreases quite fast when the temperature rises, to reach 10^6 Ω·cm at 600 °C. Most high-temperature piezoelectric crystals show better dielectric properties than LGS [47] (Figure 7.7).

That being said, it has to be underlined that LGS-based resonators constitute the state-of-the-art regarding high-temperature wireless SAW sensors, with operating temperatures in the range between 600 and 700 °C [48–50].

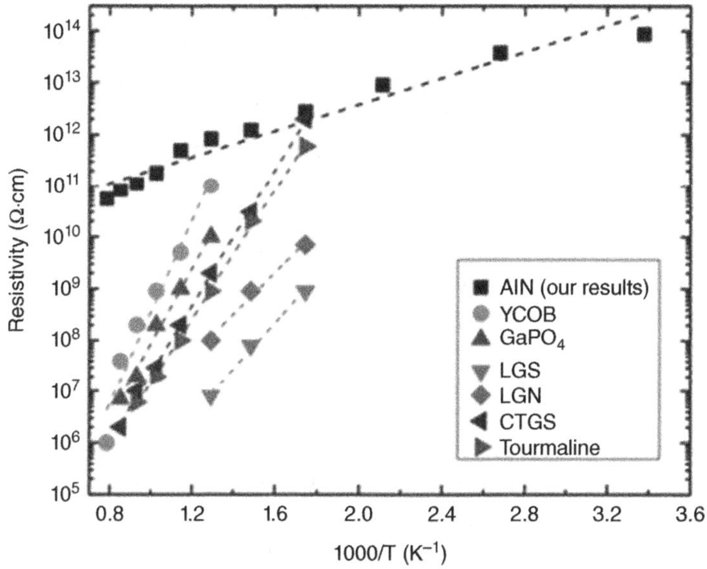

Figure 7.7 Temperature-dependent resistivity evolution of common piezoelectric substrate. *Source:* Kim et al. [47]/with permission of IEEE.

Developing wireless SAW sensors able to operate at higher temperatures would require the replacement of LGS by another piezoelectric crystal with better dielectric properties, such as gallium phosphate $GaPO_4$ [51] or YCOB [52] for instance. Indeed, the electrical resistivity of those crystals is 2–3 orders of magnitude higher than that of LGS at 800 °C [47], which would allow the achievement of SAW resonators showing high-quality factors Q compatible with remote interrogation, above 700 °C. Unfortunately, the use of these promising materials is hindered by the lack of availability of high-quality wafers of different orientations.

Another way to achieve wireless SAW sensors operating above 700 °C has been explored. It consists in replacing piezoelectric bulk substrate by stacking in which at least one layer is piezoelectric. The stacking is deposited by a PVD or CVD method on a substrate which can be in that case nonpiezoelectric. Actually, the most advanced results have been obtained with the AlN/Sapphire bilayer structure. Indeed, AlN thin films are of major interest for the targeted applications. They gather several critical properties. The melting temperature of AlN is higher than 3 000 °C. It is stable under high vacuum up to 1 040 °C [53], but this limitation is pushed back around 1 200 °C under N_2/H_2 atmosphere [54]. It undergoes oxidation under air from 700 °C. However, this phenomenon is very limited as the thickness of the formed Al_2O_3 overlayer is less than 20 nm, whereas it is 10 nm at its native state [29]. Moreover, AlN keeps its piezoelectricity up to at least 1150 °C [33]. However, the greatest strength of AlN for high-temperature SAW applications lies in its dielectric properties. This semiconductor material has a very large bandgap of 6.2 eV, making AlN extremely resistive even at high temperature. Thus, the resistivity of AlN at 700 °C is more than 10^{11} $\Omega\cdot$cm, six orders of magnitude larger than that of LGS, and three orders more than that of $GaPO_4$ or YCOB (Figure 7.7). The resistivity of AlN remains higher than 10^{10} $\Omega\cdot$cm at 1000 °C. In other words, the dielectric properties of AlN make it theoretically possible to achieve wireless SAW sensors able to be operated up to 1 000 °C. Besides, the electromechanical properties of AlN are also very interesting. Indeed, AlN is a very hard material which consequently shows high acoustic velocities: Rayleigh waves propagating at the surface of (002) AlN have a velocity around 5 600 m/s. Thus, when combined with a substrate showing high acoustic velocities as well (diamond, sapphire, ...), AlN thin films are particularly suitable to make SAW sensors operating in the 2.45 GHz ISM band, even more as AlN propagation losses in the GHz range are close to those of $LiNbO_3$ [55]. These outstanding electromechanical properties can be obtained only with AlN thin films with excellent crystalline quality, in which the impurities quantity is as low as possible. This condition requires to grow AlN thin films on a substrate with lattice parameters compatible with AlN ones so that hetero-epitaxy can be realized. Sapphire is particularly suitable for that aim, even more so as this substrate is particularly stable at high temperature [56]. Finally, it has to be noted that K^2 of AlN is relatively limited ($\approx$0.3%), making it unsuitable for the R-DL technology. But AlN-based resonators constitute a unique and promising solution to achieve SAW sensors able to operate at temperatures higher than 700 °C (especially under low-O environments) and at high frequencies, namely in the 2.45 GHz band.

The first studies on the AlN/Sapphire bilayer structure for high-temperature SAW applications were published a decade ago [57]. In particular, Aubert et al. measured a stable signal on a SAW delay line based on the AlN/Sapphire structure up to 1 050 °C, for more than 40 hours [58] (Figure 7.8). One can observe that the insertion losses evolve very little with the temperature, which is likely related to the compensation of the propagation losses increase by that of the K^2 of the structure [59]. More recent studies fully confirmed the potential of the structure, showing that SAW resonators with quality factors higher than 8 000 at 433 MHz can be achieved on it [60]. Moreover, the factor of merit $Q.K^2$ of the structure, an essential parameter for remote interrogation, was shown to be higher than 2, and very stable up to 400 °C at least. Based on these very promising results, future studies will have to focus on the wireless interrogation of SAW resonator sensors based on AlN/Sapphire, in order to definitively establish the high potential of this structure, in particular for applications above 700 °C.

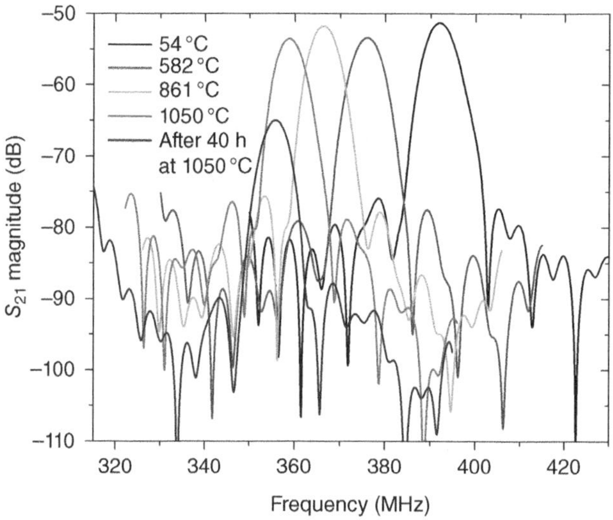

Figure 7.8 AlN/Sapphire-based delay-line response up to 1 140 °C. *Source:* Adapted from Aubert et al. [58].

7.6.2 Metallic Material for Electrodes

The materials constituting the electrodes of SAW devices designed for applications in harsh environments should gather several properties:

- a low electrical resistivity, close to that of Al or Cu, that is, of some $\mu\Omega \cdot cm$; limiting the losses by Joule effect;
- a high melting temperature, which leads to a high Tamman temperature, over which thin films transform into a collection of separate beads, which is called agglomeration phenomena [61];
- a high resistance to oxidation, especially for applications taking place at high temperature under air atmosphere;
- a low density, close to that of Al, allowing the generation of propagating SAW at high frequencies, for applications in the 915 MHz or 2.45 GHz ISM bands.

Consequently, till the end of the first decade of the twenties, Pt electrodes were massively used for high-temperature SAW applications [62]. Indeed, Pt shows an outstanding resistance to oxidation, a reasonable electrical resistivity of $10.6\,\mu\Omega \cdot cm$, and a rather high melting temperature of 1 768 °C, corresponding to a Tamman temperature of 750 °C. The only drawback of Pt is related to its very large density (21.5). Such a high-value leads, when the frequency increases and thus the wavelength decreases, to the rise of the reflection coefficient of the IDT fingers, and consequently hinders the SAW emission outside the IDT. This phenomenon may lead to the use of very thin Pt electrode fingers (around 50 nm at 2.45 GHz), which are too resistive and too brittle for harsh environments applications. That is the reason why Al electrodes are used to achieve SAW R-DL sensors operating in the 2.45 GHz ISM band.

In 2007, Pereira da Cunha et al. suggested new classes of materials to make electrodes for high-temperature SAW applications above 700 °C: Pt-based alloys and nanocomposites [63]. In such materials, agglomeration phenomena are significantly hindered since Pt diffusion coefficients are much smaller than in pure Pt films [64].

Thus, Pereira da Cunha et al. measured successfully a SAW signal on a device made of a LGS substrate and Pt-10%Rh/ZrO_2 nanocomposite thin film electrodes for 5½ months at 800 °C under air atmosphere [65]. The main drawback of these films is related to their relatively high electrical resistivity, around 150 μΩ·cm [66]. Some other compositions were tested by this group in further studies. The best results were obtained with Pt-Ni/Pt-Zr multilayer thin films, which showed great stability for several days at 1 000 °C under air [67].

Another way to achieve thin film electrodes able to withstand temperatures higher than 700 °C in air consists of replacing Pt with another noble metal with intrinsic lower diffusion coefficients, thus a higher melting temperature. In this regard, iridium constitutes an excellent candidate with a melting temperature of 2 440 °C, leading to a Tamman temperature close to 1 100 °C. Moreover, the electrical resistivity of Ir is particularly low, with a value of 5.1 μΩ·cm. Thus, Aubert et al. measured a SAW signal based on a LGS substrate and Ir electrodes up to 1 140 °C, under low-vacuum environment ($P = 5.10^{-5}$ mbar) [43]. Unfortunately, Ir is more sensitive to oxidation than Pt. It undergoes oxidation in air from a temperature between 600 and 700 °C, the phenomenon accelerating at 800 °C [68]. However, the formed oxide, IrO_2, is an excellent electrical conductor, with a metallic behavior (the electrical resistivity of IrO_2 increases with the temperature). Thus, the resistivity of IrO_2 thin films can reach values as low as 37 μΩ·cm [69]. Unfortunately, nor is IrO_2 stable above 900 °C as it oxidizes into volatile IrO_3: thus Ir thin film electrodes sublimate around 900 °C under air atmosphere. However, as shown by Osamura and Abe in the case of bulk spark plug electrodes, the resistance of Ir to high-temperature oxidation and sublimation can be greatly improved by alloying Ir with a small part of Rh [70]. This phenomenon has been confirmed in the case of thin films by Taguett et al. [71, 72]. In these studies, the authors observe the great potential of 120 nm-thick Ir–Rh thin films, with an Ir atomic percentage of at least 50%, for high-temperature SAW applications in air atmosphere. Actually, such films oxidize completely after 4 days at 800 °C, transforming into a conductive $Ir_xRh_{1-x}O_2$ film, surrounded by an insulating Rh_2O_3 layer. The thickness of the latter decreases with the initial Rh atomic percentage: it is 50 nm for $Ir_{0.5}Rh_{0.5}$ films, but only 5 nm in the case of

$Ir_{0.85}Rh_{0.15}$. Consequently, the electrical resistivity of Ir–Rh films after 4 days at 800 °C in air decreases with the initial Rh atomic percentage: 185 μΩ · cm for $Ir_{0.5}Rh_{0.5}$ films, whereas it is only 100 μΩ · cm for $Ir_{0.85}Rh_{0.15}$. It has to be noted that this value is still larger than the resistivity of pure IrO_2 films (around 40 μΩ · cm). Actually, the superiority of $Ir_xRh_{1-x}O_2$ films appears at temperatures higher than 800 °C, or during a very long exposure at 800 °C (more than one week). In these extreme conditions, the sublimation process of IrO_2 films occurs gradually, whereas $Ir_xRh_{1-x}O_2$ films are much more stable. A very good compromise seems to be reached with $Ir_{0.7}Rh_{0.3}O_2$ films. Films richer in Rh are too resistive for the targeted application, whereas films richer in Ir are less stable, especially above 850 °C. $Ir_{0.7}Rh_{0.3}O_2$ films show a great stability, for at least several days, at temperatures as high as 900 °C in air atmosphere. Finally, conductive oxide films have a great advantage over metal-based films to make electrodes for high-temperature SAW applications. They have a much larger chemical affinity with substrates since the latter are generally oxides themselves. Consequently, agglomeration phenomena seem nonexistent in oxide films since the driving force of agglomeration is the surface tension between the film and the substrate.

7.7 Antennas for Harsh Environments

Another key element for wireless SAW sensors in harsh environment is the antenna. The radiating element, in this context, has its own set of requirements.

- the conducting part has to be a good conductor, with typically a conductivity > 106 S/m as a bare minimum and retain this property at high temperatures. Here, platinum or, less expensive, stainless steels/Inconel superalloys are materials of choice.
- the conductors have to present a thickness at least in the range of the skin depth and preferably over three to five times. At the ISM frequencies considered for SAW sensors, typically in the 433 to 2500 MHz region, this means multimicrons conductors, which can be challenging to achieve with thin film deposition techniques.
- if the antenna is not a fully metallic construct, a dielectric substrate is present. It has to be low loss (loss tangent $\delta < 0.01$ even at high temperatures) and present relatively stable dielectric properties with the increasing temperature to avoid an excessive frequency drift. Here, ceramics like pure alumina are to be considered [73].
- last but not least, the antenna needs to be integrated with the SAW sensors and its package, which can be challenging due to the heterogeneity of materials, various corrosive and (thermo-)mechanical stresses and the potential

limitation of the interconnect techniques. Mechanical clamping [74], soldering/brazing [49], and wire-bonding [75] all present challenges at high temperatures. Capacitive coupling [76] can provide a solution without direction connection but with parasitic effect. Proximity coupling between a feed structure and an antenna could provide an elegant solution [77].

A few examples of high-temperature SAW-antenna systems can be found in the literature, using various design and fabrication techniques: metallic helicoidal and PIFA antennas are used in [49], stainless steel dipole in [78], electroless Pt on alumina folded dipole [79], platinum paste antenna on ceramics [50, 76]. A comprehensive study of antenna and interconnect performance under high-temperature conditions remains very much needed.

7.8 Packaging for Harsh and Severe Environments

7.8.1 Packaging for Saw Devices

SAW devices are known for their sensitivity to surface perturbations. For many applications, the use of packaging is, as a consequence, critical for a correct operation. In telecommunications SAW/BAW devices, there has been a constant progression toward ever-small chip size and wafer-level packaging [80]. However most "standard" packaging techniques are not suitable for harsh environments.

For extreme environments, metallic [50, 81] and ceramic [74, 78, 81] packages have been developed. Klein et al. proposed a metallic package based on Kovar (Fe–Ni–Co alloy) and Da Cunha developed one based on Inconel alloy [50] with specific sealing techniques. Ceramic packaging based on platinum HTCC (High Temperature Co-Fired Ceramic) [81] is particularly attractive because of the resilience of the materials involved and the fact that it is an industrial technique. Despite some development by Kyocera [82], Pt-ceramic does not appear to be widely available commercially.

7.8.2 Packageless Solution

Many challenges remain for harsh environment packaging. Herein, we discuss a new solution, called "packageless", which is based on additional layers over the IDTs.

To protect the device from the external environment (chemical attack, accidental scratches), a simple—and probably the most intuitive—approach is the use of a passivating deposition above the electrodes [66]. However, this layer remains very thin, making the acoustic wave propagation still at the surface of the device and consequently sensitive to all external perturbations.

Bhattacharjee et al. went further with this idea of additional layers and have shown in 2007 two theories to achieve full packageless structures [83]. The first one is the Isolated Layer Acoustic Wave (ILAW) where the high impedance substrate is covered by a relatively thick low acoustical impedance layer (SiO_2 or Pyrex), itself covered with a thin high acoustical impedance layer. The significant difference in the acoustic properties at the interface limits the wave propagation inside the upper layer. By alternating thin layers of high and low acoustical impedance thus forming a Bragg mirror structure in the vertical direction (Figure 7.9a), the wave confinement can even be enhanced.

The second one is the Waveguiding Layer Acoustic Wave (WLAW) which can be seen as the acoustic equivalent of the optical fiber (Figure 7.9b). Indeed, in this three-layer structure, the two outer layers (including the substrate) have a high acoustic impedance, while the middle layer has a low acoustic impedance. The velocity gradient generated by the impedance difference is thus responsible for the wave propagation inside the structure. More precisely, the thicknesses of the two upper layers must be judiciously chosen to ensure the wave confinement, i.e. so that the surface displacement is no longer significant. Without a surface propagation, the wave cannot be disturbed by external perturbations (dust, droplet), thus explaining the so-called "packageless" device.

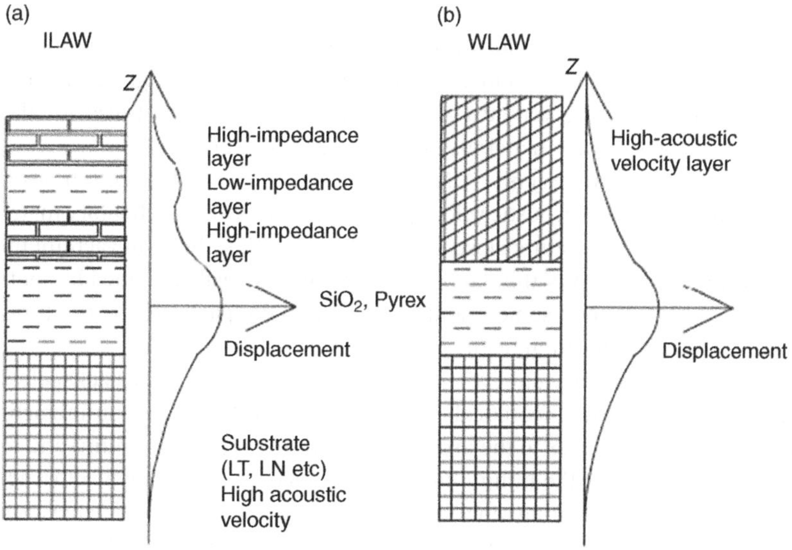

Figure 7.9 Cross-section illustration of ILAW (a) and WLAW (b) layered structures to achieve packageless devices. *Source:* Bhattacharjee et al. [83]/with permission of IEEE.

In addition to their interest in the wave confinement, these top layers can improve the acoustic properties of the overall structure depending on the targeted application. For example:

- **Zero TCF (temperature coefficient of frequency):** coming from a compensation between a negative TCF layer and a positive TCF layer [84]
- **High coupling coefficient K^2:** useful for delay lines structures [85]
- **High quality factor Q:** useful for wireless measurements

The thicknesses choices must therefore result from a trade-off between a satisfying wave confinement and a performance optimization.

Due to the wave confinement inside the three-layer structure, packageless devices are not suitable for surface detection such as chemical sensors (gases, humidity). However, they perfectly fit with physical bulk parameters that propagate through the structure (temperature, strain) or that can be remotely controlled (magnetic field).

Compared to the encapsulation, which is also a mechanical protection, this packageless solution does not withstand violent destruction. Packageless devices are nevertheless the best in terms of miniaturization leading to a use for unreachable environments or with harsh conditions (pressure, temperature). Indeed, layers in the order of the micrometer are only needed. Shen et al. have recently achieved a packageless filter with an ultra-small size of $0.695 \times 0.490 \times 0.122\,mm^3$ by using a $SiN\textit{x}/SiO_2$/IDTs(Cu)/15°Y–X $LiNbO_3$ multilayered structure [86].

The material choice must obviously be made according to the acoustical impedance but also according to the environment constraints of the targeted applications [85]. This is, by the way, true for the additional layers but also for the electrodes and the substrate materials as explained in Section 7.6.

In case of harsh environments, AlN is a material of choice for the top layer. Indeed, it has a very high acoustic velocity (5 600 m/s). It is also a refractory material with a melting temperature close to 3 200 °C and with stability in vacuum up to 1 040 °C [53]. Finally, AlN thin layers can be relatively easily obtained by sputtering [87]. Table 7.2 gives some examples of potentially packageless structures and their intended application field.

Naumenko and Nicolay went even further in the concept of SAW packageless devices. Indeed, they have theoretically shown a Rayleigh-type wave confinement in a bilayer structure thanks to the use of dense electrodes (like platinum) and to the specific properties of the lithium niobate Y°+128 cut [94]. This type of structure is particularly interesting since it involves only one thin layer deposition (here AlN) which should be easier to optimize in terms of acoustic performances. The wave confinement has been verified experimentally and measurements were done up to 300 °C [95], making this bilayer structure particularly interesting for

Table 7.2 Examples of packageless structures.

Concept	Structure	Targeted application	References
ILAW	HfO_2/Simax/IDTs(Au/Ti)/15°Y–X $LiNbO_3$	Communication systems, resonators	Zhgoon et al. [88]
WLAW	SiN_x/SiO_2/IDTs(Cu)/15°Y–X $LiNbO_3$	Communication systems	Shen et al. [86]
WLAW	AlN/ZnO/IDTs(Al)/Diamond	/	Elmazria et al. [2]
WLAW	Al_2O_3/ZnO/IDTs(Al)/$LiNbO_3$	Biomedical devices	Hage-Ali et al. [89]
WLAW	AlN/ZnO/IDTs(Au/Ti)/$LiNbO_3$	On-body temperature sensors	Floer et al. [90]
WLAW	AlN/IDTs(Al)/ZnO/Si	/	Legrani et al. [3]
WLAW	AlN/IDTs(Pt)/AlN/Sapphire	High temperature environments	Legrani et al. [91]
WLAW	Al_2O_3/IDTs(Al)/GaN/Sapphire	High temperature environments	Zhgoon et al. [92]
WLAW	AlN/IDTs(Al)/GaN/Sapphire	High temperature environments	Bartoli et al. [93]
WLAW	Si_3N_4/IDTs(Pt)/GaN/Sapphire	High temperature environments	Naumenko [85]

high-temperature applications. Based on the same idea, a study with other materials like Cu or Au for the electrodes and ScAlN for the top layer has already been investigated [96].

7.9 Conclusion and Outlooks

SAW sensor technology has great potential for measurements under harsh environments, especially in wireless configurations. Indeed, when combined with an antenna, SAW sensors can be wirelessly interrogated without embedded power sources and electronics, which are the limiting parts of conventional wireless sensors in such conditions. Moreover, SAW sensor technology offers the RFID functionality in the R-DL configuration: R-DL SAW sensors can be unambiguously identified through their coded time-resolved response.

Many efforts have been dedicated to the development of SAW sensors for harsh environments, especially high temperatures, in the last two decades. Many of these efforts have been focused on the materials constituting the sensor. Numerous piezoelectric substrates or stackings have been studied. Three of them are of particular interest. Lithium niobate because of its large K^2 and low propagation losses enabling the achievement of R-DL sensors operating in the 2.45 GHz ISM band. This substrate seems however limited to 400–500 °C in the congruent composition because of the segregation process. Higher temperatures could be reached with stoichiometric $LiNbO_3$ substrates, more stable at high temperatures. Langasite crystals have also been thoroughly studied because of their exceptional stability at high temperatures up to 1000°C, at least. Their limited K^2 does not allow the realization of R-DL sensors, but LGS-based SAW resonators constitute the state-of-the-art regarding high-temperature wireless SAW sensors, with operating temperatures in the range between 600 and 700 °C. This limitation is related to the relatively poor dielectric properties of LGS crystals at high temperatures. Higher temperatures could be reached with the AlN/Sapphire bilayer structure which combines great stability at high temperatures and outstanding dielectric properties. Regarding thin film electrodes, advanced solutions already exist. In particular, Pt-based nanocomposites or multilayers show great stability above 800 °C for months. Such compositions, very dense, are however not suitable for R-DL sensors that operate at high frequencies, that is, in the large 915 MHz or 2.45 GHz ISM bands. Electrodes with a density close to that of the substrate are required in that case, otherwise, the thickness of the electrodes should be very limited to compensate for the difference in acoustic impedance between the electrodes and the substrate.

Based on all these developments, SAW devices for harsh environments are currently on the market with companies like Transens, Sensideon, or SenSanna. Their field of application can refer to:

- aerospace, where the measurement of torque in engines is used to improve safety, pilot control and engine reliability [97].
- electric motors and drives, where a reliable torque and temperature measurement from the motor rotor is needed to improve both motor efficiency and safety, which is particularly suitable for the growing electric vehicle market [97].
- off-highway machines like handling or agriculture machine, where the remote measurement of torque or temperature can lead to fully autonomous unit and reduce unnecessary breakdowns and costly downtime [97].
- motorsport, where the torque in race vehicle can be measured for monitoring and regulatory compliance purposes [97].
- industrial robotics, where a measurement over a joint with a full rotation of 360° can be achieved [98].
- metallurgy industry, where high temperature need to be controlled like in a ladle in steel works, in forging dies or also in continuous furnaces [99].
- power lines, where the current or the tension needs to be monitored [100].

Many more applications can be targeted for the future, especially if higher temperatures of stability can be reached for SAW sensors. This will require to dedicate much more effort to unconventional but highly promising materials, like stoichiometric lithium niobate substrates or new thin film electrode materials that gather high conductivity, high melting temperature, great resistance to oxidation, and low density. This last property excludes the use of Pt-based materials. First studies on NiAl alloy, for instance, are of great interest. Of course, it will also be necessary to work hard on the peripheral parts of the system that are also placed under harsh environments, namely the antenna and even more the packaging that has been hardly studied up to now.

References

1 Reindl, L., Ruppel, C.C.W., Kirmayr, A. et al. (2001). Radio-requestable passive SAW water-content sensor. *IEEE Transactions on Microwave Theory and Techniques* 49 (4): 803–808.

2 Elmazria, O., Zhgoon, S., Le Brizoual, L. et al. (2009). AlN/ZnO/diamond structure combining isolated and surface acoustic waves. *Applied Physics Letters* 95 (23): 233503.

3 Legrani, O., Elmazria, O., Zhgoon, S. et al. (2013). Packageless AlN/ZnO/Si structure for SAW devices applications. *IEEE Sensors Journal* 13 (2): 487–491.

4 Plessky, V.P. and Reindl, L.M. (2010). Review on SAW RFID tags. *IEEE Transactions on Ultrasonics, Ferroelectrics, and Frequency Control* 57 (3): 654–668.

5 Floer, C., Hage-Ali, S., Nicolay, P. et al. (2020). SAW RFID devices using connected IDTs as an alternative to conventional reflectors for harsh environments. *IEEE Transactions on Ultrasonics, Ferroelectrics, and Frequency Control* 67 (6): 1267–1274.

6 SAW COMPONENTS Dresden [Internet]. SAW COMPONENTS Dresden GmbH, Dresden Germany. https://www.sawcomponents.de/en/ (accessed 28 November 2022).

7 SENSeOR [Internet]. SENSeOR - WIKA Group, Mougins France. https://senseor.com/ (accessed 28 November 2022).

8 Transense [Internet].Transense Technologies plc, Bicester, Oxfordshire, UK. https://www.transense.com/saw-technology/technology (accessed 28 November 2022)

9 Fachberger, R., Bruckner, G., Hauser, R., et al. (2004). Properties of radio frequency Rayleigh waves on langasite at elevated temperatures. *IEEE Ultrasonics Symposium, 2004*, vol. 2, Montreal, QC, Canada (23–27 August 2004). IEEE. 1223–1226.

10 Reindl, L., Scholl, G., Ostertag, T. et al. (1998). Theory and application of passive SAW radio transponders as sensors. *IEEE Transactions on Ultrasonics, Ferroelectrics, and Frequency Control* 45 (5): 1281–1292.

11 White, R.M. and Voltmer, F.W. (1965). Direct piezoelectric coupling to surface elastic waves. *Applied Physics Letters* 7 (12): 314–316.

12 Campbell, C. (1998). *Surface Acoustic Wave Devices for Mobile and Wireless Communications*. San Diego: Academic Press.

13 Zhgoon, S., Shvetsov, A., Ancev, I. et al. (2015). SAW temperature sensor on Quartz. *IEEE Transactions on Ultrasonics, Ferroelectrics, and Frequency Control* 62 (6): 1066–1075.

14 Bergman, J.G., Ashkin, A., Ballman, A.A. et al. (1968). Curie temperature, birefringence, and phase-matching temperature variations in $LiNbO_3$ as a function of melt stoichiometry. *Applied Physics Letters* 12 (3): 92–94.

15 Li, Y.Q., Zheng, Y.Q., Tu, X.N. et al. (2014). The high temperature resistivity of lithium niobate and related crystals. *Proceedings of the 2014 Symposium on Piezoelectricity, Acoustic Waves, and Device Applications*, Beijing, China (30 October–02 November 2014). IEEE. 283–286.

16 Hornsteiner, J., Born, E., and Riha, E. (1997). Langasite for high temperature surface acoustic wave applications. *Physica Status Solidi A: Applications and Materials Science* 163 (1): R3–R4.

17 Nicolay, P. (2007). Les capteurs à ondes élastiques de surface : applications pour la mesure des basses pressions et des hautes températures. PhD thesis. Université Henri Poincaré – Nancy 1. https://hal.univ-lorraine.fr/tel-01748241.

18 Ieki, H. and Kadota, M. (1999). ZnO thin films for high frequency SAW devices. *1999 IEEE Ultrasonics Symposium Proceedings International Symposium (Cat No99CH37027)*, vol. 1, Tahoe, NV, USA (17–20 October 1999). IEEE. 281–289.

19 Assouar, M.B. (2001). Etude de dispositifs à ondes acoustiques de surface (SAW) à structure multicouche nitrure d'aluminium diamant : croissance de matériaux en couches minces et technologie de réalisation. PhD thesis. Université Henri Poincaré – Nancy 1. https://hal.univ-lorraine.fr/tel-01746509.

20 Justice, J., Lee, K., and Korakakis, D. (2012). Harmonic surface acoustic waves on gallium nitride thin films. *IEEE Transactions on Ultrasonics, Ferroelectrics, and Frequency Control* 59 (8): 1806–1811.

21 Caliendo, C. (2008). Theoretical and experimental investigation of gigahertz-band, temperature-compensated electromechanical coupling configurations based on AlN films. *Applied Physics Letters* 92 (3): 033505.

22 Caliendo, C. (2003). Gigahertz-band electroacoustic devices based on AlN thick films sputtered on Al_2O_3 at low temperature. *Applied Physics Letters* 83 (23): 4851–4853.

23 Lurz, F., Ostertag, T., Scheiner, B. et al. (2018). Reader architectures for wireless surface acoustic wave sensors. *Sensors* 18 (6): 1734.

24 Shvetsov, A., Zhgoon, S., Antcev, I. et al. (2016). Quartz orientations for optimal power efficiency in wireless SAW temperature sensors. *2016 European Frequency and Time Forum (EFTF)*, York, UK (4–7 April 2016). IEEE. 1–4.

25 Lakin, K.M. (1974). Electrode resistance effects in interdigital transducers. *IEEE Transactions on Microwave Theory and Techniques* 22 (4): 418–424.

26 Friedt, J.M., Saintenoy, A., Baron, T. et al. (2013). High-overtone bulk acoustic resonator as passive sensor acting as buried cooperative target interrogated by Ground Penetrating RADAR. *2013 Joint European Frequency and Time Forum & International Frequency Control Symposium (EFTF/IFC)*, Prague, Czech Republic (21–25 July 2013). IEEE. 466–469.

27 Nicolay, P., Bruckner, G., Chambon, H., and Ortner, M. SAW RFID sensors and devices for industrial applications, a short review. *ECSA-5 Symposium 2018 (keynote lecture) [Internet]* (15–30 November, 2018). MDPI. https://www.youtube.com/watch?v=IqOKdRe9ecc&t=1s

28 Mishra, H., Hehn, M., Hage-Ali, S. et al. (2020). Microstructured multilayered surface-acoustic-wave device for multifunctional sensing. *Physical Review Applied* 14 (1): 014053.

29 Aubert, T., Elmazria, O., Assouar, B. et al. (2012). Investigations on AlN/sapphire piezoelectric bilayer structure for high-temperature SAW applications. *IEEE Transactions on Ultrasonics, Ferroelectrics, and Frequency Control* 59 (5): 999–1005.

30 Sveshnikov, B.V. and Shitvov, A.P. (2018). Multitransducer SAW device architecture for passive wireless sensor tags. *IEEE Transactions on Ultrasonics, Ferroelectrics, and Frequency Control* 65 (11): 2176–2183.

31 Haines, J., Cambon, O., Keen, D.A. et al. (2002). Structural disorder and loss of piezoelectric properties in α-quartz at high temperature. *Applied Physics Letters* 81 (16): 2968–2970.

32 Damjanovic, D. (1998). Materials for high temperature piezoelectric transducers. *Current Opinion in Solid State & Materials Science* 3 (5): 469–473.

33 Turner, R.C., Fuierer, P.A., Newnham, R.E., and Shrout, T.R. (1994). Materials for high temperature acoustic and vibration sensors: a review. *Applied Acoustics* 41 (4): 299–324.

34 Svaasand, L.O., Eriksrud, M., Nakken, G., and Grande, A.P. (1974). Solid-solution range of $LiNbO_3$. *Journal of Crystal Growth* 22 (3): 230–232.

35 Born, E., Hornsteiner, J., Metzger, T., and Riha, E. (2000). Diffusion of niobium in congruent lithium niobate. *Physica Status Solidi A: Applications and Materials Science* 177 (2): 393–400.

36 Streque, J., Aubert, T., Kokanyan, N. et al. (2019). Stoichiometric lithium niobate crystals: towards identifiable wireless surface acoustic wave sensors operable up to 600 °C. *IEEE Sensors Letters* 3 (4): 1–4.

37 Fachberger, R., Bruckner, G., Knoll, G. et al. (2004). Applicability of $LiNbO_3$ langasite and $GaPO_4$ in high temperature SAW sensors operating at radio frequencies. *IEEE Transactions on Ultrasonics, Ferroelectrics, and Frequency Control* 51 (11): 1427–1431.

38 Fachberger, R., Bruckner, G., Hauser, R., and Reindl, L. (2006). Wireless SAW based high-temperature measurement systems. *2006 IEEE International Frequency Control Symposium and Exposition*, Miami, FL, USA. IEEE. 358–367.

39 Hornsteiner, J., Born, E., Fischerauer, G., and Riha, E. (1998). Surface acoustic wave sensors for high-temperature applications. *Proceedings of the 1998 IEEE International Frequency Control Symposium (Cat No98CH36165)*, Pasadena, CA, USA (29 May 1998). IEEE. 615–620.

40 Mill, B.V. and Pisarevsky, Y.V. (2000). Langasite-type materials: from discovery to present state. *Proceedings of the 2000 IEEE/EIA International Frequency Control Symposium and Exhibition (Cat No00CH37052)*, Kansas City, MO, USA (9 June 2000). IEEE. 133–144.

41 Fritze, H. (2006). High temperature piezoelectric materials: defect chemistry and electro-mechanical properties. *Journal of Electroceramics* 17 (2): 625–630.

42 da Cunha, M.P., Moonlight, T., Lad, R. et al. (2008). High temperature sensing technology for applications up to 1000°C. *2008 IEEE Sensors*, Lecce, Italy (26–29 October 2008). IEEE. 752–755.

43 Aubert, T., Bardong, J., Elmazria, O. et al. (2012). Iridium interdigital transducers for high-temperature surface acoustic wave applications. *IEEE Transactions on Ultrasonics, Ferroelectrics, and Frequency Control* 59 (2): 194–197.

44 Jen, S. and Bobkowski, R. (2000). Black lithium niobate SAW device fabrication and performance evaluation. *2000 IEEE Ultrasonics Symposium Proceedings An International Symposium (Cat No00CH37121)*, vol. 1, 269–273.

45 Bardong, J., Bruckner, G., Kraft, M., and Fachberger, R. (2009). Influence of packaging atmospheres on the durability of high-temperature SAW sensors. *2009 IEEE International Ultrasonics Symposium*, Rome, Italy (20–23 September). IEEE. 1680–1683.

46 Aubert, T., Kokanyan, N., and Elmazria, O. (2021). Langasite as piezoelectric substrate for sensors in harsh environments: investigation of surface degradation under high-temperature air atmosphere. *Sensors* 21 (17): 5978.

47 Kim, T., Kim, J., Dalmau, R. et al. (2015). High-temperature electromechanical characterization of AlN single crystals. *IEEE Transactions on Ultrasonics, Ferroelectrics, and Frequency Control* 62 (10): 1880–1887.

48 François, B., Richter, D., Fritze, H. et al. (2012). Wireless and passive sensors for high temperature measurements. *Third International Conference on Sensor Device Technologies and Applications (SENSORDEVICES 2012)*. IARIA: Rome, Italy. 46–51. https://hal.archives-ouvertes.fr/hal-00767695.

49 Wall, B., Gruenwald, R., Klein, M., and Bruckner, G. (2015). A 600°C wireless and passive temperature sensor based on langasite SAW-resonators. *Proceedings of the Association for Sensors and Measurement* 2015: 390–395.

50 Pereira da Cunha, M. (2017). Wireless microwave acoustic sensor system for condition monitoring in power plant environments [Internet]. University of

Maine, Orono, ME (United States), Report No.: DOE-UMaine-FE0007379. https://www.osti.gov/biblio/1406890.

51 Krempl, P., Schleinzer, G., and Wallno, W. (1997). Gallium phosphate, $GaPO_4$: a new piezoelectric crystal material for high-temperature sensorics. *Sensors and Actuators A: Physical* 61 (1): 361–363.

52 Zhang, S., Fei, Y., Chai, B.H.T. et al. (2008). Characterization of piezoelectric single crystal $YCa_4O(BO_3)_3$ for high temperature applications. *Applied Physics Letters* 92 (20): 202905.

53 Ambacher, O., Brandt, M.S., Dimitrov, R. et al. (1996). Thermal stability and desorption of Group III nitrides prepared by metal organic chemical vapor deposition. *Journal of Vacuum Science & Technology, B: Microelectronics and Nanometer Structures--Processing, Measurement, and Phenomena* 14 (6): 3532–3542.

54 Lin, C.Y. and Lu, F.H. (2008). Oxidation behavior of AlN films at high temperature under controlled atmosphere. *Journal of the European Ceramic Society* 28 (3): 691–698.

55 Uehara, K., Yang, C.M., Shibata, T. et al. (2004). Fabrication of 5-GHz-band SAW filter with atomically-flat-surface AlN on sapphire. *IEEE Ultrasonics Symposium, 2004*, vol. 1, Montreal, QC, Canada (23–27 August 2004). IEEE. 203–206.

56 Harrop, P.J. and Creamer, R.H. (1963). The high-temperature electrical conductivity of single-crystal alumina. *British Journal of Applied Physics* 14 (6): 335.

57 Aubert, T., Elmazria, O., Assouar, B. et al. (2010). Surface acoustic wave devices based on AlN/sapphire structure for high temperature applications. *Applied Physics Letters* 96 (20): 203503.

58 Aubert, T., Bardong, J., Legrani, O. et al. (2013). In situ high-temperature characterization of AlN-based surface acoustic wave devices. *Journal of Applied Physics* 114 (1): 014505.

59 Blampain, E., Elmazria, O., Aubert, T. et al. (2013). AlN/sapphire: promising structure for high temperature and high frequency SAW devices. *IEEE Sensors Journal* 13 (12): 4607–4612.

60 Streque, J., Camus, J., Laroche, T. et al. (2020). Design and characterization of High-Q SAW resonators based on the AlN/sapphire structure intended for high-temperature wireless sensor applications. *IEEE Sensors Journal* 20 (13): 6985–6991.

61 Aubert, T., Elmazria, O., Assouar, B. et al. (2011). Behavior of platinum/tantalum as interdigital transducers for SAW devices in high-temperature environments. *IEEE Transactions on Ultrasonics, Ferroelectrics, and Frequency Control* 58 (3): 603–610.

62 Aubert, T., Sarry, F., Elmazria, O. et al. (2008). Extraction of COM parameters on Pt/LGS for high temperature SAW sensor. *2008 IEEE Ultrasonics Symposium*, Beijing, China (2–5 November 2008). IEEE. 820–823.

63 Pereira da Cunha, M., Moonlight, T., Lad, R. et al. (2007). Enabling very high temperature acoustic wave devices for sensor and frequency control applications.

2007 IEEE Ultrasonics Symposium Proceedings, New York, NY, USA (28–31 October 2007). IEEE. 2107–2110.

64 Selman, G.L., Day, J.G., and Bourne, A.A. (1974). Dispersion strengthened platinum. *Platinum Metals Review* 18 (2): 46–57.

65 Pereira da Cunha, M., Lad, R.J., Moonlight, T. et al. (2008). High temperature stability of langasite surface acoustic wave devices. *2008 IEEE Ultrasonics Symposium*, Beijing, China (2–5 November 2008). IEEE. 205–208.

66 Moulzolf, S.C., Frankel, D.J., Bernhardt, G.P. et al. (2011). Thin film electrodes and passivation coatings for harsh environment microwave acoustic sensors. *Smart Sensors, Actuators, and MEMS V*. SPIE, 54–62. https://www.spiedigitallibrary.org/conference-proceedings-of-spie/8066/806606/Thin-film-electrodes-and-passivation-coatings-for-harsh-environment-microwave/10.1117/12.886602.full.

67 Pereira da Cunha, M., Maskay, A., Lad, R.J. et al. (2015). Pt–Ni/Pt–Zr electrodes for stable SAW resonator operation during repeated temperature cycling up to 1000°C. *2015 IEEE International Ultrasonics Symposium (IUS)*, Taipei, Taiwan (21–24 October 2015). IEEE. 1–4.

68 Richter, D., Sakharov, S., Forsén, E. et al. (2011). Thin film electrodes for high temperature surface acoustic wave devices. *Process Engineering* 25: 168–171.

69 Liu, Y., Masumoto, H., and Goto, T. (2004). Electrical and optical properties of IrO_2 thin films prepared by laser-ablation. *Materials Transactions* 45 (10): 3023–3027.

70 Osamura, H. and Abe, N. (1999). Development of new iridium alloy for spark plug electrodes [Internet]. Warrendale, PA: SAE International, 1999 Mar [cited 2022 Nov 24]. Report No.: 1999-01–0796. https://www.sae.org/publications/technical-papers/content/1999-01-0796/.

71 Taguett, A., Aubert, T., Lomello, M. et al. (2016). Ir-Rh thin films as high-temperature electrodes for surface acoustic wave sensor applications. *Sensors and Actuators A: Physical* 243: 35–42.

72 Taguett, A., Aubert, T., Elmazria, O. et al. (2017). Comparison between Ir, $Ir_{0.85}Rh_{0.15}$ and $Ir_{0.7}Rh_{0.3}$ thin films as electrodes for surface acoustic waves applications above 800°C in air atmosphere. *Sensors and Actuators A: Physical* 266: 211–218.

73 Kubina, B., Schüßler, M., Mandel, C. et al. (2013). Wireless high-temperature sensing with a chipless tag based on a dielectric resonator antenna. *2013 IEEE Sensors*, Baltimore, MD, USA (3–6 November 2013). IEEE. 1–4.

74 Ballandras, S., Raveski, A., Bernard, F. et al. (2018). Antenna connection, particularly for surface guided elastic wave transducers [Internet]. WO2018178941 Patent. https://patentscope.wipo.int/search/en/detail.jsf?docId=WO2018178941.

75 Ernst, D., Brachmann, E., Menzel, S., and Bock, K. (2018). Wire bonding of surface acoustic wave (SAW) sensors for high temperature applications. *2018 7th Electronic System-Integration Technology Conference (ESTC)*, Dresden, Germany (18–21 September 2018). IEEE. 1–5.

76 Moulzolf, S.C., Behanan, R., Pollard, T. et al. (2013). Capacitively coupled IDT for high temperature SAW devices. *2013 IEEE International Ultrasonics Symposium (IUS)*, Prague, Czech Republic (21–25 July 2013). IEEE. 255–258.

77 Makdissy, T., Elmazria, O., and Hage-Ali, S. (2021). Interconnect-free DRA-SAW RFID sensing system for high temperature monitoring. *2021 IEEE International Symposium on Antennas and Propagation and USNC-URSI Radio Science Meeting (APS/URSI)*, Singapore, Singapore (4–10 December 2021). IEEE. 393–394.

78 François, B., Sakharov, S., Droit, C. et al. (2012). Wireless temperature measurements above 500°c using surface acoustic wave sensors. *Process Engineering* 47: 1227–1230.

79 Brachmann, E., Seifert, M., Neumann, N. et al. (2019). Electroless-deposited platinum antennas for wireless surface acoustic wave sensors. *Materials* 12 (7): 1002.

80 Hagelauer, A., Fattinger, G., Ruppel, C.C.W. et al. (2018). Microwave acoustic wave devices: recent advances on architectures, modeling, materials, and packaging. *IEEE Transactions on Microwave Theory and Techniques* 66 (10): 4548–4562.

81 Klein, M., Wall, B., Gruenwald, R., et al. (2015). Design, assembly and reliability of a hermetic package for a 600°C wireless temperature sensor. *2015 IEEE 65th Electronic Components and Technology Conference (ECTC)*, San Diego, CA, USA (26–29 May 2015). IEEE. 1351–1356.

82 Manners, D. (2014). Kyocera pushes HTCC packaging to 1000°C. Electronics Weekly. https://www.electronicsweekly.com/news/business/manufacturing/kyocera-pushes-htcc-packaging-1000c-2014-09/.

83 Bhattacharjee, K., Shvetsov, A., and Zhgoon, S. (2007). Packageless SAW devices with isolated layer acoustic waves (ILAW) and waveguiding layer acoustic waves (WLAW). *2007 IEEE International Frequency Control Symposium Joint with the 21st European Frequency and Time Forum*, Geneva, Switzerland (29 May–1 June 2007). IEEE. 135–140.

84 Shen, J., Fu, S., Li, Q. et al. (2018). Simulation of temperature compensated waveguiding layer acoustic wave devices. *Journal of Physics D: Applied Physics* 52 (7): 075105.

85 Naumenko, N. (2017). Selection of materials for multilayered structures to be used in packageless sensors. *2017 IEEE International Ultrasonics Symposium (IUS)*, Washington, DC, USA (6-9 September 2017). IEEE. 1–4.

86 Shen, J., Fu, S., Su, R. et al. (2021). A multilayered structure for packageless acoustic-wave devices with ultra-small sizes. *Journal of Microelectromechanical Systems* 30 (4): 589–596.

87 Dubois, M.A. and Muralt, P. (2001). Stress and piezoelectric properties of aluminum nitride thin films deposited onto metal electrodes by pulsed direct current reactive sputtering. *Journal of Applied Physics* 89 (11): 6389–6395.

88 Zhgoon, S., Shvetsov, A., and Bhattacharjee, K. (2011). Spurious mode suppression in multiple layer HfO_2/Simax acoustical mirror structure for ILAW on 15°-YX $LiNbO_3$. *2011 IEEE International Ultrasonics Symposium*, Orlando, FL, USA (18–21 October 2011). IEEE. 547–550.

89 Hage-Ali, S., Elmazria, O., Pierson, G. et al. (2016). Packageless acoustic wave sensors for wireless body-centric applications. *2016 IEEE Sensors*, Orlando, FL, USA (30 October–3 November 2016). IEEE. 1–3.

90 Floer, C., Hage-Ali, S., Zhgoon, S. et al. (2018). AlN/ZnO/$LiNbO_3$ packageless structure as a low-profile sensor for potential on-body applications. *IEEE Transactions on Ultrasonics, Ferroelectrics, and Frequency Control* 65 (10): 1925–1932.

91 Legrani, O., Aubert, T., Elmazria, O. et al. (2016). AlN/IDT/AlN/Sapphire SAW heterostructure for high-temperature applications. *IEEE Transactions on Ultrasonics, Ferroelectrics, and Frequency Control* 63 (6): 898–906.

92 Zhgoon, S., Legrani, O., Elmazria, O. et al. (2015). Potential of Al_2O_3/GaN/ Sapphire layered structure for high temperature SAW sensors. *2015 Symposium on Piezoelectricity, Acoustic Waves, and Device Applications (SPAWDA)*, Jinan, China (30 October–2 November 2015). IEEE. 106–110.

93 Bartoli, F., Aubert, T., Moutaouekkil, M. et al. (2018). AlN/GaN/Sapphire heterostructure for high-temperature packageless acoustic wave devices. *Sensors and Actuators, A: Physical* 283: 9–16.

94 Naumenko, N. and Nicolay, P. (2017). AlN/Pt/LN structure for SAW sensors capable of operating at high temperature. *Applied Physics Letters* 111 (7): 073507.

95 Floer, C., Elmazria, O., Naumenko, N. et al. (2021). AlN/Pt/LN-Y_{128} packageless acoustic wave temperature sensor. *IEEE Transactions on Ultrasonics, Ferroelectrics, and Frequency Control* 68 (6): 2315–2318.

96 Nicolay, P, Naumenko, N, Aubert, T, and Chambon, H. (2018). Towards an AlScN-based packageless acoustic wave sensor with RFID capabilities, for applications above 350°C. *IUS 2018 [Internet]*. https://www.researchgate.net/publication/333059164_Towards_an_AlScN-based_packageless_acoustic_wave_sensor_with_RFID_capabilities_for_applications_above_350C.

97 Transense_website [Internet]. Transense Technologies plc, Bicester, Oxfordshire, UK. https://www.transense.com/saw-technology/applications (accessed 28 November 2022).

98 Vimeo_website [Internet]. "Surface Acoustic Wave (SAW) Technology in Robotics," Video from Transense company. https://vimeo.com/745796839 (accessed 28 November 2022).

99 Sensideon_website [Internet]. sensideon GmBH, Wels, Austria. https://www.sensideon.com/en/tracking-monitoring/ (accessed 28 November 2022).

100 Sensanna_website [Internet]. SenSanna Inc, Hanover, Maryland, USA. https://www.sensanna.com/ (accessed 28 November 2022).

8

Wireless Sensing for Harsh and Severe Environments Based on Saw Sensors

Manuel Monedero[1], *Robert Staraj*[2], *and Philippe Le Thuc*[2]

[1] *NXP Semiconductors GmbH, Munich, Germany*
[2] *Université Côte d'Azur, LEAT Laboratoire Electronique Antennes, et Télécommunications, Sophia Antipolis, France*

8.1 Introduction

One of the most common industrial applications for sensors is temperature measurement. Indeed, the need to measure in real time the temperature of moving or static devices has led to the development of a new generation of sensors capable of converting a temperature variation into an electrical signal by different principles of physics, depending on the technology used. These can be wired sensors, but when their number becomes large, or simply because it is not possible to use them for certain applications, wireless sensor systems are preferred. Then, the sensor requires to be associated with an antenna for the transfer of information. In some applications, these systems can be placed in harsh environments from an application point of view, but also from an electromagnetic point of view. In addition, the space available is often reduced, which implies the need to miniaturize the antenna associated with the sensor. The performances of electrically small antennas are however inversely proportional to their size, their characterization is more difficult, and they can be more sensitive to their environment. It is, therefore, necessary to take all these factors into account in the design and modeling of these systems.

In this chapter, the various technologies used in wireless temperature sensors are first described. We will focus more particularly on SAW sensors adapted to wireless systems. The measurement of the device made up of the antenna and the SAW sensor is described. The phenomenon known as "frequency pulling" created

Wireless Identification and Sensing Systems for Harsh and Severe Environments,
First Edition. Edited by Smail Tedjini and Valentina Palazzi.

Published 2025 by John Wiley & Sons, Inc.

by the association of two resonators (i.e. the antenna and the SAW sensor) is also investigated. Finally, we will present a modeling of the complete system making it possible to find the conditions to which the impedance of the antennas associated with the sensor must respond to avoid this phenomenon.

8.2 State of the Art

This part is dedicated to the state of the art of wireless temperature sensors. In general, these systems comprise a unit of measurement to transform physical quantities into an electrical signal (the sensor), transmit-receive modules for wireless transmission and a computer processing unit with data storage.

Two main categories of wireless sensors exist those that require additional power (active or semi-passive sensors) and those that do not (passive sensors).

8.2.1 Active Wireless Sensors

Active sensors have a great advantage in terms of interrogation distance. Indeed, they use energy from an external source, like a battery. But it is precisely this external source that reduces their lifespan and requires periodic maintenance. In some cases, once the sensor is placed, it is no longer accessible for its maintenance. In addition, in most industrial applications, measuring temperatures above 120 °C, or any other extreme condition such as severe vibrations or humidity, can cause damage to the circuit and/or the battery integrated in the sensor.

Except for the systems using optical pyrometry, most wireless sensors use radio frequencies (RF) and are generally part of a network (Wireless Sensor Network WSN) composed of mono or multi-sensor modules attached to a central unit by a RF link. All the sensors can operate wirelessly if they are either associated with a traditional battery-operated communication module, or to an energy harvesting system. Many of them are based on RFID technology which allows the identification of the RFID chip to which an external sensor has been added allowing the measurement of one or more physical quantities. We then speak of "augmented RFID sensor tags" [1, 2].

8.2.2 Passive Wireless Sensors

Passive wireless sensors do not have an external power source, instead, they are powered by the interrogator unit with electromagnetic waves. They can be classified into two categories: sensors using RFID technology and those using SAW technology. Many cases of RFID applications can be found in the other chapters of this book; hence, we will focus here on SAW sensors working in the 434 MHz ISM band, other bands are also covered in this book.

SAW sensors are therefore completely passive and do not need any discrete electronic component, which allows them, among other things, to measure very high temperatures (up to 600 °C) [3]. They can also be implanted in environments with severe propagation conditions, as in the case of the human body. To be remotely powered, these sensors are associated with antennas, which ensure the conversion of electromagnetic energy into electrical energy (and vice versa during the response). They are designed with piezoelectric materials, also called crystals, which have the property of being able to transform electrical energy into mechanical energy, that is, electrical wave into an acoustic wave, and vice versa.

As the physical and mechanical properties of the crystal depend on several variables, such as temperature or pressure, these sensors allow to measure various physical quantities without contact in numerous environments and applications. In the literature, many uses can be found, among which:

In the industrial field: temperature measurement in a fixed [4] or rotating [5] object
In the automotive field: measurement of temperature [6], tyre pressure [7], and torque [8]
In the agricultural field: measuring the moisture content of the soil [9]
In the medical field: the measurement of different temperatures including brain temperature [10], chemical composition [11], and blood pressure [12].

Other sensors are able to make these measurements but SAW-based solutions are particularly competitive for measuring high temperatures, for example, metallurgy manufacturing processes. The low weight and reduced size of these sensors also allow them to be placed on rotating objects or machines because they generate very little inertia. Since these sensors are also very resistant to strong electromagnetic fields, they can be found inside electrical motors. Moreover, due to their passivity, they can be placed in explosive environments (ATEX regulations) for the functional control of equipment in the chemical and petrochemical fields. SAW sensors thus allow to carry out measurements in real time and are particularly suitable for harsh environments. A SAW sensor is based on detecting the changes in the propagation of an acoustic wave, its speed and amplitude being modified in case of a disturbance. This change in speed can be measured from the resonance frequency of the sensor and thus be correlated with a temperature value or any other corresponding physical quantity. Two types of SAW devices exist: delay lines and resonators.

8.3 Surface Acoustic Wave Sensors

In the case of a SAW sensor, an electrode placed on a piezoelectric material transforms an electrical signal into an acoustic wave using inverse piezoelectric effect. The physical dimensions of the piezoelectric material determine its resonant frequency. The lowest possible resonant frequency is derived from the widest dimension. The quality factors obtained with these materials are very high. For example, the quality factor of a quartz crystal is 10000 at a frequency of 30 MHz. A high-quality factor ensures greater stability and better performance in terms of accuracy.

The transformation operation is carried out by means of transducers, called Interdigital Transducer (IDT), which are metal electrodes produced by depositing metal on the piezoelectric crystal. The simplest transducer is composed of two electrodes having the form of an interdigital comb. The width of the metallization and the metallization period determine the frequency of the desired acoustic wave. One of the most attractive properties of this technology is the fact that the transducers can be placed freely along any propagation axis of the crystal, without any constraint on the distance between the transducers. The principle of operation is based on the fact that the speed of propagation of the SAW on the piezoelectric crystal depends on the density and the coefficients of elasticity of the material. The mechanical and physical properties of the crystal can vary depending on the environment, temperature, pressure, etc., which makes it possible to measure these physical quantities due to changes in the resonance frequency or propagation time in the crystal.

The main piezoelectric materials used in wireless SAW sensors are Quartz (SiO_2), PZT (Lead Titanium–Zirconate) ceramics, Lithium Tantalate ($LiTaO_3$), Lithium Niobate ($LiNbO_3$), Zinc Oxide (ZnO), Aluminum Nitride (ALN), and Langasite ($La_3Ga_5SiO_{14}$) [13].

8.3.1 The Different Types of SAW Sensors

The difference between the SAW sensor types depends on the number of transducers (or port). For a two-port device (Figure 8.1a), the input transducer ensures the conversion of the electrical signal into an acoustic wave and the output transducer allows to convert the acoustic wave after propagation in the material, into an electrical signal. A one-port device has a single transducer (which serves as both an input and an output port) and electrodes that form reflective planes (Figure 8.1b). These reflecting planes can either create a total reflection of the surface waves and produce a resonant cavity characterized by its resonant frequency (case of resonator-type sensors) or partially reflect the energy emitted by the transducer (Figure 8.1c) characterized by a delay (case of "delay line" type sensors).

8.3.2 Resonator SAW Sensors

A resonator is made up of reflectors located on either side of the transducer. The excitation is done by the RF signal coming from an interrogation unit picked up by the antenna of the sensor. During the conversion of the electrical signal

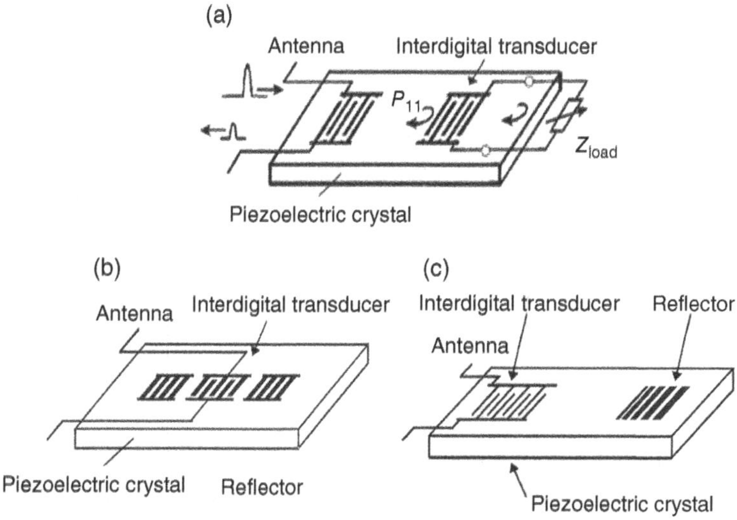

Figure 8.1 (a) 2 ports SAW device. *Source:* Adapted from Kalinin et al. [8]. (b) Resonator wireless SAW sensor. (c) Reflective delay line wireless SAW sensor.

provided by the antenna, the surface waves propagate on both sides of the transducer. Each reflector totally reflects the energy emitted by the transducer. Between the reflecting planes is created a standing wave. The sensor thus behaves like a resonant cavity characterized by its resonant frequency.

Then, the resonator transmits to the antenna a signal at its resonant frequency. This later varies according to the environment and therefore reflects a variation of the physical phenomenon to be measured. This signal radiated by the antenna of the sensor is reemitted toward the interrogator. The response of a resonator-type SAW sensor is a time-damped decaying exponential (Figure 8.2).

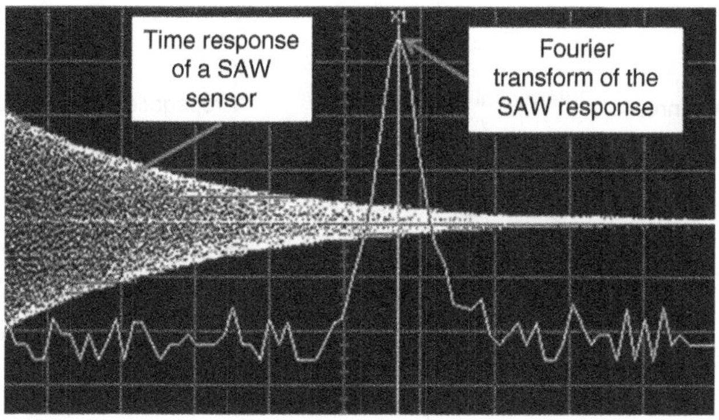

Figure 8.2 Time and frequency response of a SAW resonator sensor.

8.3.3 Delay Line SAW Sensors

A delay line-type SAW sensor consists of a single transducer and several (minimum 3) reflector bars. The excitation is also done by the RF signal coming from an interrogation unit. The electrical signal is translated into an acoustic wave using the IDTs. The acoustic wave is partially reflected on the reflector bars found in the direction of propagation. The different acoustic wave bursts are converted into an electrical signal and then into an electromagnetic wave. The time response of a delay line SAW sensor is a series of time-shifted pulses, identical in shape to the transmitted signal but attenuated. The measurement of the physical quantity is obtained with the calculation of the time difference between the retransmitted signals.

8.3.4 Two Resonators-One-Port Sensors

Buff et al. have shown that the frequency response of a SAW sensor using only one resonator is disturbed by the propagation channel [14]. Indeed, a variation in the distance between the antennas modifies the natural resonance frequency of the sensor, which results in a significant error in the physical quantity measured in the case of sensors of the resonator type. To limit this phenomenon, they proposed to use two resonators to obtain a differential structure (Figure 8.3a).

These two resonators must have a different resonant frequency, and one of them (resonator $n°2$) must be placed along an axis of the crystal which is almost insensitive to the physical quantity to measure (Figure 8.3b). This resonator serves as a reference. The resonator $n°1$ is oriented along an axis sensitive to the physical quantity to be measured. Thus, this later remains constant, the two resonators are affected by the same shift with respect to the propagation channel, and the difference between resonance frequencies will remain the same. The measured physical quantity will then be calculated from the difference between the resonance frequency of the reference resonator (resonator $n°2$) and the one of the resonator

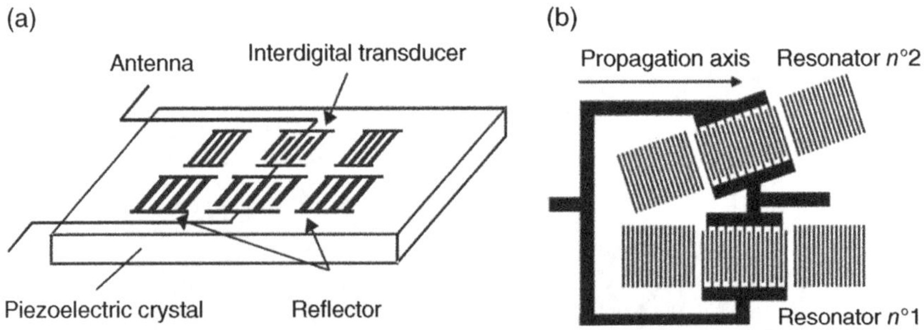

Figure 8.3 (a) SAW sensor based on a structure with two series-connected resonators, (b) SEAS10 temperature sensor structure.

sensitive to it (resonator n°1). The differential structure also allows to limit the effects of aging. Moreover, the frequency difference of the resonators is quasi-linear as a function of the temperature since the non-linear residues can be corrected in a first phase of calibration of the sensors. The SEAS10 temperature sensor developed by the SENSeOR company, for example, is based on a one-port resonator-type SAW sensor and has two resonators to perform the differential measurement. The main advantage of resonator-type sensors compared to those of the delay line type lies in the level of insertion losses. Indeed, the insertion losses are of the order of 2 dB for a resonator-type sensor, whereas they can rise up to 30 dB for delay line-type sensors. Therefore, the interrogation distances of resonator-type sensors are greater than those of delay line-type sensors. Their main drawback appears when several sensors must operate in the same volume. The identification of the information returned by each of them is based on the sharing of the total available frequency band. This amounts to reducing the frequency excursion relative to the quantity to be measured and therefore a reduction in terms of precision. However, this type of sensor is rarely deployed in large numbers, but rather in a unique way in the applications already mentioned (high temperatures, etc.).

In the following paragraphs, we will focus on SAW sensor technology dedicated to the 434 MHz ISM band. Other frequency bands are covered in other chapters of this book. Similarly, we will only discuss temperature measurements.

8.3.5 Temperature Effect on the Resonance Frequencies

The variation of the resonance frequency as a function of temperature is described by the following equation (Eq. (8.1)), in which, f_0 is the resonant frequency at the reference temperature T_0 (by convention 25 °C), CTF_1 is the temperature coefficient of the first-order frequency in ppm/°C, and CTF_2 is the temperature coefficient of the second-order frequency in ppm/°C^2:

$$f(T) = f_0\left(1 + \mathrm{CTF}_1\left(T - T_0\right) + \mathrm{CTF}_2\left(T - T_0\right)^2\right) \tag{8.1}$$

The resonant frequency of each resonator is given by the previous equation but with different *CTF* coefficients. The difference in frequency can then be expressed as follows (Eq. (8.2)):

$$\Delta f(T) = \delta_0 + S_T \cdot T + \varepsilon \cdot T^2 \tag{8.2}$$

with:

$$\begin{aligned}
\delta_0 &= \Delta f_0 + \left(f_{0/2} \cdot \mathrm{CTF}_{1/2} - f_{0/1} \cdot \mathrm{CTF}_{1/1}\right)T_0 + \left(f_{0/2} \cdot \mathrm{CTF}_{2/2} - f_{0/1} \cdot \mathrm{CTF}_{2/1}\right)T_0^2 \\
S_T &= f_{0/1} \cdot \mathrm{CTF}_{1/1} - f_{0/2} \cdot \mathrm{CTF}_{1/2} + 2\left(f_{0/2} \cdot \mathrm{CTF}_{2/2} - f_{0/1} \cdot \mathrm{CTF}_{2/1}\right)T_0 \\
\varepsilon &= f_{0/1} \cdot \mathrm{CTF}_{2/1} - f_{0/2} \cdot \mathrm{CTF}_{2/2}
\end{aligned}$$

The frequency variation versus the temperature (Eq. (8.1)) is of parabolic type. The temperature at which the frequency difference is maximum is called the "turn over" temperature. The temperature can be calculated by solving the quadratic equation when Δf is known. The solution of the equation gives two solutions, but due to the design of the sensor, the solution to be retained is the one that is found where the slope of the parabola is negative, which visually corresponds to the temperature to the right of the "turn over" temperature for Δf. The solution of the equation therefore gives (Eq. (8.3)):

$$T = \frac{-S_T + \sqrt{S_T^2 - 4\varepsilon\left(\delta_0 - \Delta f\right)}}{2\varepsilon} \tag{8.3}$$

The sensor taken as an example (SEAS10) is able to measure temperatures between −20 and 160 °C. The following figures show the variation of resonance frequencies and the difference between these two resonant frequencies as a function of temperature (Figure 8.4). This last one varies almost linearly as a function of the temperature with a slope coefficient of 2.5 kHz/°C.

8.4 Remote Interrogation System for Surface Acoustic Wave Sensors Based on Differential Mode

Interrogation/listening type systems are generally based on the principle of monostatic radar: the central unit works as a T_X/R_X (interrogation and reception) module. The complete system is made up of the unit associated with its transmitting antenna, and a receiving antenna associated with the SAW sensor. In this context, many interrogation units operate in the ISM (Industrial-Scientific-Medical) band which in Europe extends from 433.050 to 434.790 MHz. The interrogation unit sends a wave and listens for the sensor response over the entire ISM band with a constant frequency step. The maximum radiated transmission power in the band is limited to a value of 10 dBm, i.e. 10 mW. Each resonance, visible by a peak (Figure 8.5), corresponds to a maximum of received power. If the sensors are detected, both peaks are visible. The visible quantization is due to the successive exponential discharges of the two resonators. The amplitude of the peaks is identical, the two resonators have the same quality coefficient.

8.4.1 Operation of the Interrogation System in Transmission Mode (T_x)

The interrogation unit emits a time pulse of duration τ (between 20 and 40 μs) modulated by a signal of frequency f_i. The time signal is the product of a "gate" function of duration τ_{Tx} with a cosine of period T_i corresponding to the frequency f_i of the carrier belonging to the ISM band (Figure 8.6). During an interrogation

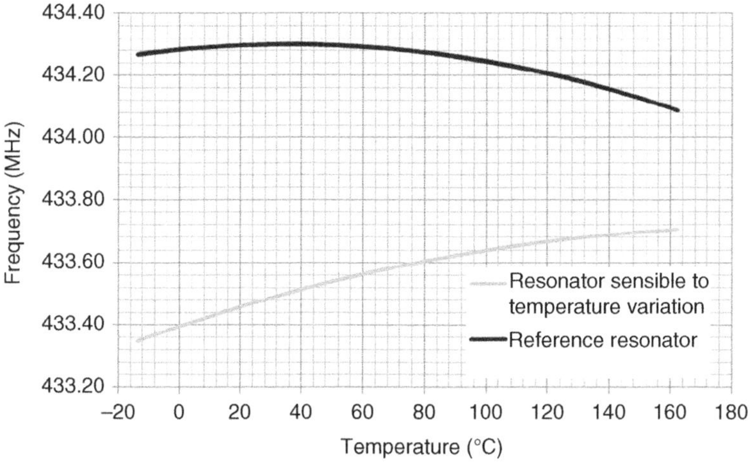

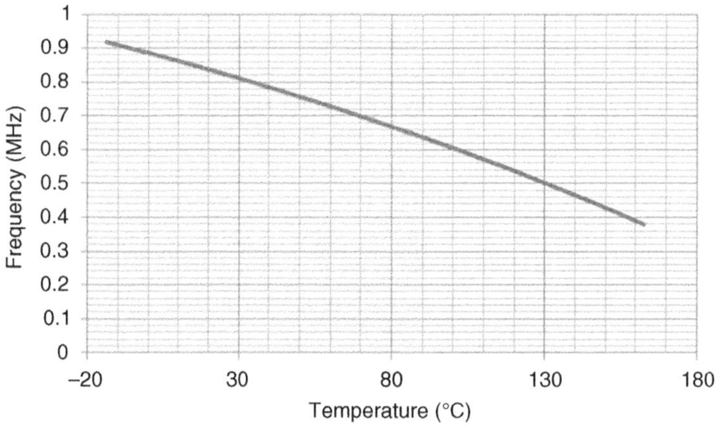

Figure 8.4 Frequency evolution of the two resonators and the difference in resonance frequencies as a function of $T°$.

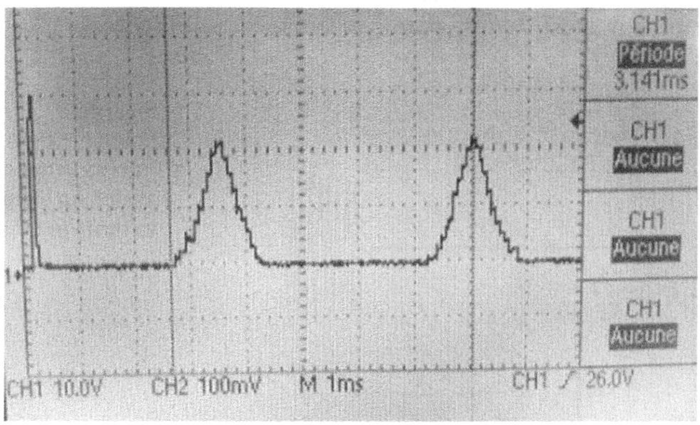

Figure 8.5 Time response of the time response of a SAW sensor with two resonators.

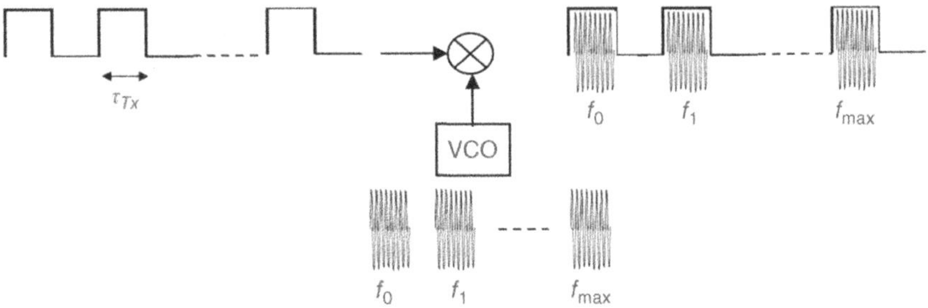

Figure 8.6 Interrogation signal: rectangular signal modulated by a sinusoidal one.

cycle, the transmitter performs interrogation for signals of 128 different frequencies in the ISM band, which will be converted into an electromagnetic wave by the transmitting antenna and sent to the SAW sensor module.

A part of this emitted wave will be collected by the antenna associated with the SAW sensor, which transforms this electromagnetic signal into an electrical one which, in turn, will be converted into an acoustic wave by the inverse piezoelectric effect by the transducer. As long as the transmitter signal is not zero, an acoustic wave will be created. It is stored in the cavity formed between the two reflector planes of the resonators. When the frequency of the emission signal received by the resonator is close to its resonant frequency, the sensor resonates during the charging period. When the frequency of the emitted signal is equal to the resonance frequency of the sensor, a standing wave is created in the cavity between the two reflecting planes. This resonance frequency is proportional to the speed of propagation of the surface wave in the piezoelectric crystal and therefore depends on the temperature of the resonator. After a time $\tau_{SAW} < \tau_{Tx}$, a steady state is established.

8.4.2 Operation of the Interrogation System in Reception Mode (R_x)

When the interrogation unit stops transmitting the signal, it switches to the receiving mode after a switching time τ_{com} necessary to limit interference in the receiving chain and enters a listening period of duration τ_{Rx} (Figure 8.7).

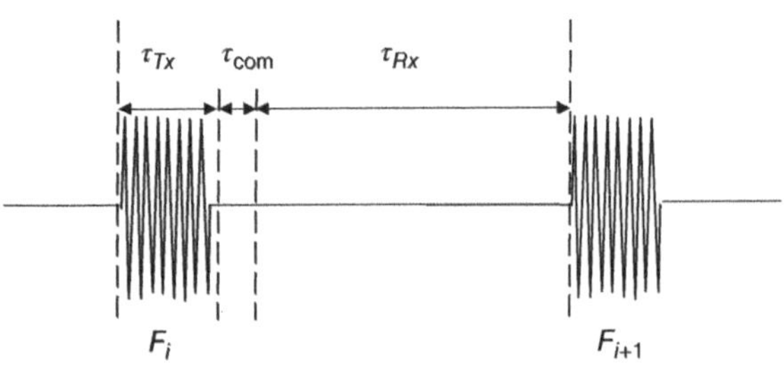

Figure 8.7 Timing diagram of the interrogation unit for switching from transmitter to receiver mode.

Once the emission is finished, the SAW sensor, switches to "discharge" mode, meaning all the mechanical energy of the SAW is transformed into an electrical signal by direct piezoelectric effect. The signal is a damped oscillation of time constant $\tau_{discharge}$, proportional to the quality factor of the resonator, and frequency equal to that of the received transmission signal. This signal will itself be transformed into an electromagnetic wave by the sensor's antenna. Some of the reemitted electromagnetic energy will be collected by the antenna of the central unit which measures the power level associated with the transmission frequency.

This procedure is repeated for all frequencies of the ISM. A guard band of 250 kHz on both sides is necessary to limit the transmission in power, the useful band is reduced to (433.300–434.540 MHz) with a first emission burst emitted at 433.300 MHz and a last at 434.540 MHz.

When the carrier frequency is different from the resonant frequency of the resonator, the sensor re-emits only a very low power. As the carrier frequency approaches the resonant frequency of the resonator, the sensor re-emits an increasingly powerful signal. The resonant frequency is calculated by using an interpolation from the point for which the maximum power is received and from the previous and next point (Figure 8.8).

However, a physical problem called "frequency pulling" for which the company SENSEoR has managed to provide a solution [15], exists in the case of these sensors. Indeed, tests carried out with a dipole antenna device printed on

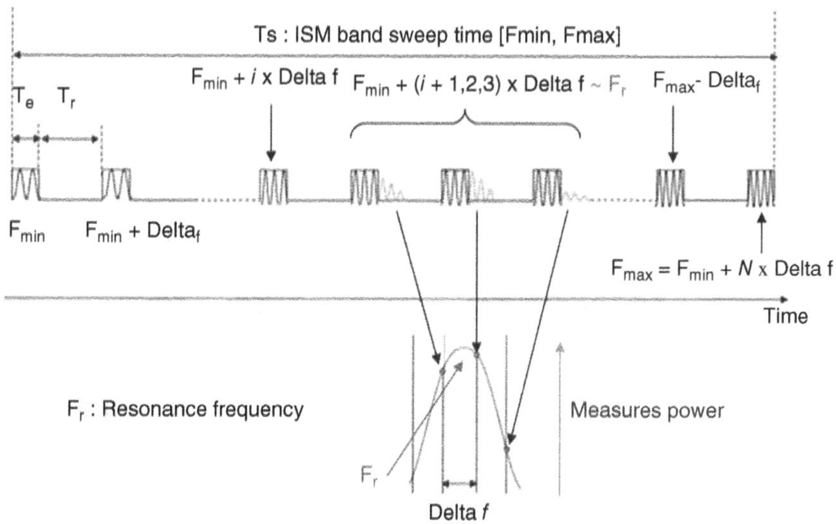

Figure 8.8 Measurement method for resonance frequency. *Source:* Monedero [16]/with permission of CNRS.

FR4/SENSEoR SEAS10 sensor placed on different supports show that the measured temperature can vary (up to 3.35 °C) depending on the support on which the antenna rests. But the level of the signal received by the interrogation unit can also change, depending on the configuration of the measurement. When the antenna is located in free space, the signal level is indeed higher than in other configurations. For a better understanding, we will describe the phenomenon in the case of this dipole antenna printed on FR4 substrate, which can be found more precisely in the reference [16]. The resonant frequency of a dipole antenna is proportional to its length, and it resonates when this physical length is equal to half wavelength, i.e. approximately 35 cm at 434 MHz. When this same dipole is placed on a Formica support ($\varepsilon_r = 4.8$), its resonance frequency decreases by 3–5%. This shift due to the operating frequencies of the sensor, which remain unchanged generates an increase in losses by mismatch which may explain a decrease in the strength of the signal received. However, the fact that the measured temperature is not constant when the antenna is placed on different materials necessarily implies that the difference between the two frequencies of the SAW sensor resonator varies. This phenomenon, called "frequency pulling", is known in the field of SAW filters. One method of correction is to adjust the frequencies of a filter by exhibiting a parallel or series impedance of a SAW device [17]. Antenna structures with a ground plane are less sensitive to a change in environment and their use could be considered. The disadvantage of this solution is that it requires the use of a large ground plane (three times the wavelength) and that this is not necessarily available in all the desired applications. In addition, the input impedance of the antenna can drop drastically when a metal plane is in its vicinity, which can immediately affect the frequency draw more strongly. A better understanding of the phenomenon requires a global modeling of the system. The difficulty encountered in this case is the in-situ measurement of the whole device sensor/antenna which cannot be done with classical methods and which is nevertheless necessary to validate the model.

8.5 Miniature Antenna/Saw Sensors Characterization

A very small size is generally sought to make the integration of the sensor/antenna device as minimally invasive as possible. If the sensor is generally small, it is not the same for the antenna. As mentioned previously, in the ISM band at 434 MHz, the basic antenna which is the dipole has an optimal theoretical length of approximately 35 cm. Of course, many miniaturization techniques can be found in the literature, applied to the case of antennas for SAW sensors or not. Thereby, the miniaturization of the antenna takes on even more importance, even if it is accompanied by a loss of efficiency more critical for totally passive sensors. However, the characterization of miniature antennas is particularly sensitive because the measuring cable or the support participates in the radiation. Under these conditions, it becomes impossible to isolate the radiation of the antenna from that of the cable and also to determine the proper impedance of the antenna. In the case of antennas associated with SAW sensors, the combination of these two resonators creates a mutual disturbance of operation, which further complicates the experimental characterization. The so-called classic measurement methods (time and frequency measurements), but also more innovative methods allowing to replace or eliminate the cable (RF-optical conversion), are therefore poorly adapted to the SAW/antenna sensor case. Indeed, a near-field, non-contact measurement technique of an antenna in a given environment with any connected load (the SAW sensor) is in this case more particularly desirable [18]. Indeed a wireless characterization of the impedance of the SAW sensor could correct fluctuations in measurements brought only by variations in the propagation channel.

Traditional cable characterization is therefore not appropriate in the particular case of SAW/antenna sensors. Another possible technique is the Radar Cross Section (RCS) method whose precursor is Professor Wiesbeck who used this technique to determine the gain, impedance and reflection coefficient of large antennas compared with the wavelength, obtained from the measurements of the antenna-charged by three different impedances (short circuit, matched load and open circuit) [19, 20]. This method, called impedance modulation, was also exported to the RFID domain by Nikitin and Rao, in order to be able to characterize RFID antennas and especially to be able to calculate the gain of antennas associated with an RFID chip [21, 22] in the UHF frequency band from 800 to 950 MHz. Vicencio and Romeu, from Universitat Politècnica de Catalunya, have also used the RCS method for antennas in the UHF band from 400 to 470 MHz [23]. The description of this method can be found in Ref. [16]. However, the notion of RCS is normally valid only in the far-field region. Used in near field for the case of SAW sensors, the results obtained are of poor quality because the equations established in the far field are developed in the hypothesis of a given field polarization. However, in the near field region, it is difficult to speak of linear polarization. Therefore, to be able to characterize a near-field antenna by this method, we would have to know all the terms of the Sinclair matrix which increases the number of equations and unknowns to find. The resolution of this new system would result in an addition of additional measurements of the RCS of the Antenna Under Test (AUT) using this time, not three but by nine loads. This approach can be simplified using a transmission line model for the RF channel. Indeed, when this quadrupole is modeled by a transmission line, the results obtained in the measurements are quite poor in ten cases of SAW sensors, due to the presence of the measurement cable which makes the equivalent matrix non-symmetrical [16].

Another possible modeling method associated to the impedance modulation technique is the "equivalent circuit" one. The results obtained with this method are very promising, as long as we remain in the analysis of the two-port network. This makes it possible to model the coupling between the antennas and the propagation channel which will be explained in the next paragraph.

8.6 Global Modelization of a Wireless Saw Sensor Interrogation System

8.6.1 Theoretical Sensor Models

The modelization of a SAW device can be done with "analytical" models based on piezoelectricity equations, or simplified models based on one-dimensional modeling of piezoelectricity. Mason's and Butterworth-Van Dyke's (BVDs) models, originally made to study the behavior of resonators operating at low frequencies, are derived from this principle [24, 25].

Mason's model associates for each layer of material (piezoelectric substrate, electrodes, etc.) an equivalent scheme based on acoustic wave propagation equations such as Hooke's law and Christoffel's equations [26]. This type of model allows to consider all the materials of the sensor and to observe harmonic resonances higher in frequencies that are not visible with the BVDmodel. However, the BVD model can be used to predict the behavior of the resonator in the vicinity of its resonance. The BVD model is used because it is sufficient to characterize a SAW resonator intended for RF link measurements in the relatively narrow ISM band. It is composed of a static capacitance C_0, modeling the capacitance created by the two electrodes deposited on the piezoelectric material, in parallel with a series L_iC_i circuit, representing the acoustic behavior of the resonator. A resistor R_i models the mechanical losses of the piezoelectric layer (Figure 8.9).

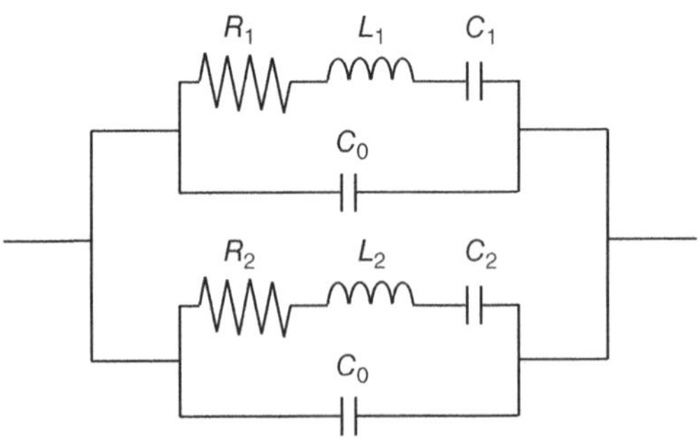

Figure 8.9 BVD model of a SAW sensor with two parallel resonators.

8.6.2 Radio Link Modeling

Maxwell's equations are the basis for the study of antennas and propagation phenomena. From a system of equations established in harmonic regime, it is possible not only to establish an equivalent circuit of an antenna in the far field but also an equivalent circuit modeling the coupling between two antennas in both near and far field. In this paragraph, the different equivalent models of the "circuit" type of an antenna in transmitting and receiving modes are presented.

8.6.3 Impedance Matrix [Z]

As demonstrated by Silver [27] and Balanis [28], when a transmitting antenna and a receiving antenna are considered as an isolated system, it can be modeled by a two-port network whose current-voltage relations are as follows:

$$\begin{cases} U_1 = Z_{11}I_1 + Z_{12}I_2 \\ U_2 = Z_{21}I_1 + Z_{22}I_2 \end{cases} \tag{8.4}$$

Where, Z_{11} is the transmitting antenna impedance, Z_{22} is the receiving antenna impedance, Z_{12} and Z_{21} are the mutual impedance between the antennas.

The terms Z_{12} and Z_{21} represent the coupling between the antennas. Indeed, if the transmitting antenna is considered an isolated system, the following relation can be written:

$$U_1 = Z_{11}I_1 \tag{8.5}$$

Hence, the input impedance of the dipole is defined by:

$$Z_{in} = \frac{U_1}{I_1} = Z_{11} \tag{8.6}$$

However, when a second antenna is added to the system, the input impedance of the system is written:

$$Z_{in} = \frac{U_1}{I_1} = Z_{11} + Z_{12}\frac{I_2}{I_1} \tag{8.7}$$

According to the Reciprocity Theorem, the coupling between the antennas is the same whether the antenna is transmitting or receiving, so $Z_{12} = Z_{21} = Z_m$.

The equivalent impedance matrix for an RF link is therefore:

$$\begin{bmatrix} U_1 \\ U_2 \end{bmatrix} = \begin{bmatrix} Z_{11} & Z_m \\ Z_m & Z_{22} \end{bmatrix} \begin{bmatrix} I_1 \\ I_2 \end{bmatrix} \tag{8.8}$$

8.6.4 Equivalent Circuit

In the previous paragraph, the impedance matrix of an RF link between two antennas was determined. The current-voltage relations are given by the following system:

$$\begin{cases} U_1 = Z_{11}I_1 + Z_m I_2 \\ U_2 = Z_m I_1 + Z_{22} I_2 \end{cases} \tag{8.9}$$

These equations can be modified as follows:

$$\begin{cases} U_1 = Z_{11}I_1 + Z_m I_2 + Z_m I_1 - Z_m I_1 \\ U_2 = Z_m I_1 + Z_{22} I_2 + Z_m I_2 - Z_m I_2 \end{cases} \Rightarrow \begin{cases} U_1 = (Z_{11} - Z_m) I_1 + Z_m (I_1 + I_2) \\ U_2 = Z_m (I_1 + I_2) + (Z_{22} - Z_m) I_2 \end{cases} \tag{8.10}$$

These equations can be represented by a T network (Figure 8.10). It is also possible to convert it into an equivalent π network by applying Y–Δ transform. In the rest of the chapter, only the T network will be used.

8.6.5 Mutual Coupling Between Antennas Modeling

As seen before, a two antennas system can be modeled by a T network that considers the coupling between the two antennas by the mutual impedance Z_{12} between the two antennas (Figure 8.10).

This model is used for mainly two reasons. The first one is the reliability of electromagnetic results. Indeed, with this type of model, the use of impedances calculated by electromagnetic simulation software is possible. This avoids using different models in reception and transmission. The second reason is that this model appears to be accurate not only in the far field but especially in the near field region.

The modeling of the interrogator in receiver mode is done by making an equivalent Thevenin generator seen by the SAW sensor in which, the impedance and the equivalent Thevenin voltage are given by the following expressions, determined

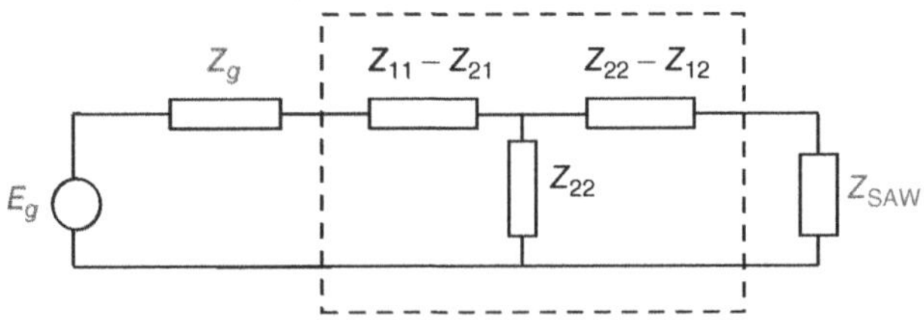

Figure 8.10 Equivalent T network of a two-antenna system including the electrical equivalent circuit of the interrogation system and the load.

from the emission model of the interrogation unit connected to the input of the equivalent circuit (Figure 8.10).

$$\begin{cases} \overline{E_T} = \dfrac{Z_{12}}{Z_{11}+Z_g}\overline{E_g} \\ Z_T = Z_{22} - \dfrac{Z_{12}^2}{Z_{11}+Z_g} \end{cases} \tag{8.11}$$

8.6.6 Theoretical Aspects of the Model

The expression of the transferred power from the equivalent Thevenin circuit to the SAW sensor can be written as:

$$P_{\text{transferred}} = \left| \frac{E_T(\omega)}{Z_T(\omega)+Z_{\text{SAW}}(\omega)} \right|^2 \text{Re}\left(Z_{\text{SAW}}(\omega)\right) \tag{8.12}$$

To determine the maximum of this equation, it is necessary to calculate its derivative and find its cancellation points.

An obvious solution is the cancellation of the real and imaginary parts of the sensor for the same pulse, but two pulsations must be considered here. To illustrate the solution of this equation, different cases were tested.

Figure 8.11 shows different examples of received power curves, obtained for different values of sensor antenna impedance. The one with the highest peaks meets the needs of the system. Both selectivity and power of the peaks are high.

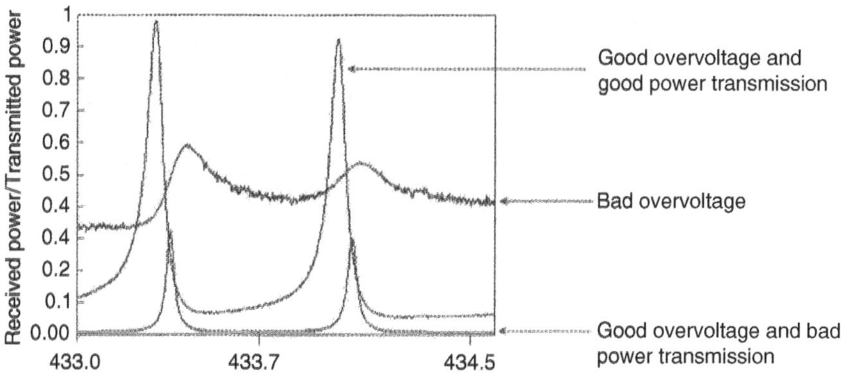

Figure 8.11 Power transmitted by the SAW for different values of sensor antenna impedance. *Source:* Monedero and Chommeloux [15]/with permission of SENSeOR/U.S. Patents.

The second curve in terms of peak amplitude shows a bad overvoltage so a little selectivity. The curve with the lowest peaks in terms of levels shows that the overvoltage and therefore the selectivity are good but the small amplitude of the peaks proves that the power transmission was badly done.

The optimization of these systems based on differential SAW sensor comprising at least two resonators can therefore be summarized by the following process:

1) The determination for a given sensor of its impedance Z_{SAW} by the presented techniques,
2) The definition by electromagnetic simulations of a set of power curves P_{SAW} in the frequency band of interest for a given reader antenna and a set of sensor antennas that may be of different types,
3) The selection of the solution associated to the curve that meets the following two criteria: The curve must have two peaks of maximum power, representative of the operation of a differential and these pics must have a width at mid-height less than a threshold value (typically less than or equal to 100 kHz).

This method could also be used for volume acoustic wave (BAW) sensors.

8.7 Conclusion

In this chapter, a review of the ISM 434 MHz SAW sensor systems was presented. This chapter aimed to offer a global vision of this technology and to underline the main advances in SAW sensor tags (Antenna and SAW resonator) modelization. Indeed, this chapter describes a possible global modeling method of a wireless sensor interrogation system using SAW technology, to allow a better integration of the antenna/sensor tag in the interrogation chain. This model is intended to establish a "system approach" in which each module, transmitting antenna or receiving antenna could be reworked or reoptimized according to the applications to enhance the range of the system.

A short state of the art of the different sensor technologies was first presented. Among all the possible solutions, wireless, passive sensors based on SAW technology allow reliable measurements of various physical quantities such as temperature, pressure, etc. Their main advantage lies in the fact that, combined with a radiating element, they can be interrogated remotely without the need for an external source of energy.

SAW sensors were described in detail. Their principle is based on the fact that a change in the physical quantity intended to be measured leads to a change in the resonance frequencies of the SAW sensor. The value of the physical quantity is calculated from the variation of the resonance frequency, and more precisely from the difference between this frequency and a reference one in the case of differential measurement. The principle of interrogation of this type of sensor was also presented and the phenomenon of "frequency pulling" that occurs when the antenna environment changes was highlighted.

As it is, therefore, the variation of the impedance of these antennas that causes this phenomenon, the existing wireless methods of measuring antennas have been briefly described. The objective could be not only to measure the characteristics of a wireless antenna, but also to determine remotely, the intrinsic impedance of the SAW sensor in the presence of the other resonator that is the antenna, but also possibly to arrive at a method of interrogation in which one could compensate for the propagation channel. Two methods were introduced a little bit more in detail: the first one based on the RCS and the other one on the "equivalent circuit" of a two-antenna system. However, the characterization in near field with the method based on RCS measurement requires three times more measurements than in far field. The results obtained by the "equivalent circuit" method have shown that it can be usefully used to precisely model the coupling between antennas.

Based on these results, a complete system model that can serve as a valid working basis for both near and far fields was described. In this system model, each resonator of the SAW sensor is modeled by an equivalent BVD circuit. This approach can be used to develop the theoretical aspect of the complete model, but the use of direct acquisitions of the sensors S-parameters as load would be preferable. The propagation channel and antenna coupling can be modeled with the "equivalent circuit" also described. For dual resonator SAW sensors associated directly with an antenna, this solution allows to determine the behavior of the complete system for a given application and to predetermine the optimal antenna impedance configuration to present to the sensor. This can be useful for any new type of application-specific antenna that will be chosen to be connected to the SAW sensor.

In conclusion, this chapter could serve as a design guide for SAW sensor tags based on resonators. The readers can find a complete development of the different parts in [16].

References

1 Tedjini, S., Andia-Vera, G., Zurita, M., et al. (2016). Augmented RFID tags. *IEEE Topical Conference on Wireless Sensors and Sensor Networks (WiSNet)*, Austin, TX, USA (24–27 January 2016), pp. 67–70. doi: https://doi.org/10.1109/WISNET.2016.7444324.

2 K. Zannas, H. El Matbouly, Y. Duroc, and S. Tedjini, "From identification to sensing: augmented RFID tags", in *Wireless Power Transmission for Sustainable Electronics*, N. Borges Carvalho, A. Georgiadis 2020, Wiley. pp. 223–246. https://doi.org/10.1002/9781119578598.ch8

3 Floer, C., Hage-Ali, S., Nicolay, P. et al. (2020). SAW RFID devices using connected IDTs as an alternative to conventional reflectors for harsh environments. *IEEE Trans. Ultrason. Ferroelectr. Freq. Control* 67 (6): 1267–1274. https://doi.org/10.1109/TUFFC.2019.2943310.

4 Reindl, L. and Shrena, I. (2004). Wireless measurement of temperature using surface acoustic waves sensors. *IEEE Trans. Ultrason. Ferroelectr. Freq. Control* 51(11): 1457–1463. https://doi.org/10.1109/TUFFC.2004.1367486.

5 Wolff, U., Schmidt, F., Scholl, G., and Magori, V. (1996). Radio accessible SAW sensors for non-contact measurement of torque and temperature. In: *IEEE Ultrasonics Symposium*, San Antonio, CA, USA (November 1996), vol. vol. 1, 359–362.

6 Tourette, S., Collin, G., Le Thuc, P. et al. (2009). Association of a small meandered PIFA and a SAW passive sensor for monitoring the inner temperature of a car exhaust header. In: *IEEE International Workshop on Antenna Technology (IWAT)*, Santa Monica, CA, USA (2–4 March 2009), pp. 1–4. doi: 10.1109/IWAT.2009.4906888.

7 Dixon, B., Kalinin, V., Beckley, J., and Lohr, R. (2006). A second generation in-car tire pressure monitoring system based on wireless passive SAW sensors. In: *IEEE International Frequency Control Symposium and Exposition*, Miami, FL, USA (4–7 June 2006), 374–380. doi: 10.1109/FREQ.2006.275414.

8 Kalinin, V., Bown, G., Beckley, J., and Lohr, R. (2004). Pulsed interrogation of the SAW torque sensor for electrical power assisted steering. In: *IEEE Ultrasonics Symposium*, Montreal, QC, Canada (23–27 August 2004), vol. vol. 3, 1577–1580. doi: 10.1109/ULTSYM.2004.1418121.

9 Reindl, L., Ruppel, C.C., Kirmayr, A. et al. (2001). Radio-requestable passive SAW water-content sensor. *IEEE Trans. Microwave Theory Tech.* 49 (4): 803–808.

10 Gopalsami, N., Osorio, I., Kulikov, S. et al. (2007). SAW microsensor brain implant for prediction and monitoring of seizures. *IEEE Sensors J.* 7 (7): 977–982.

11 Tsortos, A., Papadakisa, G., and Gizeli, E. (2008). Shear acoustic wave biosensor for detecting DNA intrinsic viscosity and conformation: a study with QCM-D. In: *IEEE International Frequency Control Symposium*, Honolulu, HI, USA (19–21 May 2008), 346–349. doi: 10.1109/FREQ.2008.4623017.

12 G. Collin, A. Chami, C. Luxey, P. Le Thuc, and R. Staraj, "Small electrical antenna for SAW sensor biotelemetry", *Microwave Opt. Technol. Lett.*, vol. 51, n°10, pp. 2286–2293, 2009, https://doi.org/10.1002/mop.24622

13 Mayer, E., Shrena, I., Eisele, D. et al. (2009). Characterization of Langasite as a material for SAW based RFID and sensing systems at high temperatures. In: *IEEE MTT-S International Microwave Workshop on Wireless Sensing, Local Positioning and RFID (IMWS)*, Cavtat, Croatia (24–25 September 2009), pp. 1–4. doi: 10.1109/IMWS2.2009.5307882.

14 Buff, W., Klett, S., Rusko, M. et al. (1998). Passive remote sensing for temperature and pressure using SAW resonator devices. *IEEE Trans. Ultrason. Ferroelectr. Freq. Control* 45 (5): 1388–1392.

15 Monedero, M. and Chommeloux, L. (2018). Method for optimizing the design of a device comprising interrogation means and a remotely-interrogatable passive sensor. US Patent Number 2018/0335350 (22 November 2018).

16 Monedero, M. (2012). Modélisation globale de systèmes d'interrogation sans fil de capteurs basés sur l'association éléments rayonnants/résonateur SAW. PhD dissertation. University Nice Sophia Antipolis.

17 Cross, P.S. and Shreve, W.R. (1982). Frequency trimming of SAW resonators sonics and ultrasonics. *IEEE Trans. Sonics Ultrason.* 29 (4): 231–234.

18 Monedero, M., Le Thuc, P., Seguenot, E., and Staraj, R. (2012). Near-field contactless return loss measurement of a 434 MHz dipole antenna-SAW sensor system. In: *6th European Conference on Antennas and Propagation (EUCAP)*, Prague, Czech Republic (March 2012), 2166–2169. doi: 10.1109/EuCAP.2012.6206341.

19 Heidrich, E. and Wiesbeck, W. (1989). Features of advanced polarimetric RCS-antenna measurements. In: *Proceedings of IEEE Antennas and Propagation Society International Symposium and URSI National Radio Science Meeting*, San Jose, CA, USA (26–30 June 1989), vol. vol. II, 1026–1029.

20 Wiesbeck, W. and Heidrich, E. (1998). Wide-band multiport antenna characterization by polarimetric RCS measurements. *IEEE Trans. Antennas Propag.* 46 (3): 341–350.

21 Nikitin, P.V. and Rao, K.V.S. (2006). Theory and measurement of backscattering from RFID tags. *IEEE Antennas Propag. Mag.* 48 (6): 212–218.

22 Pursula, P., Hirvonen, M., Jaakkola, K., and Varpula, T. (2007). Antenna effective aperture measurement with backscattering modulation. *IEEE Trans. Antennas Propag.* 55 (10): 2836–2843.

23 Vicencio, O. and Romeu, J. (2008). Application of RCS measurement method for characterization of small antennas. In: *IEEE AP-S International Symposium*, San Diego, CA, USA (5–11 July 2008), 1–4. doi: 10.1109/APS.2008.4618966.

24 Berlincourt, D.A., Curran, D.R., and Jaffe, H. (1964). Piezoelectric and piezomagnetic materials and their function in transducers. In: *Physical Acoustics*, vol. vol. 1 (ed. W.P. Masson), 169–270. Academic Press. doi.org/10.1016/B978-1-4832-2857-0.50009-5.

25 Rosebaum, J.F. (1988). *Bulk Acoustic Wave Theory and Devices*. Boston: Artech House.

26 Dieulesaint, E. and Royer, D. (1999). *Propagation et génération des ondes élastiques*. Techniques de l'Ingénieur.

27 Silver, S. (1949). *Microwave Antenna Theory and Design*, 45–50. New York: MIT Radiation Lab. Series, McGraw-Hill.

28 Balanis, C. (2005). *Antenna Theory: Analysis and Design*, 468–469. Wiley-Interscience.

9

Microwave Encoders for Motion Control and Chipless-RFID Applications

Ferran Martín, Ferran Paredes, and Amirhossein Karami-Horestani

CIMITEC, Departament d'Enginyeria Electrònica, Universitat Autònoma de Barcelona, Bellaterra, Spain

9.1 Introduction

Optical encoders constitute a well-known, and profusely utilized, technology that can be found in many industrial systems [1–3]. In particular, the so-called optical rotary (or angular) encoders are used as high-resolution angular displacement and velocity sensors for the accurate control of motion systems, such as elevators, conveyor belts, etc. Optical encoders, either linear or angular, exhibit very high accuracy with a relatively moderate cost, but such encoders are not the preferred option in harsh environments subjected to pollution, dirtiness, or grease, such as those found in many industrial scenarios. The reason is that optical encoders are implemented by means of apertures arranged circularly (rotary encoders) or linearly (linear encoders) in the moving part of the system, made of an opaque material (e.g. a metal disc in many rotary encoders). System functionality is based on the fact that any time an aperture lies in the optical path between the optical source (a laser beam) and the detector (a photodetector), a signal in the form of an electric pulse is generated in the detector (see a sketch of the working principle of optical rotary encoders in Figure 9.1). Thus, according to this working principle, it is obvious that the presence of pollution and dirtiness in the apertures may jeopardize encoder functionality. Nevertheless, it is clear that the angular velocity can be retrieved from the time lapse between adjacent pulses, whereas the number of cumulative pulses determines the encoder position. Actually, this functionality corresponds to the so-called incremental encoders. In such encoders, the key

Wireless Identification and Sensing Systems for Harsh and Severe Environments,
First Edition. Edited by Smail Tedjini and Valentina Palazzi.

Published 2025 by John Wiley & Sons, Inc.

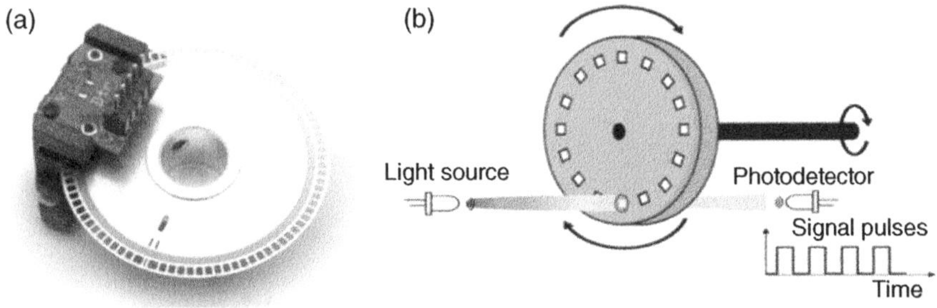

Figure 9.1 Photograph (a) and sketch showing the working principle (b) of an incremental optical rotary encoder. *Source:* Ferran Martín et al. 2020 [4]/Reproduced with permission from Springer Nature.

figure of merit is the number of pulses per revolution (PPR), for optical rotary encoders, or the number of pulses per unit length, for linear encoders. Position resolution (either angular or linear) is directly related to such figure of merit (e.g. in rotary encoders the angular resolution is simply 360°/PPR). By contrast, in the so-called absolute encoders, the position is determined by a unique ID code. Consequently, it is not necessary to define a reference (REF) position in such encoders, and, therefore, after a system reset, the absolute position can be retrieved. In absolute encoders, the position resolution is related to the number of bits of the encoder.

The previous limitative aspect of optical encoders (operation in harsh environments subjected to pollution, dirtiness, and grease) can be circumvented by using magnetic fields or microwave signals, rather than optical signals, for the determination of the position and velocity of the movable part of the encoder. The reason is that magnetic fields and microwave signals are robust against the effects of pollution and dirtiness. Thus, magnetic encoders, inductive encoders, and microwave encoders, constitute an interesting alternative to optical encoders in certain applications. Nevertheless, magnetic and inductive encoders are based on magnets or coils [5, 6], thereby representing certain complexity, as compared to the so-called microwave encoders, the subject of the present chapter. There are also various position sensors based on the Hall effect [7–13].

Microwave encoders can be applied to the measurement of linear and angular displacements and velocities [14–25], similar to the optical and magnetic counterparts, and are therefore of interest in motion control applications in diverse industrial scenarios. Nevertheless, microwave encoders can also be applied to the implementation of chipless radiofrequency identification (chipless-RFID) systems based on near field and sequential bit reading [26–40]. Such chipless-RFID systems operate in time domain, using a time-division multiplexing scheme, as it will be later shown, and their main relevant feature is the fact that the data capacity is only limited by the size of the encoder. Thus, chipless-RFID systems based on microwave encoders can compete against chip-based RFID systems in terms of data storage capacity. Nevertheless, tag (or encoder) reading proceeds by proximity and alignment, thereby representing a drawback in applications where the tag and the reader should be placed at distant positions. Indeed, the canonical application of these time-domain near-field chipless-RFID systems based on microwave encoders is anti-counterfeiting and security (e.g. authentication of premium products, secure paper, etc.), where the read distance is not an issue (or a critical requirement) [4].

The chapter is organized as follows. In Section 9.2, the working principle of microwave encoders, including both the incremental and the absolute types, is presented in detail. In Section 9.3, an example of a microwave encoder system devoted to motion control, using an absolute encoder, is presented. A chipless-RFID application is highlighted in Section 9.4. In Section 9.5, a hybrid approach that exploits also encoding in the frequency domain, in order to boost up the data density, is presented. Finally, Section 9.6 concludes the chapter by emphasizing the main relevant aspects.

9.2 Working Principle of Microwave Encoders and Case Example

The working principle of microwave encoders is actually very similar to the one of optical encoders, sketched in Figure 9.1. The difference concerns the fact that microwave signals are used for sensing the velocity and position of the encoder, and, rather than apertures, the encoder is made of metallic or dielectric inclusions. In the former case, electrically small metallic resonators, strips, or patches arranged in a chain (encoder chain) are etched or printed on a dielectric substrate. When the inclusions are made of a dielectric material, such inclusions should be embedded in a substrate with a higher or lower dielectric constant, so that the inclusions can be detected by virtue of the change in the permittivity. To detect the inclusions (either metallic or dielectric), when the encoder is in relative motion with regard to the static part (called stator in the nomenclature of position sensors and reader in the argot of RFID), a transmission line-based structure fed by, at least, a harmonic signal, is used.

Let us first consider the working principle of the incremental microwave encoders, and let us particularize to the case of a rotary encoder (this does not represent a loss of generality) [15]. The inclusions should be periodically distributed along the periphery of the rotor, made of a dielectric material, forming a circular chain, as Figure 9.2 illustrates. In order to detect the inclusions when they pass over the sensitive part of the stator, or reader, such sensitive part consists

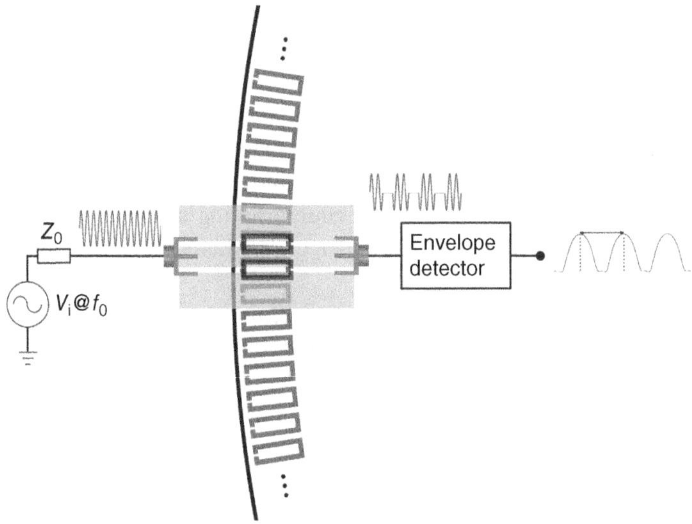

Figure 9.2 Sketch illustrating the working principle of an incremental microwave rotary encoder. *Source:* Reproduced with permission from Mata-Contreras et al. [15]; copyright 2017 IEEE.

of a transmission line-based structure (typically a transmission line loaded with a resonator), whose transmission coefficient varies when an encoder inclusion is on top of it. The specific type of line and loading element (if it is present) depends on the type of inclusion. Nevertheless, the idea is that, as far as the transmission coefficient of the line is affected by the presence of an inclusion on top of it, there should be a frequency range where the magnitude of the transmission coefficient varies significantly. Thus, by tuning the frequency of the harmonic signal feeding the line to a value within that range, it is expected that the output signal is amplitude modulated (AM). Therefore, the envelope of such signal, which can be retrieved by means of an AM detector, should contain as many pulses per time interval as number of inclusion crossings over the sensitive part of the stator in that interval. Since the distance between adjacent inclusions (or chain period) is well-known, it is obvious that the instantaneous velocity of the encoder is determined by the time lapse between two adjacent pulses of the envelope function, whereas the cumulative number of pulses determines the angular position from a predefined REF position [15]. A sketch showing the working principle of an incremental encoder is depicted in Figure 9.2, where, as mentioned, the inclusions are rectangular-shaped split-ring resonators (SRRs). In this particular incremental encoder, the transmission line of the stator (the sensitive element) is a coplanar waveguide (CPW) loaded with a pair of SRRs, oppositely oriented to the SRRs of the encoder chain, and etched in the back substrate side,

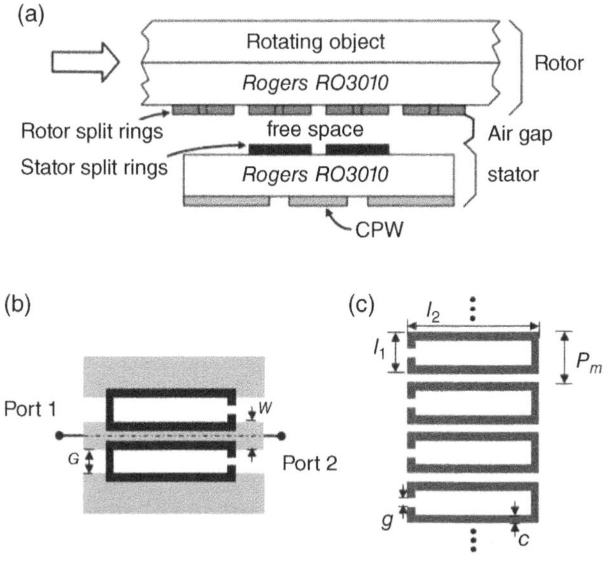

Figure 9.3 Cross section of the microwave encoder based on a CPW loaded with a pair of SRRs (a), and layout of the (b) stator and (c) rotor (zoom view). Dimensions (in mm) are: $W = 1.3$, $G = 0.9$, $P_m = 2$, $l_1 = 1.6$, $l_2 = 6.2$, $c = 0.4$, and $g = 0.2$. The substrates have relative permittivity $\varepsilon_r = 11.2$, thickness $h = 0.635$ mm (stator) and $h = 1.27$ mm (rotor), and loss tangent $\tan\delta = 0.0023$.

below the slots of the CPW, see Figure 9.3. With this arrangement, the variation of the transmission coefficient at the frequency of the pair of vertically aligned SRRs (forming the so-called broadside coupled SRR, BC-SRR) when the encoder chain crosses the sensitive part of the stator, is significant, since inter-resonator coupling, as well as multiple couplings between the line and the encoder SRRs are not relevant at this frequency. It was also demonstrated in [15] that system functionality is even better by tuning the frequency of the feeding signal slightly above the intrinsic resonance frequency of the BC-SRR, since by this means, two pulses, rather than one, per period of the encoder chain, are generated.

To gain more insight into system functionality, Figure 9.4 depicts the simulated transmission coefficient for different relative displacements between the rotor and stator (specific dimensions are indicated in the caption of Figure 9.3, corresponding to the considered case example). The so-called Ref displacement corresponds to a perfect alignment between the pair of SRRs of the stator, and a pair of SRRs of the rotor. The other responses correspond to displacements of a fraction of the chain period, P_m. For the indicated frequencies in Figure 9.4, designated as f_0, f_1, and f_3, the transmission coefficients as a function of the rotor displacement are depicted in Figure 9.5, where it can be appreciated that for the

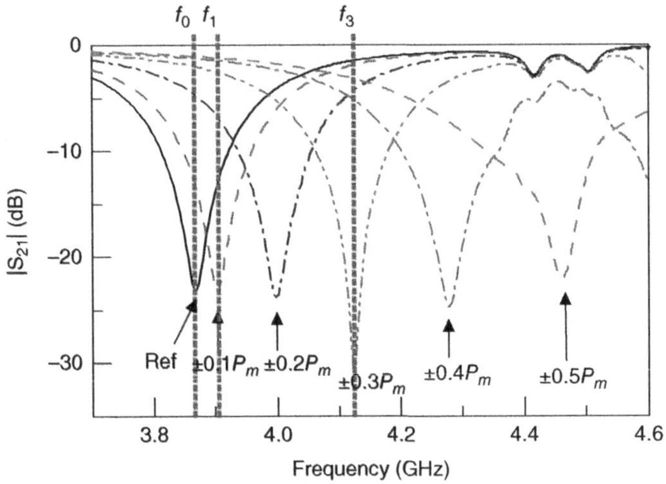

Figure 9.4 Electromagnetic simulation of the transmission coefficient for different relative displacements between the stator and the rotor. The considered air gap, or distance between the stator and rotor, is 0.5 mm. *Source:* Reproduced from Mata-Contreras et al. [15]/with permission of IEEE.

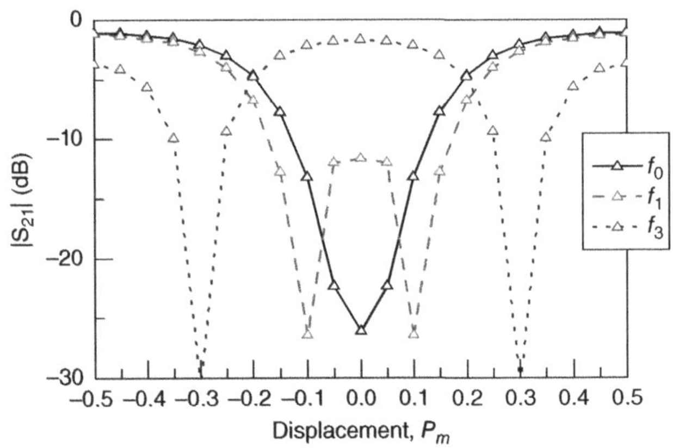

Figure 9.5 Attenuation as a function of the rotor displacement at the indicated frequencies. *Source:* Reproduced from Mata-Contreras et al. [15]/with permission of IEEE.

frequency f_0, a single peak per chain period is expected, whereas, for the frequencies f_1 and f_3, two peaks per chain period should result. It is important to mention that in the simulated responses of Figures 9.4 and 9.5, the considered vertical distance between the rotor and stator, called air gap, was set to 0.5 mm. However, such results should vary by modifying the air gap space, as discussed

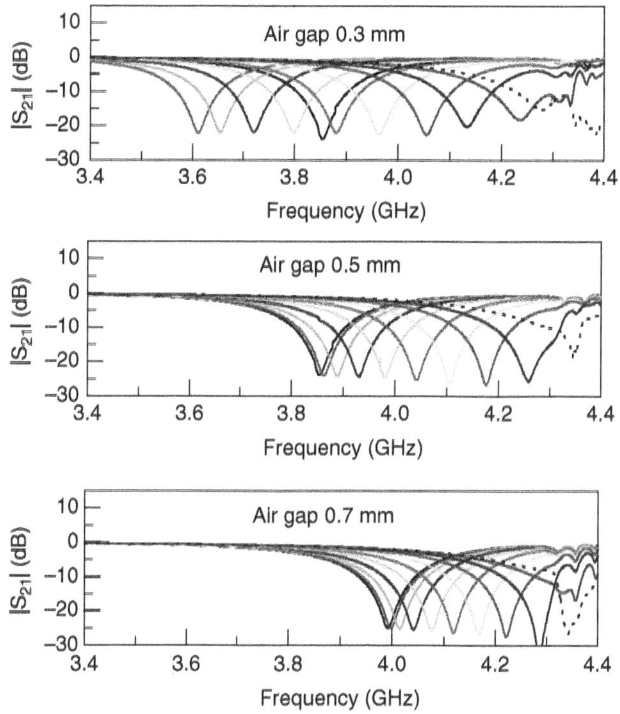

Figure 9.6 Electromagnetic simulation of the transmission coefficient for different relative displacements (from 0 to 1 mm in steps of 0.1 mm) between the stator and the rotor, parameterized by the air gap separation. Note that such relative displacements correspond to variation between $0P_m$ and $P_m/2$ in steps of $0.05P_m$. Dashed lines indicate the curves corresponding to $P_m/2$. *Source:* Reproduced from Mata-Contreras et al. [15]/ with permission of IEEE.

in [15]. This means that by setting the frequency to a certain value providing two peaks per chain period, it suffices to enhance the air gap in order to achieve an envelope function with only one peak per chain period, as Figure 9.6 illustrates (note that the response progressively displace to the right as the air gap space increases).

The encoder system was fabricated, and its functionality was demonstrated by obtaining the envelope function for nominal angular velocities of the rotor of 60 rpm. The carrier frequency was tuned to $f_c = 4.2$ GHz. The experimental setup and the measured envelope functions for different air gap distances are depicted in Figure 9.7. It can be appreciated that, depending on the air gap separation, either one or two pulses (or dips) per chain period are inferred, in agreement with the previous analysis. The details of the experimental setup are given in the caption of Figure 9.7. This case example has been useful to illustrate the working

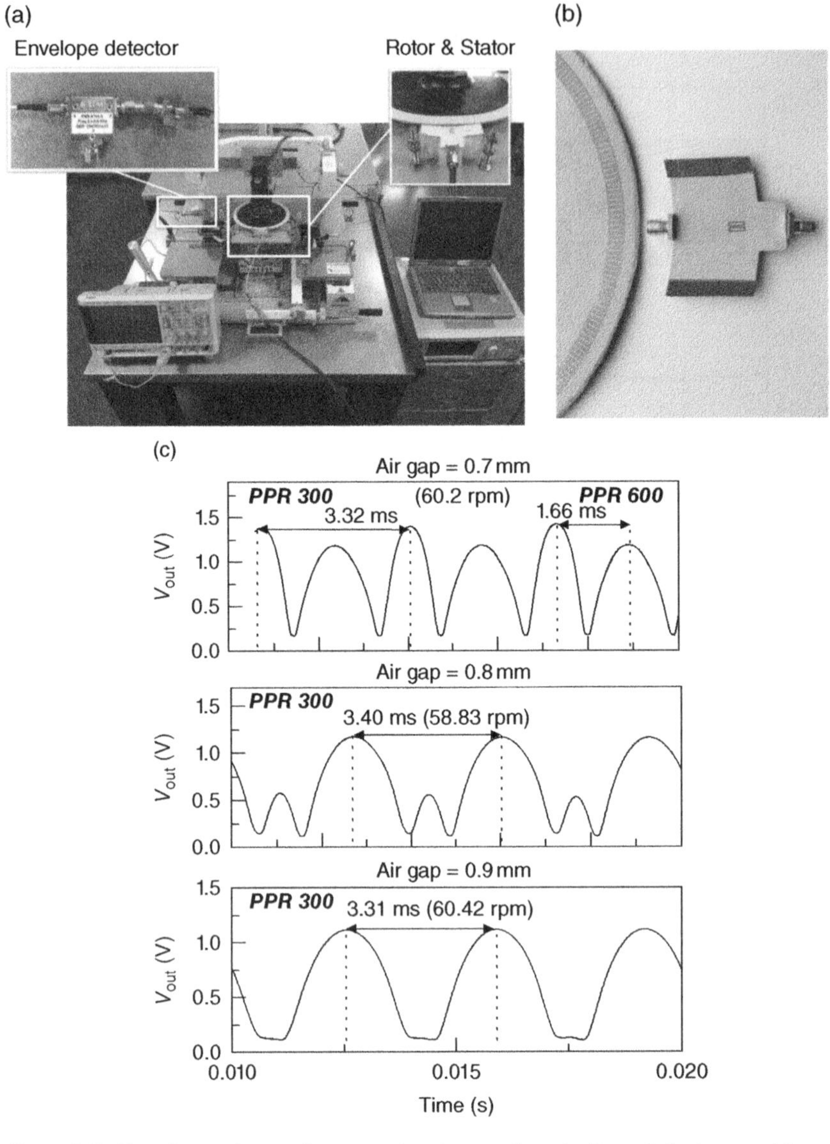

Figure 9.7 Experimental setup for measuring the angular velocity and displacement by means of rotors based on the edge configuration (a) photographs of the fabricated rotor, based on a single chain of SRRs, and stator (b), and envelope of the modulated carrier signal for different air gaps (c). For the smaller air gap, the amplitude of the pulses is similar, and the angular velocity can be obtained from the time distance between adjacent pulses, or between higher amplitude pulses, as indicated. The envelope detector was implemented by means of an RC filter consisting of an active probe (Agilent N2795A, with resistance $R = 1\,\text{M}\Omega$ and capacitance $C = 1\,\text{pF}$), and the Avago Technologies HSMS-2860 diode. The source impedance, the CPW characteristic impedance, and the circulator impedance were set to $Z_0 = 50\,\Omega$. The harmonic signal was injected by means of the Agilent N5182A MXG signal generator. An isolator, implemented by means of the circulator ATM ATc1-2 model, was cascaded between the output of the CPW and the envelope detector, in order to prevent from mismatching reflections from the diode. The oscilloscope Agilent 3054A was used to visualize the envelope signals in time domain. *Source:* Mata-Contreras et al. [15]/from IEEE.

principle of incremental microwave (rotary) encoders. As indicated, the main limitation of such encoders is the fact that the angular position is determined from the cumulative number of pulses from a certain REF position, and, consequently, after a system reset, the rotor should be returned to that position. The solution to this limitative aspect is the so-called absolute-type encoder, where a different identification (ID) code is assigned to each position. Absolute optical encoders are well-known. However, the implementation of absolute microwave encoders is not so simple. Let us next discuss this aspect.

In the incremental microwave encoder system of Figure 9.7, it is possible to generate a code along the periphery of the rotor by simply skipping certain SRRs (inclusions) of the circular chain. For example, the presence of an inclusion at the predefined positions in the chain can be associated to the "1" logic state, whereas the "0" is identified with the absence of an inclusion (or with a detuned inclusion, e.g. cut). Apart from the fact that reading such encoder with the considered reader would not be easy (because the CPW of the reader, the sensing element, is loaded with a pair of SRRs, as discussed), with this approach, only one bit per position results, which is, naturally, insufficient for an absolute encoder. Note that the different number of positions of the encoder determines the number of required bits. Let us assume that the number of different positions is N. Such number of positions depends on the ratio between the length of the encoder and the period of the chain. Let us suppose without loss of generality that the encoder is linear and that the length and period are L and p, respectively, i.e. $N = L/p$. Thus, the number of bits necessary to unequivocally determine the position, N_b, must satisfy

$$N_b \geq \log_2\left(\frac{L}{p}\right) \tag{9.1}$$

For high-resolution encoders, with periods much smaller than the encoder length, typically required in operative environments, N_b is typically above 10 bits. However, it is not possible to consider 10 parallel inclusion chains, or even more if $N_b > 10$, since reading would require as many harmonic signals as number of chains, which does not seem realistic.

To solve the previous limitation, the "quasi-absolute" microwave encoder concept was introduced in [21]. As the encoder moves, the absolute position is determined by simply reading the bit corresponding to the predefined position of the inclusion (either present or absent) crossing the sensitive part of the reader. Such bit plus the previous $N_b - 1$ bits provide a unique N_b-bit subcode that univocally identifies the encoder position. For that purpose, the complete ID code of the chain along the whole encoder must be chosen according to the so-called De Bruijn sequence [41]. This guarantees that any N_b-bit subcode does not repeat for the total set of different positions of the encoder. Note that the system must incorporate a table with the position assigned to the different N_b-bit subcode sequences, and this is necessary for both motion directions. For this reason, it is necessary to provide a means to discriminate the encoder direction. Note that after a system reset, or after a change in the direction of motion of the encoder chain, it must displace N_b positions in order to read the N_b bits of the subcode necessary to identify the absolute position from the table. This "after-reset uncertainty" can be tolerated in many applications. However, strictly speaking, a system like this cannot be categorized as absolute encoder, and for that reason, the designation "quasi-absolute encoder" has been coined.

In order to read the chain encoded following the De Bruijn sequence (which can be designated as position chain), a clock signal providing the instants of time for reading is necessary. Such clock signals can be generated by means of a parallel chain with all the inclusions present at their predefined positions. Note that such chain, which can be designated as clock chain, provides also the instantaneous velocity (thus, velocity chain is a reasonable alternative designation). Obviously, the predefined inclusion locations of the clock and position chains must be situated at the same axial positions of the encoder (in linear encoders), a necessary condition for synchronously reading the position chain. It should also be mentioned that, by properly designing the reader (to be discussed in the next section), the clock chain can also be used to determine the motion direction, thereby avoiding the use of a third chain for that purpose. Figure 9.8 depicts a sketch of a quasi-absolute encoder based on two inclusion chains consisting of rectangular metallic patches. Let us next review in detail in the next section an example of a synchronous quasi-absolute microwave encoder.

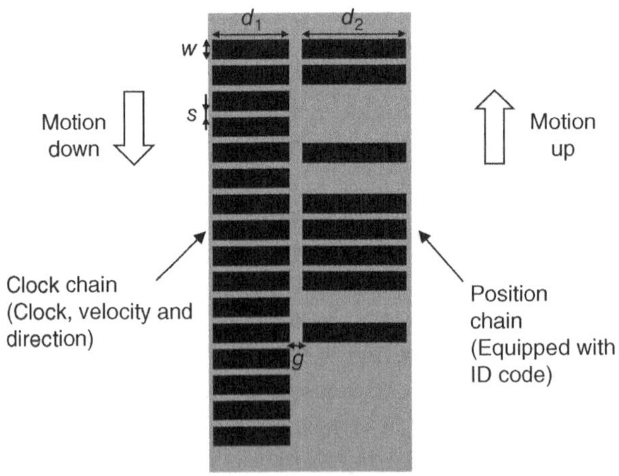

Figure 9.8 Sketch of a quasi-absolute microwave encoder based on metallic patches and relevant dimensions.

9.3 Quasi-Absolute Synchronous Encoders

Since not all the inclusions are present in the encoded position chain, quasi-absolute encoders should be equipped with a clock chain, that is, should be synchronous. Let us report in this section an illustrative example of a quasi-absolute encoder, which can be applied to motion control [25]. Indeed, the encoder was implemented by means of rectangular metallic patches (similar to Figure 9.8) screen-printed on a rubber material, particularly on a piece of a commercial elevator belt (nevertheless, direct printing of the encoder chains on different moving elements such as conveyor belts, can be envisaged, as well). The encoder is depicted in Figure 9.9a, whereas the considered reader is depicted in Figure 9.9b and c. With the considered encoder length and period (see caption of Figure 9.9), 48 different positions can be discerned, and the number of bits of the subcode sequence is $N_b = 6$ (i.e. any subset of 6 adjacent bits of the position chain does not repeat). For the implementation of the encoder, the metallic inclusions were screen-printed using the *Norcote ELG* conductive silver ink. Then the ink was UV cured by a halogen light (5 s at 500 W), and the printed belt was finally introduced in an air flow oven (10 minutes at 130 °C) to make thermal curing.

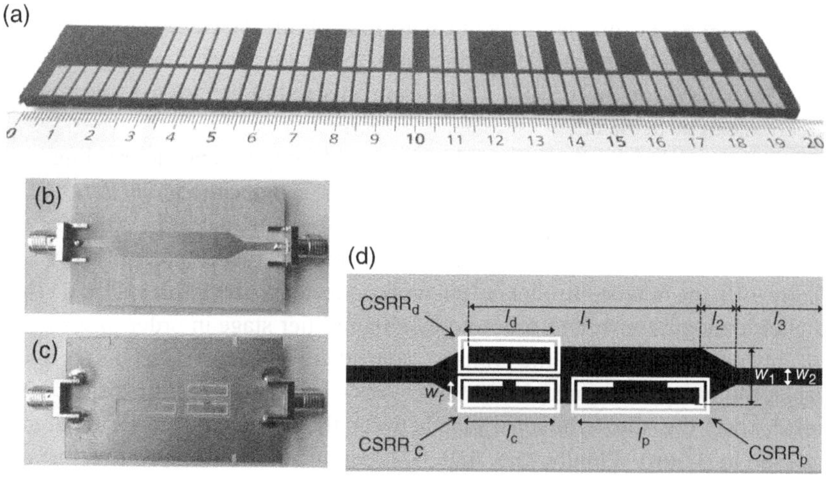

Figure 9.9 Photograph of the fabricated encoder, with patches screen-printed on a piece of commercial elevator belt (a) and photograph of the reader top (b) and bottom (c) faces. In reference to Figure 9.8, encoder dimensions are: $d_1 = 11.5$ mm; $d_2 = 15.9$ mm, $w = 3$ mm, $s = 1$ mm, and $g = 1.9$ mm. Moreover, the length, width and thickness of the encoder is 20, 3 and 3 cm, respectively. In reference to the reader layout (d), reader dimensions are $l_1 = 26.6$ mm; $l_2 = 3.8$ mm, $l_3 = 10$ mm, $w_1 = 6.4$ mm, $w_2 = 1.9$ mm, $l_c = l_d = 10.5$ mm; $l_p = 14.5$ mm, $w_r = 2.9$ mm. CSRR slots width is $c = 0.5$ mm, and ring splits are $s_d = 0.4$ mm, $s_c = 1.6$ mm, and $s_p = 6.2$ mm.

The reader is a microstrip line loaded with three rectangular complementary split-ring resonators (CSRRs) [42] etched in the ground plane and tuned to different frequencies. The resonator designated as $CSRR_p$ is used to determine the ID code (thus the position chain should be displaced above this resonator). For reading the ID code, a harmonic (carrier) signal tuned to the resonance frequency of the $CSRR_p$, $f_{0,p}$, should be injected to the line. When a patch is on top of the $CSRR_p$, the resonator is detuned and, consequently, the transmission coefficient at $f_{0,p}$ changes drastically (increases). Therefore, the input signal is AM modulated at the output port, and the envelope function contains the ID code (with as many peaks as patches in the position chain). The other two resonators, designated as $CSRR_c$ and $CSRR_d$, are both aligned with the clock chain. When a patch of that chain is located below the $CSRR_c$ (tuned to $f_{0,c}$), the transmission coefficient at $f_{0,c}$ varies. Thus, by injecting a harmonic signal tuned to that frequency, such signal should be AM modulated at the output port, the envelope function being the clock signal necessary for synchronously obtaining the ID code. The clock signal also provides the encoder velocity, from the time lapse between adjacent peaks, since the clock-chain period is well-known. To determine the motion direction, a redundant signal is generated from the AM modulation of a third injected harmonic signal tuned to the frequency of the resonator $CSRR_d$, $f_{0,d}$. By encoder motion, the clock chain generates an AM signal at the output port of the line with a period (envelope function) identical to one of the clock signals. The only difference is a lag or a lead depending on whether the clock chain patches first cross the $CSRR_c$ or the $CSRR_d$. Thus, from this redundant signal, the motion direction can be deduced.

Figure 9.10 depicts the system architecture and setup for encoder reading [25]. For the generation of the three required harmonic signals, the *VCO HMC391LP4* was used. The control voltage of such component was managed by means of the *ATmega328* microcontroller (*Arduino* development platform). As shown in Figure 9.10, the microcontroller output was connected to the 12-bit *MCP4725 DAC* (digital/analogic converter), followed by an amplifier stage in order to cover the full operation range of the voltage controller oscillator (VCO). A linear displacement system (model *STM 23Q-3AN*) was employed to displace the encoder over the stator (the separation between the stator and encoder, or air gap, was set to 1 mm). Finally, the *ADL5511* commercial envelope detector was connected between the output port of the stator and the input port of the microcontroller. The procedure to generate the signals and process the data is as follows. First, the microcontroller sets a specific control voltage V_1 (so that the VCO generates a tone at $f_{0,p}$, the carrier frequency of the position chain) and reads the output signal of the envelope (AM) detector. After the sample period, $t_s = 1.25$ ms, the control voltage is changed to V_2 (corresponding to the unmodulated tone $f_{0,d}$, carrier frequency of the direction chain) and the procedure is repeated (the microcontroller retrieves the signal given by the AM detector) until the end

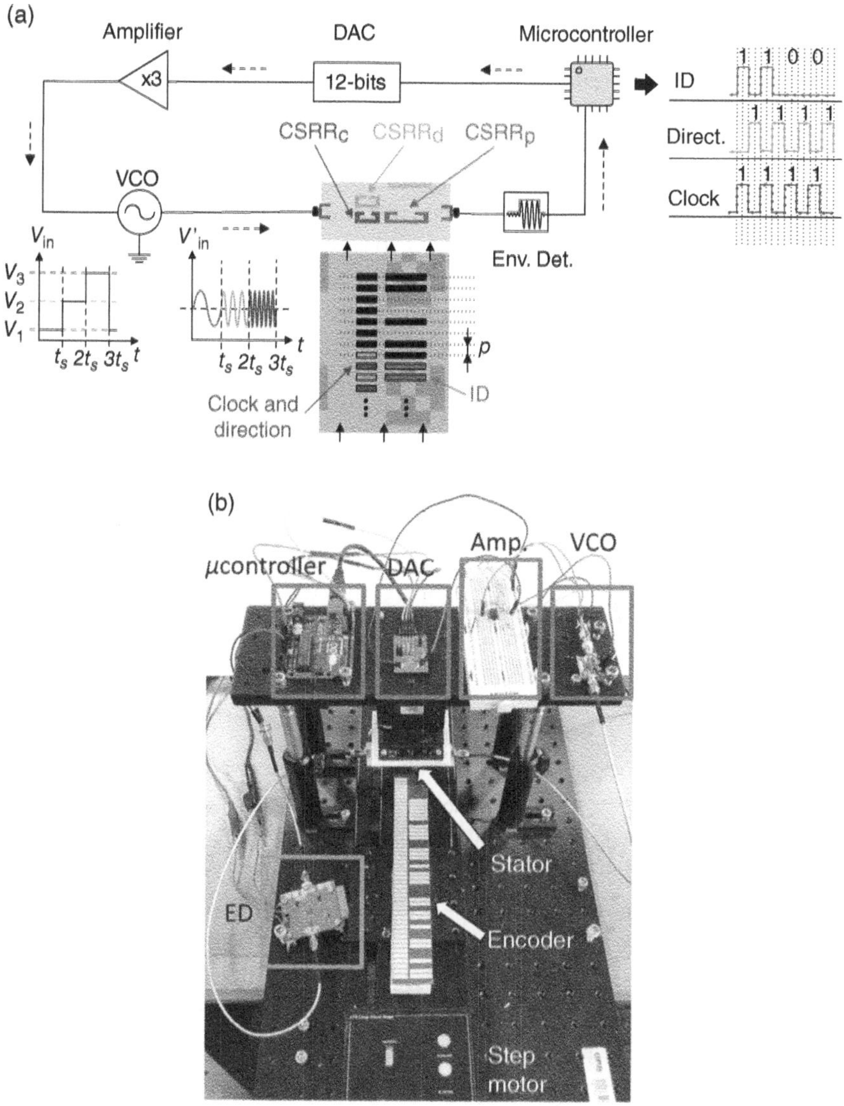

Figure 9.10 Sketch (a) and photograph (b) of the quasi-absolute encoder system with synchronous reading based on an advanced architecture with signal generation and processing by means of a microcontroller. *Source:* Ferran Paredes et al. [25]/MDPI/CC BY 4.0.

of the sample period, $2t_s$. Finally, at $2t_s$, the voltage is set to V_3 (with the VCO giving the tone at $f_{0,c}$, carrier frequency of the clock chain), and this voltage is kept until $t = 3t_s$. This sequence is constantly repeated while the encoder is in motion, and the three AM modulated signals (envelope functions) are obtained.

To validate the system, the encoder was displaced at a velocity of $v = 40$ mm/s in the upwards direction [25]. The measured signal at the output port of the envelope (AM) detector, which contains the three AM demodulated signals (envelope functions), is depicted in Figure 9.11a, whereas Figure 9.11b shows the separated envelope functions (achieved by the microcontroller). There is a good synchronism between the clock and the ID code signals, and the motion direction signal is delayed with regard to the clock signal (this indicates that the encoder moves upwards). The separation between peaks is 100 ms, in coherence with the nominal encoder velocity of 40 mm/s (the period of the metallic inclusions is $p = 4$ mm). Finally, the peaks in the ID code signal perfectly correlate with the presence of rectangular patches in the position chain (see Figure 9.9).

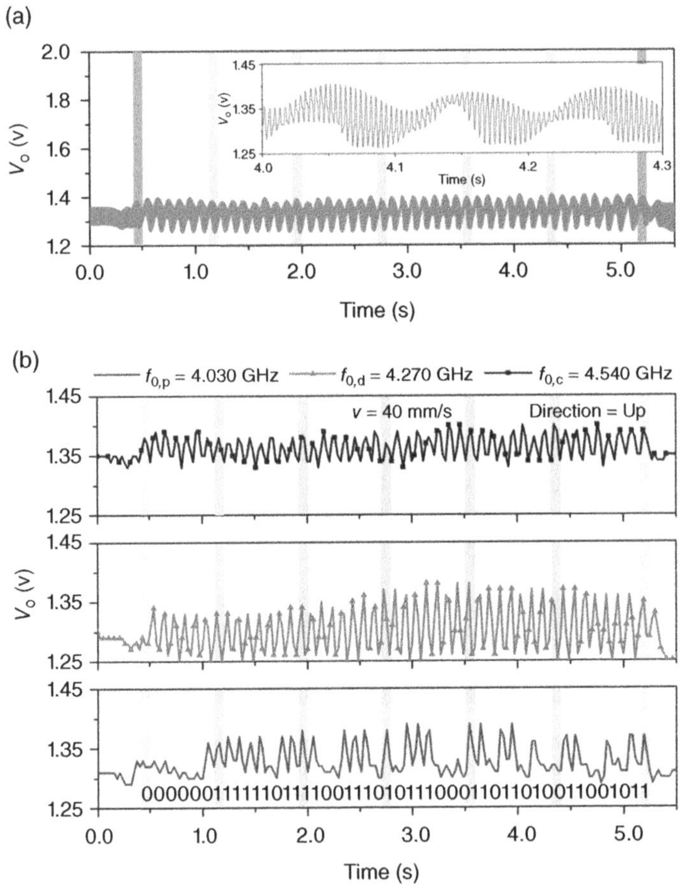

Figure 9.11 (a) Measured signal at the output port of the envelope (AM) detector; (b) measured envelope functions for the clock, direction, and position signals.

The previous results validate system functionality with the quasi-absolute encoder implemented in a commercial elevator belt. The screen-printed rubber encoder of Figure 9.9a supports significant tensile stress and torsion, but tolerance against wear out, for example, as consequence of pulley action in elevator systems or conveyor belts, still needs to be demonstrated. Let us also indicate that due to mechanical vibration of the belts during motion, the reader/stator should be closely located to one of the pulleys that the considered systems typically require. The belt movement in the radial direction of the pulley is negligible, and, by this means, the air gap, or distance between the reader and the encoder, can be maintained within very restrictive limits, an important aspect for the correct functionality of the sensor.

9.4 Chipless-RFID Application

In quasi-absolute encoders devoted to motion sensing, synchronism is necessary since the instants of time for retrieving the information relative to the position chain should be dictated by a clock signal. By contrast, in chipless-RFID, synchronism is not necessary provided the relative velocity between the tag (encoder) and the reader is well-known. Nevertheless, in manual systems, where such velocity can vary, even during tag motion over the reader, synchronous systems are required. In this section, an encoder system devoted to chipless-RFID applications is reported [30]. The encoder contains a single chain of inclusions, SRRs, and the system is non-synchronous since it is assumed that the displacement velocity of the tag is well-known.

The active part of the reader is a transmission line fed by a harmonic signal. Particularly, the structure under consideration is a microstrip line loaded with a SRR in bandpass configuration, as depicted in Figure 9.12a [27]. This structure exhibits two singular frequencies: (i) a reflection zero frequency (where the signal between the input and the output port is transmitted), and (ii) a transmission zero frequency. Accordingly, a large excursion of the transmission coefficient is achieved when a resonant element identical to the SRR of the line is aligned and oppositely oriented to it. The net effect is an overall displacement of the transmission coefficient to the left, as depicted in Figure 9.12b. By properly tuning the resonator dimensions as well as the vertical distance between the line and the oppositely oriented resonator (air gap), it is possible to optimize the variation of the transmission coefficient at the frequency of maximum transmission of the unloaded SRR-loaded line, f_c, as seen in Figure 9.12b. Thus, the tags were implemented as chains of SRRs, printed on a different substrate (paper in [27]), and tag reading proceeds by mechanically displacing the tag chain over the SRR of the reader line, with the SRRs of the tag oppositely oriented and aligned with regard to the SRR of the line. In practice, it is

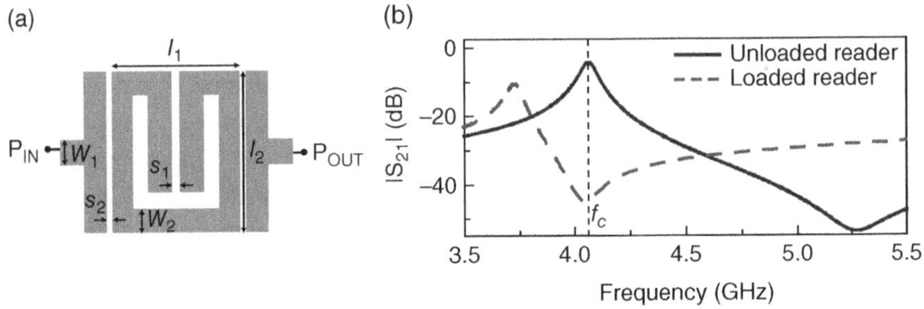

Figure 9.12 Layout of the SRR-loaded microstrip line of the reader (a), and transmission coefficients with and without oppositely oriented SRR on top of it (b). Dimensions are (in mm): $l_1 = 3.16, l_2 = 3.35, s_1 = 0.2, s_2 = 0.2, W_1 = 0.56$, and $W_2 = 0.5$. The considered air gap is 0.25 mm. The substrate of the line is the Rogers RO3010 with thickness $h = 0.635$ mm and dielectric constant $\varepsilon_r = 10.2$. The substrate of the tag is the Rogers RO4003C with thickness $h = 0.204$ mm and dielectric constant $\varepsilon_r = 3.55$. *Source:* Herrojo et al. [28]. Reproduced with permission from IEEE.

difficult to keep the air gap uniform, but some tolerance in the air gap distance is acceptable as long as the variation of the transmission coefficient at f_c is significant, even when the air gap deviates from the optimum value.

Besides the SRR-loaded line, the reader includes also the associated electronics for retrieving the envelope function, that is, the envelope detector, plus additional electronics necessary to generate the harmonic (interrogation) signal. Though in many prototypes of electromagnetic encoder systems, the envelope detector was implemented by means of a Schottky diode (*Avago HSMS-2860*), an active probe acting as a low-pass filter (with $R = 1\,\mathrm{M\Omega}$ and $C = 1\,\mathrm{pF}$) and an isolator to prevent reflections from the diode, in the prototype reported in [27], and reproduced here, an integrated circuit (*Analog Devices ADL5511*), able to provide the envelope function (hence reducing cost), was alternatively used. The envelope signal was then driven to a data acquisition card (*National Instruments myRIO*) connected to a computer, where the sampled data could be viewed. The harmonic feeding signal was generated by means of the *Agilent E44338C* function generator. An important part of the reader is the mechanical system, which is needed to displace the tag over the transmission line of the reader in close proximity to it. To this end, a printer was adapted and equipped with an adhoc guiding channel to absorb the tag (printed encoder) [27]. Good alignment and uniform air gap distance between the tag chain and the resonator of the reader were achieved with this system. With regard to the speed of reading, the system is limited by the sampling rate of the reader. However, for any practical mechanical guiding system used to displace the tag over the reader, the achievable speed did not represent an issue in this regard. The photograph of the complete reader is depicted in Figure 9.13, where the different reader parts are indicated.

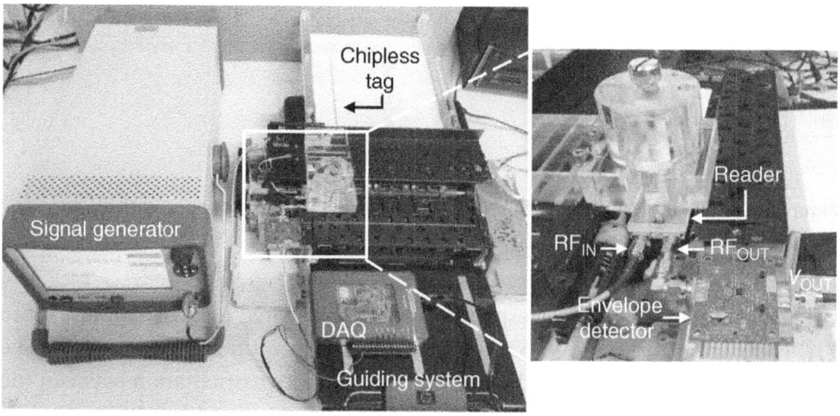

Figure 9.13 Photograph of the reader.

As indicated, the tags consist of a linear chain of SRRs identical to the one of the reader line printed on ordinary DIN A4 paper (with measured dielectric constant and loss tangent of $\varepsilon_r = 3.11$ and $\tan\delta = 0.036$, respectively) [27]. Note that the characteristics of the considered paper (tag substrate) are different than those of the considered tag substrate in Figure 9.12, but this does not represent a penalty in terms of system functionality. The tags were implemented by inkjet printing using the *Dupont™ PE410* conductive ink with measured conductivity of 7.28×10^6 S/m. One layer (with measured thickness of 2.6 μm) was demonstrated to be enough, thereby optimizing tag cost. The number of resonant elements was set to 80 for the ID code (corresponding to 80-bit tags), plus four SRRs for the header bits. Such header bits are needed since the tag can be displaced over the reader starting from the two extremes (corresponding to tag SRRs either faced up or down). The photograph of the fabricated tag with all ID bits set to the logic state "1" (i.e. with all resonators functional) is depicted in Figure 9.14. The total length and width of the tag (including the header bits) are 267.2 and 3.35 mm, respectively.

It is worth mentioning that the tags can be programmed and erased after being fabricated. This is an important aspect since fabrication costs can be further reduced, provided all-identical tags can be massively manufactured (requiring only a single mask), and then programmed in a later stage (custom level). Tag programming can be achieved by cutting those resonators with logic state "0", hence detuning them. This can be done mechanically, or by means of other systems such as, e.g. laser ablation. Moreover, tag erasing can be achieved by simply short-circuiting those previously detuned SRRs (for instance, by means of inkjet printing). Tag erasing (after programming) and reprogramming are unusual, but this possibility exists and means that a single tag can be recycled several times.

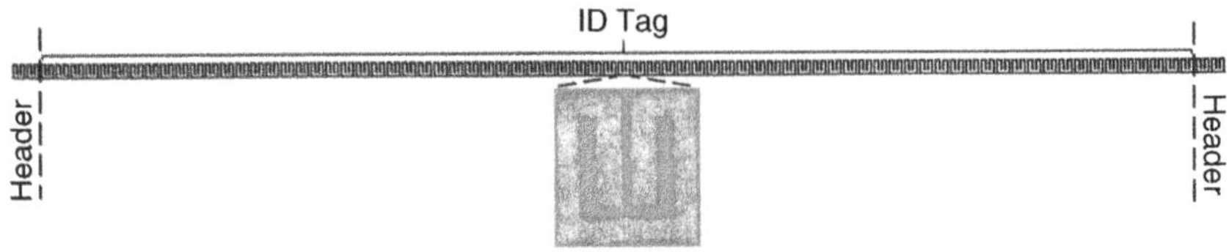

Figure 9.14 Photograph of the 80-bit tag fabricated on ordinary paper, with all bits set to "1" and with header bits.

The envelope function corresponding to the fabricated tag, with all bits set to "1," is depicted in Figure 9.15a for both the tag faced up and down. As can be seen, the code is correctly read regardless of the relative orientation between the tag and the reader. Then, in [27], the tag was programmed by cutting certain resonant elements with the code indicated in Figure 9.15b. Tag reading in this case provides the envelope functions for the tag faced up and down, also shown in Figure 9.15b, that perfectly correlate with the generated ID code. Finally, the tag was erased and reprogrammed with the ID code indicated in Figure 9.15c. Tag reading in this case also provides the correct ID code, regardless of tag orientation. Indeed, the tag was erased and reprogrammed several times, and tag reading was successful, which means that the proposed system is robust. These results point out that: (i) the number of achievable bits with this system is only limited by tag size; (ii) tag cost is very small as long as ordinary DIN A4 paper is used for tag substrate and a single layer of conductive ink suffices to correctly read the tags; and (iii) tags can be programmed and erased as many times as needed. A canonical application of this system is secure paper. By printing the ID codes directly on the document of interest, it is provided with a secure code, difficult to reproduce. Indeed, the codes can be photocopied, but tag reading requires a sophisticated system of the type shown in Figure 9.13. Moreover, it is possible to fabricate the tags through lamination, resulting in buried encoders difficult to reproduce [43].

Banknotes, certificates, corporate documents, ballots, exams, medical recipes, etc. are different types of documents that can take benefit from the proposed encoding system, providing them further security against counterfeiting.

Typically, in chipless-RFID systems, one figure of merit is the density of information per area unit. However, in the proposed chipless-RFID system, based on linear encoders, the tags exhibit an extremely long shape factor, and, consequently, the interest is to accommodate the largest possible number of bits along the tag length (typically limited by the size of the document). Thus, a significant effort has been dedicated in recent years to boost up the number of resonant elements, or inclusions, per unit length of the tag. Obviously, this can be achieved by increasing the resonance frequency of the SRRs (hence reducing

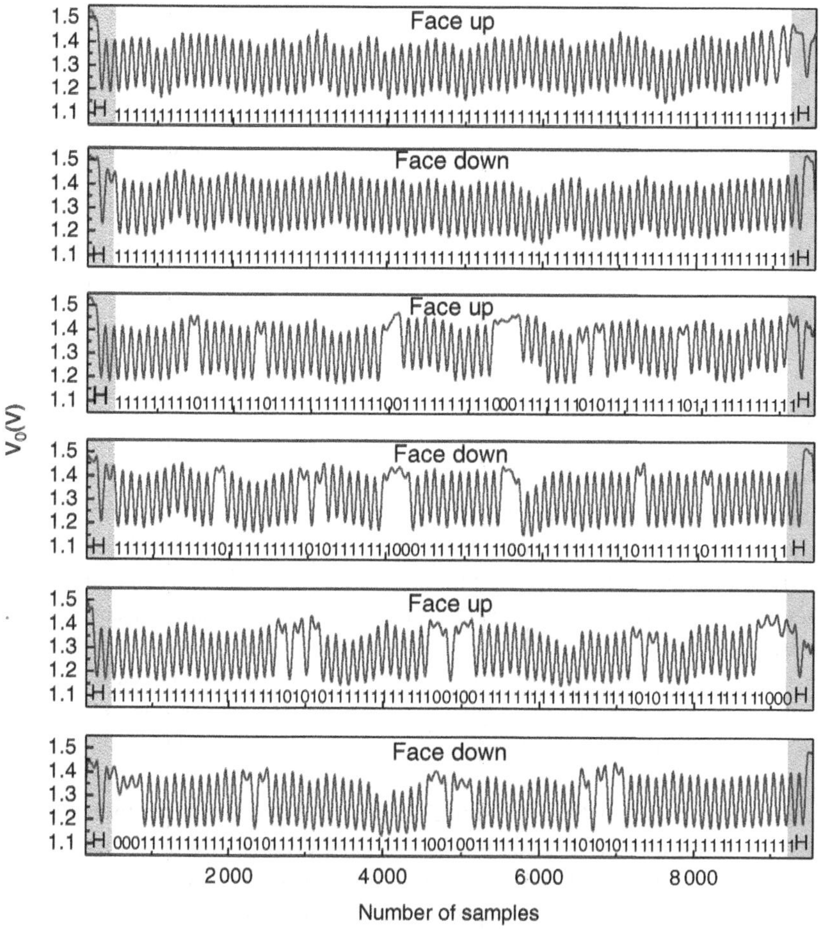

Figure 9.15 Measured envelope functions of the inkjet-printed 80-bit tag with the indicated codes. Since the envelope function has been inferred from a data acquisition system, the abscise axis is the number of samples. Note that there is no difference in the reading results between faced-up and faced-down tags. The codes are read in opposite directions for faced-up or faced-down tags, but this is indicated by the header bits, and therefore there is no confusion in this aspect.

their size), but this typically increases the cost of the associated electronics (harmonic signal generation, etc.). Therefore, novel tag inclusion topologies, very short in the longitudinal direction of the tag chain, have been proposed [19, 22, 31]. In particular, tags based on chains of linear strips transversely oriented with regard to the chain axis have been proposed in various works [19, 22]. The very high data density per unit length achieved in [19] is remarkable.

Nevertheless, it should be mentioned that in the system reported in [19], tag implementation was done in a commercial microwave substrate, rather than paper.

In the chipless-RFID system reported in this section and in other sources [26–40], tag encoding proceeds exclusively in time domain, where the absence or presence of a functional inclusions at the predefined positions determines the ID code, and the tag is read sequentially, on the basis of a time-division multiplexing scheme. In the next section, a different approach that combines encoding in time and frequency domain simultaneously is presented, as a means to enhance the data density per unit length of the tag.

9.5 Hybrid Approach

Very recently, a different type of time-domain chipless-RFID system based on sequential bit reading was proposed [38–40]. In this new system, more than a single bit per predefined position of the tag chain is achieved. For that purpose, rather than presence or absence of a functional inclusion (this corresponds to two states and, consequently, to a single bit), different-sized inclusions are considered. For example, one among four different inclusions per chain position provides two bits of information, which represents an enhancement of the data density per unit length by a factor of 2. The key aspect is to implement a reader able to discriminate the specific inclusion, and this can be achieved by virtue of the relationship between size and frequency in planar inclusions. For example, linear strips behave as resonant elements with a resonance frequency that depends on their length. Thus, inclusion detection needs not only a time-division multiplexing scheme (i.e. sequential reading of the inclusions), but also frequency multiplexing, in order to determine the inclusion through its resonance frequency. The system is called hybrid because the tags are actually encoded in time and frequency domains simultaneously. Let us report an example of this type of chipless-RFID system, presented in [40].

The inclusions are linear metallic strips transversally oriented with regard to the chain axis. An inclusion is always present at the predefined position of the chain, and this provides intrinsic synchronism to the system. However, the length of the inclusions may vary between four different values, corresponding to four different states and, consequently, to two bits of information per inclusion. Figure 9.16b depicts a typical encoder. For encoder reading, an element able to detect the different lengths of the strips is necessary. This can be achieved by means of a transmission line with a series gap (Figure 9.16a). Such a gap prevents signal transmission between the input and the output port. However, when a strip is on top of the line, and aligned with the line axis, such strip behaves similar to a

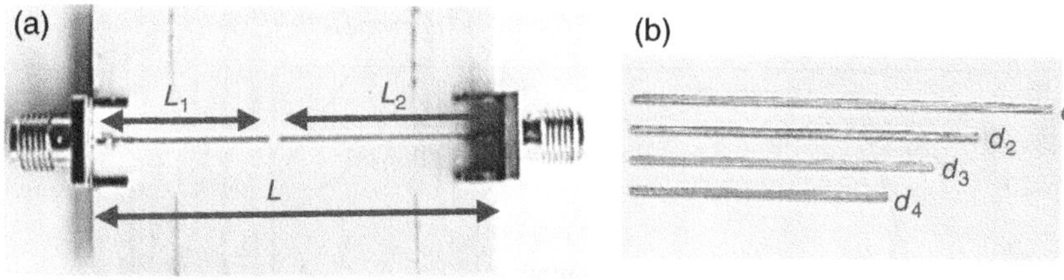

Figure 9.16 (a) Photograph of the fabricated transmission line reader, with a series gap. The width of the transmission line is 0.4 mm and other dimensions are L_1 = 22 mm, L_2 = 27 mm, and L = 50 mm. (b) photograph of a typical (8-bit) encoder. The period of the encoder is p = 2 mm and strip width is 0.5 mm. The lengths of the 4 encoder strips are d_1 = 28 mm, d_2 = 24 mm, d_3 = 20 mm, d_4 = 16 mm. The considered substrate of the encoder and reader is the Rogers 4003C with thickness h = 0.81 mm, dielectric constant ε_r = 3.55 and loss tangent tanδ = 0.0022. *Source:* Karami-Horestani et al. [40]/IEEE.

half-wavelength resonator, and a peak in the frequency response arises. Thus, by simultaneously feeding the line with four harmonic signals tuned to the resonance frequencies of the different strips, the specific strip length, when the encoder is displaced over the reader at short distance, can be detected. For each (carrier) frequency, one expects that the amplitude is modulated by encoder motion. Specifically, the amplitude of a certain carrier signal is expected to be high when the strip tuned to that frequency is on top of the line axis. Thus, by recording the envelope function of each carrier signal at the output port, the complete ID code of the encoder can be retrieved.

In a real scenario, a voltage-controlled oscillator (VCO) managed by a microcontroller, so that the frequency of the single injected signal varies periodically between the four required values (resonance frequencies of the linear strips), should be considered (similar to the system reported in Section 9.3). The AM signals are separated at the output port also by means of the microcontroller, and thus the four envelope functions can be inferred. With this approach, a single envelope detector is needed. Nevertheless, in [40], the signals were injected separately, since the purpose of that work was to demonstrate the working principle.

In order to analyze the influence of the air gap, electromagnetic simulations (by means of the *Ansys HFSS* electromagnetic solver) of the transmission coefficient for each carrier frequency as a function of the displacement for the considered 8-bit encoder of Figure 9.16b, considering the air gap as a parameter, were carried out. Note that the period of such encoder is $p = 2$ mm, corresponding to a per-unit-length bit density of 10 bit/cm, a competitive value for a synchronous encoder. The results are depicted in Figure 9.17. It can be seen that for air gaps of 0.5, 1.0, and 1.5 mm, the capacity of discrimination of the proposed system is very reasonable. Particularly, it can be appreciated that when a certain strip resonator

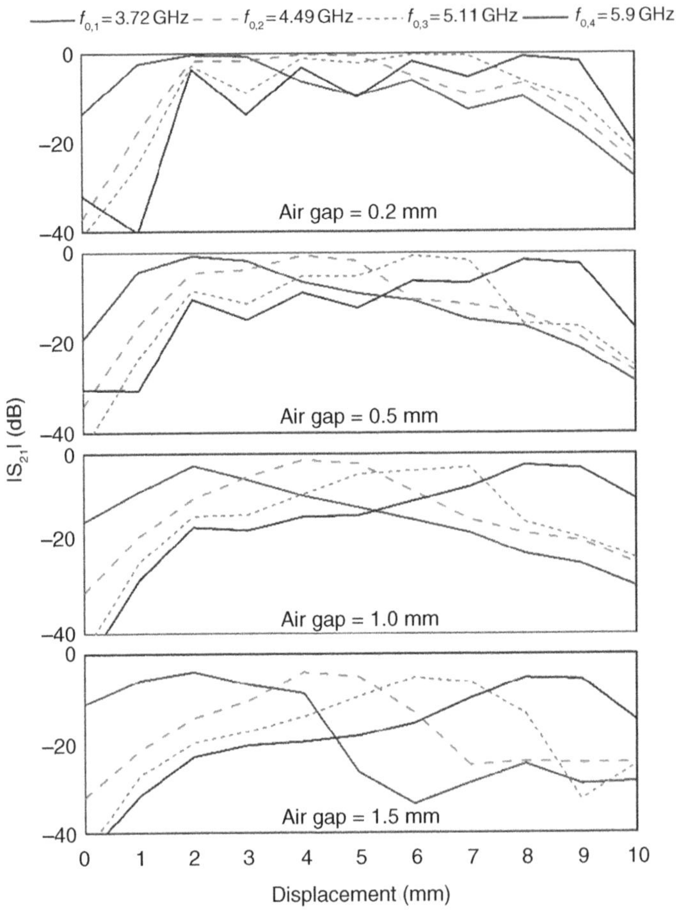

Figure 9.17 Simulation of the transmission coefficient for the resonance frequencies of the four strip resonators of the encoder, as a function of the encoder displacement, considering the air gap as a parameter. (a) Air gap 0.2 mm; (b) air gap 0.5 mm; (c) air gap 1.0 mm; (d) air gap 1.5 mm. The considered encoder is depicted in Figure 9.16b, with a period set to $p = 2$ mm, and strip width set to 0.5 mm. *Source:* Karami-Horestani et al. [40]. Reproduced with permission from IEEE.

is on top of the line axis, at the frequency of such resonator the transmission is high, whereas at the other frequencies, it decreases significantly, and this occurs for the four strip resonators. Nevertheless, for an extremely small air gap of 0.2 mm, this discrimination capability is limited. Thus, it is convenient to set the air gap separation at least to 0.5 mm.

For experimental validation, rather than the 8-bit encoder of Figure 9.16b, a 40-bit encoder consisting of a chain of 20 identical strip resonators (the length of

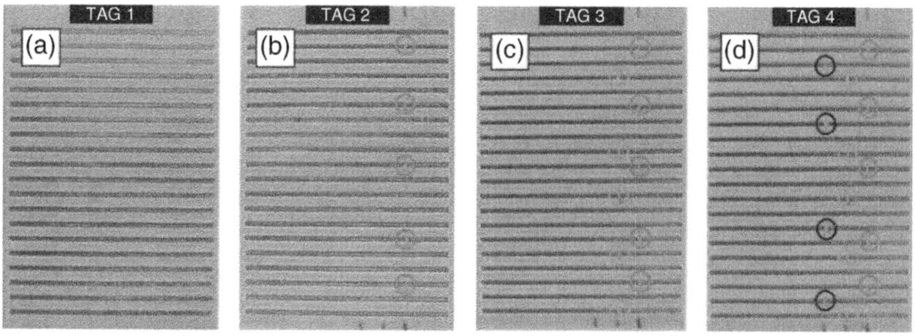

Figure 9.18 The tag before cutting and after each step of cutting. *Source:* Karami-Horestani et al. [40]/IEEE.

such strips is the longest one) was fabricated in [40]. Then, cuts were made on different resonant strips, thereby generating new ID codes by programming the encoders. Obviously, the cuts were made in positions that provide new strip resonators of one of the considered lengths. The tag before cutting and those that result after each step of cutting are shown in Figure 9.18. Figure 9.19 depicts the envelope functions of the different generated 40-bit encoders. There is a perfect correlation between the responses (envelope functions) and the strips of the different encoders. Consequently, with these results, the functionality of the system is experimentally validated.

9.6 Conclusions

In conclusion, it has been shown in this chapter that electromagnetic encoders can be applied to motion sensing and to the implementation of chipless-RFID systems with high data capacity and density. The main relevant advantage of these encoders over optical encoders as position and velocity sensors is their major robustness against the effects of pollution and dirtiness, encountered in many industrial scenarios. Moreover, as compared to magnetic encoders, electromagnetic encoders are simple since they do not include magnetic components. Within the framework of chipless-RFID, the main advantage of electromagnetic encoders over other chipless-RFID systems (typically based on the frequency domain) is their high data capacity, only limited by tag size, and density. Nevertheless, tag reading needs proximity between the tag and the reader, and alignment, and therefore the use of these encoders is not universal, but restricted to those applications where tag reading by proximity does not represent an issue (e.g. secure paper and authentication of premium products, among others).

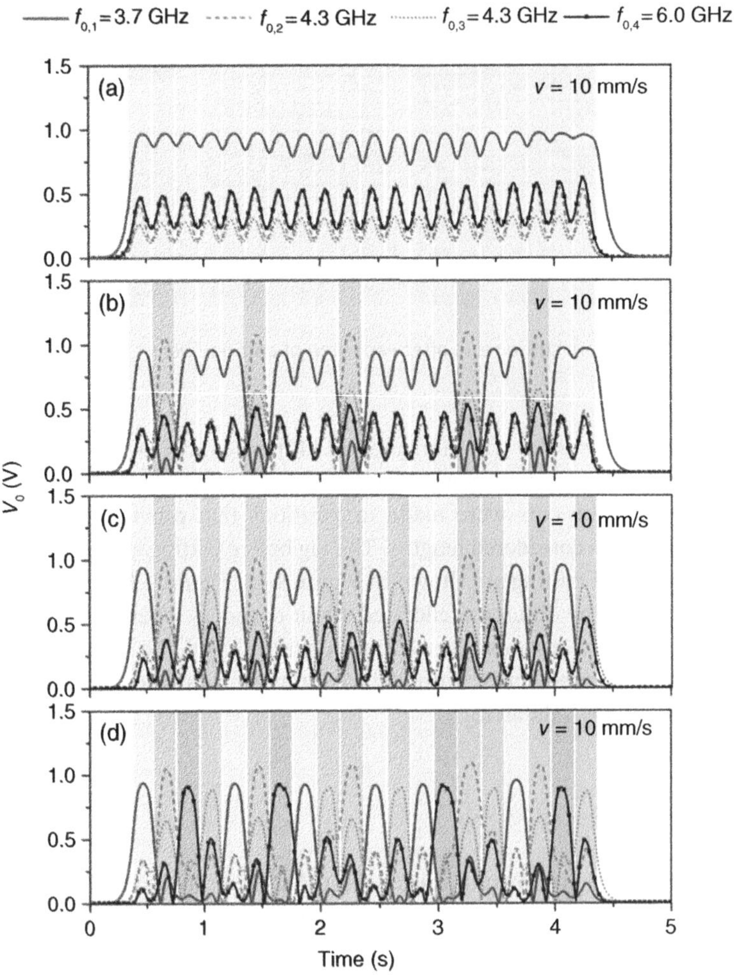

Figure 9.19 Measured envelope functions for the four frequencies corresponding to the resonances of the four strip resonators, for the encoders depicted in Figure 9.18. (a) Envelope function corresponding to TAG 1, (b) envelope function to TAG 2, (c) envelope function to TAG 3, (d) envelope function corresponding to TAG 4. The air gap was wet to 0.5 mm. *Source:* Karami-Horestani et al. [40]. Reproduced with permission from IEEE.

References

1 Petriu, E.M. (1989). Reconsidering natural binary encoding for absolute position measurement application. *IEEE Transactions on Instrumentation and Measurement* 38 (5): 1014–1016. https://doi.org/10.1109/19.39049.

2 Ueda, T., Kohsaka, F., Iino, T. et al. (1987). Optical absolute encoder using spatial filter. *Proceedings of the SPIE Photomechanics and Speckle Metrology*, San Diego, U.S.A. (17 February 1987). https://doi.org/10.1117/12.941698.

3 Eitel, E. (2014). Basics of rotary encoders: overview and new technologies. *Machine Design Magazine* (7 May), pp. 1–13.

4 Martín, F., Herrojo, C., Mata-Contreras, J. et al. (2020). *Time-Domain Signaure Barcodes for Chipless-RFID and Sensing Applications*. Cham, Switzerland: Springer https://doi.org/10.1007/978-3-030-39726-5.

5 Zhang, Z., Dong, Y., Ni, F. et al. (2015). A method for measurement of absolute angular position and application in a novel electromagnetic encoder system. *Journal Sensors* 2015 (503852): 1–10. https://doi.org/10.1155/2015/503852.

6 Zhang, Z., Ni, F., Dong, Y. et al. (2013). A novel absolute angular position sensor based on electromagnetism. *Sensors and Actuators A: Physical* 194: 196–203. https://doi.org/10.1016/j.sna.2013.01.040.

7 Jezný, J. and Curilla, M. (2013). Position measurement with hall effect sensors. *American Journal of Mechanical Engineering* 1 (7): 231–235. https://doi.org/10.12691/ajme-1-7-16.

8 Granell, P.N., Wang, G., Cañon-Bermudez, G.S. et al. (2019). Highly compliant planar Hall effect sensor with sub 200 nT sensitivity. *npj Flexible Electronics* 3 (3): 1–6. https://doi.org/10.1038/s41528-018-0046-9.

9 Lidozzi, A., Solero, L., Crescimbini, F. et al. (2007). SVM PMSM drive with low resolution hall-effect sensors. *IEEE Transactions on Power Electronics* 22 (1): 282–290. https://doi.org/10.1109/TPEL.2006.886603.

10 Scelba, G., De Donato, G., Scarcella, G. et al. (2014). Fault-tolerant rotor position and velocity estimation using binary hall-effect sensors for low-cost vector control drives. *IEEE Transactions on Industry Applications* 50 (5): 3403–3413. https://doi.org/10.1109/TIA.2014.2304616.

11 Zhang, X., Mehrtash, M., and Khamesee, M.B. (2016). Dual-axial motion control of a magnetic levitation system using hall-effect sensors. *IEEE/ASME Transactions on Mechatronics* 21 (2): 1129–1139. https://doi.org/10.1109/TMECH.2015.2479404.

12 Liu, G., Chen, B., and Song, X. (2019). High-precision speed and position estimation based on hall vector frequency tracking for PMSM with bipolar hall-effect sensors. *IEEE Sensors Journal* 19 (6): 2347–2355. https://doi.org/10.1109/JSEN.2018.2885020.

13 Rotary and Linear Motion Sensors. https://www.rls.si/eng/hilin (accessed 17 January 2022).

14 Naqui, J., and Martín, F. (2016). Application of broadside-coupled split ring resonator (BC-SRR) loaded transmission lines to the design of rotary encoders for space applications. *Proceedings of the IEEE MTT-S International Microwave Symposium (IMS)*, San Francisco, USA (22–27 May 2016). https://doi.org/10.1109/MWSYM.2016.7540017.

15 Mata-Contreras, J., Herrojo, C., and Martín, F. (2017). Application of split ring resonator (SRR) loaded transmission lines to the design of angular displacement and velocity sensors for space applications. *IEEE Transactions on Microwave Theory and Techniques* 65 (11): 4450–4460. https://doi.org/10.1109/TMTT.2017.2693981.

16 Herrojo, C., Mata-Contreras, J., Paredes, F. et al. (2017). Microwave encoders for chipless RFID and angular velocity sensors based on S-shaped split ring resonators. *IEEE Sensors Journal* 17 (15): 4805–4813. https://doi.org/10.1109/JSEN.2017.2715982.

17 Mata-Contreras, J., Herrojo, C., and Martín, F. (2018). Detecting the rotation direction in contactless angular velocity sensors implemented with rotors loaded with multiple chains of resonators. *IEEE Sensors Journal* 18 (17): 7055–7065. https://doi.org/10.1109/JSEN.2018.2853643.

18 Mata-Contreras, J., Herrojo, C., and Martín, F. (2018). Electromagnetic rotary encoders based on split ring resonators (SRR) loaded microstrip lines. *Proceedings of the IEEE MTT-S International Microwave Symposium (IMS)*, Philadelphia, USA (10–15 June 2018). https://doi.org/10.1109/MWSYM.2018.8439183.

19 Herrojo, C., Paredes, F., and Martín, F. (2019). Double-stub loaded microstrip line reader for very high data density microwave encoders. *IEEE Transactions on Microwave Theory and Techniques* 67 (9): 3527–3536. https://doi.org/10.1109/TMTT.2019.2929128.

20 Herrojo, C., Paredes, F., and Martín, F. (2020). 3D-printed all-dielectric electromagnètic encoders with synchronous reading for measuring displacements and velocities. *Sensors* 20 (17): 4837. https://doi.org/10.3390/s20174837.

21 Paredes, F., Herrojo, C., and Martín, F. (2020). Microwave encoders with synchronous reading and direction detection for motion control applications. *Proceedings of the IEEE MTT-S International Microwave Symposium (IMS)*, Los Angeles, USA (4–6 August 2020). https://doi.org/10.1109/IMS30576.2020.9224044.

22 Herrojo, C., Paredes, F., and Martín, F. (2021). Synchronism and direction detection in high-resolution/high-density electromagnetic encoders. *IEEE Sensors Journal* 21 (3): 2873–2882. https://doi.org/10.1109/JSEN.2020.3025435.

23 Paredes, F., Herrojo, C., and Martín, F. (2021). Position sensors for industrial applications based on electromagnetic encoders. *Sensors* 21 (8): 2738. https://doi.org/10.3390/s21082738.

24 Paredes, F., Herrojo, C., and Martín, F. (2021). 3D-printed quasi-absolute electromagnètic encoders for chipless-RFID and motion control applications. *Electronics* 10 (10): 1154. https://doi.org/10.3390/electronics10101154.

25 Paredes, F., Herrojo, C., Moya, A. et al. (2022). Electromagnetic encoders screen-printed on rubber belts for absolute measurement of position and velocity. *Sensors* 22 (5): 2044. https://doi.org/10.3390/s22052044.

26 Herrojo, C., Mata-Contreras, J., Paredes, F. et al. (2017). High data density and capacity in chipless radiofrequency identification (chipless-RFID) tags based on double-chains of S-shaped split ring resonators (S-SRRs). *EPJ Applied Metamaterials* 4 (8): 1–6. https://doi.org/10.1051/epjam/2017008.

27 Herrojo, C., Mata-Contreras, J., Núñez, A. et al. (2017). Near-field chipless-RFID system with high data capacity for security and authentication applications. *IEEE Transactions on Microwave Theory and Techniques* 65 (12): 5298–5308. https://doi.org/10.1109/TMTT.2017.2768029.

28 Herrojo, C., Mata-Contreras, J., Paredes, F., et al. (2018). Near-field chipless-RFID system with erasable/programmable 40-bit tags inkjet printed on paper substrates. *IEEE Microwave and Wireless Components Letters* 28 (3): 272–274. doi: https://doi.org/10.1109/LMWC.2018.2802718.

29 Herrojo, C., Mata-Contreras, J., Paredes, F. et al. (2018). Near-field chipless-RFID tags with sequential bit reading implemented in plastic substrates. *International Journal of Magnetism and Magnetic Materials* 459: 322–327. https://doi.org/10.1016/j.jmmm.2017.10.005.

30 Herrojo, C., Moras, M., Paredes, F. et al. (2018). Very low-cost 80-bit chipless-RFID tags inkjet printed on ordinary paper. *Technologies* 6: 52. https://doi.org/10.3390/technologies6020052.

31 Havlicek, J., Herrojo, C., Paredes, F. et al. (2019). Enhancing the per-unit-length data density in near-field chipless-RFID systems with sequential bit reading. *IEEE Antennas and Wireless Propagation Letters* 18 (1): 89–92. https://doi.org/10.1109/LAWP.2018.2881356.

32 Herrojo, C., Muela, F.J., Mata-Contreras, J. et al. (2019). High-density microwave encoders for motion control and near-field chipless-RFID. *IEEE Sensors Journal* 19 (10): 3673–3682. https://doi.org/10.1109/JSEN.2019.2895015.

33 Herrojo, C., Paredes, F., Mata-Contreras, J. et al. (2019). Time-domain-signature chipless RFID tags: near-field chipless-RFID systems with high data capacity. *IEEE Microwave Magazine* 20 (12): 87–101. https://doi.org/10.1109/MMM.2019.2941634.

34 Herrojo, C., Paredes, F., Bonache, J. et al. (2020). 3-D-printed high data-density electromagnetic encoders based on permittivity contrast for motion control and chipless-RFID. *IEEE Transactions on Microwave Theory and Techniques* 68 (5): 1839–1850. https://doi.org/10.1109/TMTT.2019.2963176.

35 Paredes, F., Herrojo, C., Escudé, R. et al. (2020). High data density near-field chipless-RFID tags with synchronous reading. *IEEE Journal of Radio Frequency Identification* 4 (4): 517–524. https://doi.org/10.1109/JRFID.2020.2996586.

36 Moras, M., Martínez-Domingo, C., Herrojo, C. et al. (2021). Programmable organic chipless-RFID tags inkjet printed on paper substrates. *Applied Science* 11 (17): 7832. https://doi.org/10.3390/app11177832.

37 Paredes, F., Karami-Horestani, A., and Martín, F. (2022). Strategies to enhance the data density in synchronous electromagnetic encoders. *Sensors* 22 (10): 4356. https://doi.org/10.3390/s22124356.

38 Karami-Horestani, A., Paredes, F., and Martín, F. (2022). Near-field chipless-RFID system based on hybrid time/frequency domain encoding and power splitter reader. *Proceedings of the 52nd European Microwave Conference (EuMC)*, Milan, Italy (20–25 September 2022). https://doi.org/10.23919/EuMC54642.2022.9924409.

39 Karami-Horestani, A., Paredes, F., and Martín, F. (2022). A hybrid time/frequency domain near-field chipless-RFID system. *Proceedings of the Microwave Mediterranean Symposium (MMS'22)*, Pizzo Calabro, Italy (9–13 May 2022). https://doi.org/10.1109/MMS55062.2022.9825533.

40 Karami-Horestani, A., Paredes, F., and Martín, F. (2022). Frequency-coded and programmable synchronous electromagnetic encoders based on linear strips. *IEEE Sensors Letters* 6 (8): 1–4. https://doi.org/10.1109/LSENS.2022.3193329.

41 De Bruijn, N.G. (1975). Acknowledgement of Priority to C. Flye Sainte-Marie on the counting of circular arrangements of 2n zeros and ones that show each n-letter word exactly once. *T. H.-Report 75-WSK-06,* Technol. Univ. Eindhoven, The Netherlands.

42 Falcone, F., Lopetegi, T., Baena, J.D. et al. (2004). Effective negative-/spl epsiv/ stopband microstrip lines based on complementary split ring resonators. *IEEE Microwave and Wireless Components Letters* 14 (6): 280–282. https://doi.org/10.1109/LMWC.2004.828029.

43 Herrojo, C., Paredes, F., and Martín, F. (2020). 3D-printed microwave encoders based on embedded and buried dielectric inclusions. *Proceedings of the XXXIIIrd General Assembly and Scientific Symposium of the International Union of Radio Science*, Rome, Italy (19 August to 05 September 2020). https://doi.org/10.23919/URSIGASS49373.2020.9232336.

10

Chipless RFID Technology for Operations in Harsh Environments

Simone Genovesi[1], *Filippo Costa*[1], *Michele Borgese*[2], *Francesco Alessio Dicandia*[3], *and Giuliano Manara*[1]

[1] *Dipartimento Ingegneria dell'Informazione, Università di Pisa, Pisa, Italy*
[2] *Research and Development Department, SIAE Microelettronica, Milan, Italy*
[3] *Istituto di Elettronica e di Ingegneria dell'Informazione e delle Telecomunicazioni, Consiglio Nazionale delle Ricerche (CNR), Torino, Italy*

10.1 Introduction

A transducer is a device that converts one form of energy into another one such as a battery that converts chemical energy into electrical one or a heat engine transforms chemical energy into mechanical. Transducers can be broadly divided into two main classes, actuators and sensors. An actuator usually converts an electrical signal into some kind of action, such as a mechanical activity, and its primary goal is to achieve a very efficient energy conversion. A sensor is a device that transforms a stimulus into a processable output, generally an electrical signal. The input is represented by the quantity to be sensed (i.e. the measurand), which can be related to a particular form of energy, whereas the output is an analog or digital representation of the stimulus. In the case of a sensor, efficiency does not represent the main goal and it is much more desirable to have a transducer that exhibits other features such as a linear response, a great selectivity, a high sensitivity, and good resolution. Several enabling technologies have been adopted to realize sensors but those exploiting Radio Frequency (RF) immediately provide an electrical signal as an output. A broad range of applications can benefit from sensors employing chip or chipless RFID [1] spanning from precision agriculture to wearables, from localization to food quality monitoring.

The environmental conditions upon which a sensor operates deeply affect the requirements of adopted materials and set constraints on its performance and

Wireless Identification and Sensing Systems for Harsh and Severe Environments, First Edition. Edited by Smail Tedjini and Valentina Palazzi.

Published 2025 by John Wiley & Sons, Inc.

functioning. A challenging task is represented by the deployment of sensors in a harsh environment [2, 3]. An operative scenario can be hostile to a sensor operativity and survival under different operative scenarios such as a wide range of temperature, including very high ones or abrupt excursions, severe level of pressure, high radiation dose, chemically aggressive agents, and intense vibration.

Apart from the sensing operation that may be performed in the extreme conditions another key task is represented by the data collection procedure. Sensors that depend on wired connections are prone to suffer high degradation and breakage of cable connections and could require dedicated shielding, periodic maintenance, and difficult installation. Wired sensor can be hardly connected to moving or rotating parts. On the other hand, a wireless sensor does not require any physical link and data transfer is operated without any direct connection. The wireless access to the sensor data allows more degrees of freedom in sensor placement, saves space and weight, reduces cost due to the absence of cables and offers the possibility of installing larger sensor networks by avoiding complex wiring. Another desirable feature offered by wireless sensors is represented by the possibility of designing battery-free devices. In fact, sensors based on battery support pose strict constraints on the energy source that must withstand the same extreme conditions and may require a difficult periodic replacement whereas a battery-less sensing unit guarantees a long operating lifetime and lower cost.

The research interest in sensors for harsh environment is ever-increasing, as witnessed by the growing number of papers dealing with them. This is apparent from Figure 10.1 where the number of yearly papers whose keywords include

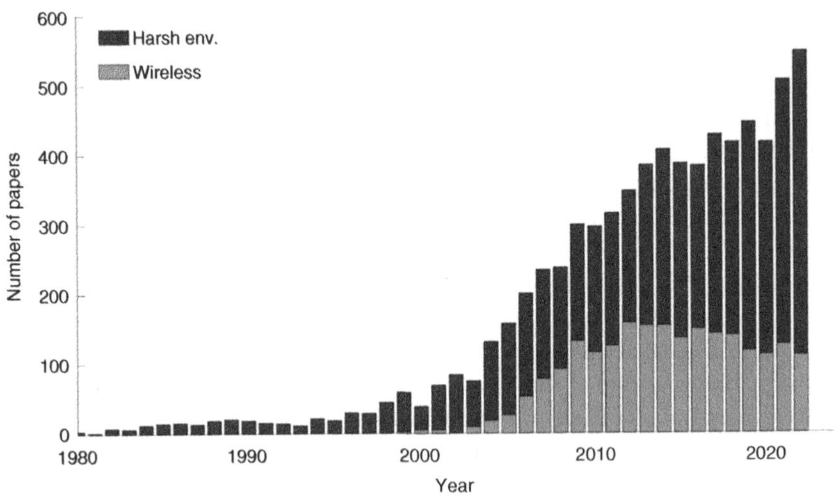

Figure 10.1 Number of papers indexed on the Scopus database 2022 according to the keywords "hostile/extreme/harsh" and "sensor" compared to those that also includes the term "wireless." Data have been collected up to December 2022. *Source:* Scopus.com.

"*harsh/hostile/extreme*" and "*sensor*" has been reported for the last 40 years. Recently, a consistent trend of papers that also include the keyword "*wireless*" has emerged that counts an average of more than 100 papers. An interesting comparison is illustrated in Figure 10.2a where the overall number of papers published and the registered patents in the last 10 years (i.e. since 2013) focused on sensor operating in harsh environments are reported. It is apparent that the number of patents largely outnumbers those of papers, but it is also important to notice that the ratio between wireless sensors and wired ones is in favor of the former in case of patents and the opposite for papers, which highlights the great interest of industry to wireless sensor solutions. Almost 15 000 patents on wireless sensor operating in hostile environments have been registered whose vast majority have been successfully filed to the United States Patent and Trademark Office (Figure 10.2b).

This chapter offers a review of the chipless RFID sensors that can operate in harsh environment. In particular, the focus is on sensors that exploit wireless connections between the reader and the sensing unit. Great attention is paid to analyze the different employed materials and all the other challenges involved in the design and manufacturing of an object that must undergo a significant stress without compromising its operativity.

The most adopted wireless sensor paradigms are presented in Section 10.2 and representative works found in the literature are provided to describe their peculiar

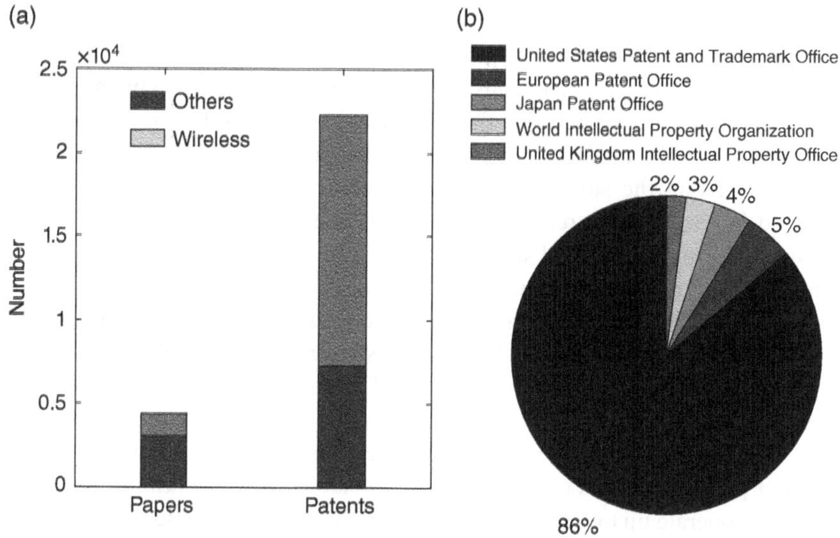

Figure 10.2 (a) Number of papers and patents of the last 10 years on sensors operating in harsh environments that are relying or not on wireless, (b) distribution among patent offices of inventions involving wireless sensors operating in harsh environments. Data have been collected up to December 2022. *Source:* Scopus.com.

features. The following sections are devoted to the wide range of applications that can benefit from the addressed technology such as sensors for space (Section 10.3), devices for supervising processes of the oil and gas industry (Section 10.4), sensors designed for the automotive sector (Section 10.5) and sensor-employed monitoring industrial tools.

10.2 Wireless Sensor Paradigms

Different solutions have been proposed to cope with the tough requirements imposed by a hostile environment, but the most adopted approaches can be grouped into three main categories, namely, that based on a surface acoustic wave (SAWs), another exploiting the near-field coupling and one relying on a far-field interrogation scheme and the collection of the sensor backscattered signal (Figure 10.3). It is worthwhile to mention that all of them are considering only completely passive wireless sensors.

10.2.1 Surface Acoustic Wave Sensors

SAW sensors are based on a suitable crystal (e.g. quartz) exhibiting the piezoelectricity effect, which means that when a voltage is applied to it, this will cause a mechanical deformation, converting electrical energy into mechanical energy. The opposite occurs when such a crystal is mechanically compressed or expanded [4]. The basic operating principle is illustrated in Figure 10.3a. The SAW tag receives an electromagnetic pulse (interrogating signal) emitted by the reader and then a transduction between the interrogated pulse and the slower acoustic wave is achieved by unidirectional interdigital transducer (IDT). The SAW pulse propagates along the surface of the piezoelectric material, and it is partially reflected by each of the reflectors. The train of reflected SAW pulses is then reconverted into an electrical signal by the IDT and retransmitted by the sensor antenna to the reader. The change in the piezoelectric substrate because of the interaction with the surrounding environment determines a change in the sensor response and this can be exploited to realize a sensor. Different coding approaches can be used to convey the sensed parameter [5, 6]. A typical application of SAW sensors is in the field of high temperature monitoring where materials that are able to undergo a significant thermal shock and an uninterrupted use are required [7, 8]. Materials such as lithium niobate (LNO) and lithium tantalate (LTO) can operate up to 300 °C before suffering from severe degradation whereas quartz can resist up to 576 °C. Gallium orthophosphate can be used up to around 900 °C whereas Langasite crystals (LGS) have remarkable properties and can operate at high temperatures up to 1400 °C [7].

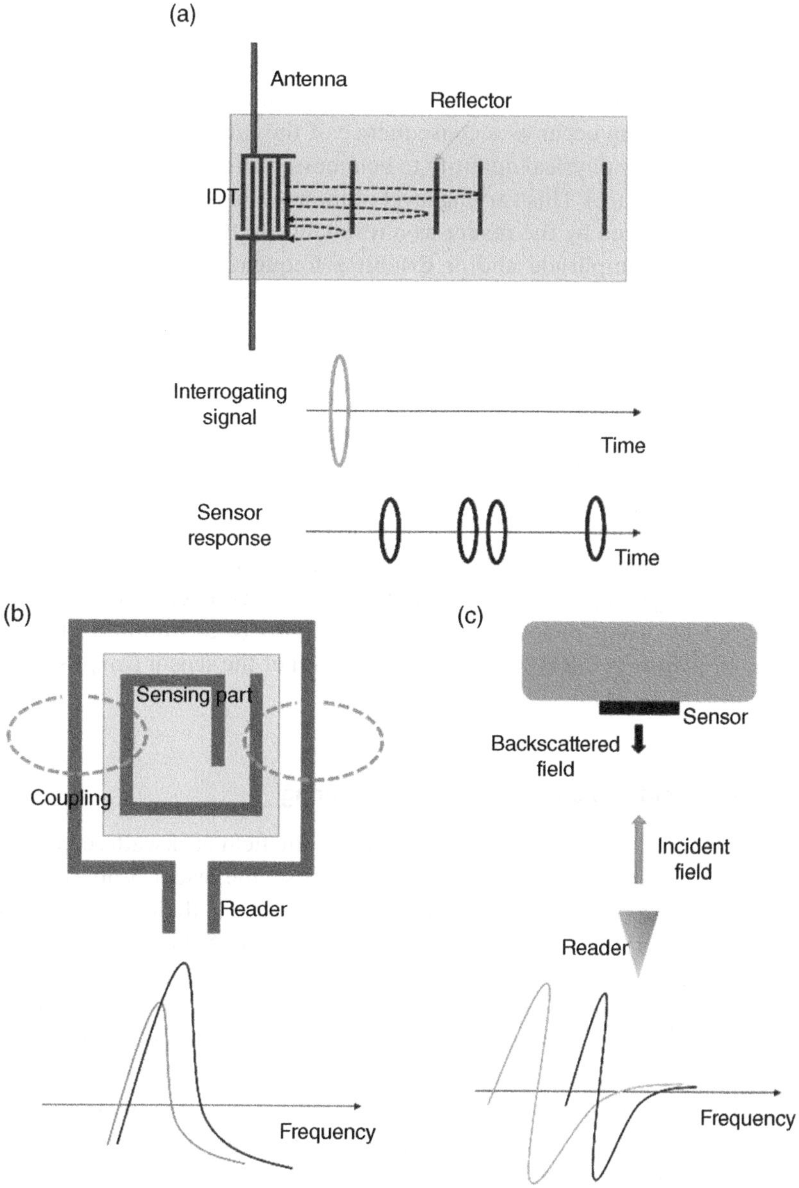

Figure 10.3 Different wireless sensor paradigms are suitable for extreme environments: (a) SAW exploitation, (b) near-field coupling, (c) far-field RF backscattering.

10.2.2 Near-Field Sensors

The near-field sensor consists of a sensing part and a reader. As illustrated in Figure 10.3b, the two parts experience a mutual coupling and changes in the level of this coupling occur as a consequence of the interaction between the sensing part and the physical quantity to be monitored (e.g. gas concentration humidity, temperature). This varying level of interaction is generally observed in the signal collected by the reader as a resonance in the frequency domain that may change amplitude and/or exhibit a frequency shift. This kind of sensor is also referred to as *LC sensor* since it may involve a change of inductance or capacitance in the sensing element equivalent circuit [8]. They generally operate at low frequency regime, mostly under 200 MHz, making them cost-effective and not interferent with other wireless communication systems. Also in this case the energy necessary for the sensing is provided by the reader and since the required amount is small this makes them a low-power consumption class of devices hence guaranteeing a long life span [9]. Also in this case a common field of application for this kind of sensor is related with scenarios where a high temperature makes impossible the adoption of standard sensing equipment [10, 11]. Temperature sensing up to 1 000 °C can be achieved by using dielectric ceramic substrates (e.g. alumina) but the adoption of platinum thick-film for the realization of the sensor can push this limit up to 1 500 °C [12].

10.2.3 Far-Field Radio Frequency Backscattering

The sensing approach based on the collection of the field backscattered by the sensing device is shown in Figure 10.3c. In this case, the reader is not in close proximity to the sensor, and it is generally at a distance sufficient to consider the sensor in its far field, or at least in the radiative near-field region, in order to consider the impinging field as a plane wave, or a reasonable approximation of it. The sensing node consists of a dielectric resonator which is generally of cylindrical shape and made of high dielectric permittivity materials with a low loss tangent [13, 14]. A possible implementation requires a reader that transmits a broadband signal toward the resonator that, at its specific resonance frequency, backscatter the received energy with some time delay that is proportional to its quality factor. Any change that occurs to the resonator as a consequence of the interaction with the environmental sensed parameters (e.g. temperature) causes a variation of the time response and, in turn, this determines a shift in the resonance peak in the frequency domain. The shift can be correlated to the monitored quantity to infer the proper calibration curve of the sensor.

10.3 Sensors for Space

As already mentioned, chipless sensors are sensors that do not rely on traditional chips to function. Instead, they rely on energy from the environment, such as light, heat, or radio waves, to operate. Therefore, they can be used in aerospace applications with some potential benefits:

- **Reduced size and weight:** Chipless sensors often do not require the use of chips, which can be relatively large and heavy. Consequently, chipless sensors are typically smaller and lighter than traditional ones, which can be beneficial for space applications where weight and size constraints are important.
- **Increased reliability:** Chipless sensors are passive and do not require a power source, which means they are less likely to fail due to power outages or other issues.
- **Reduced cost:** Chipless sensors may be less expensive to produce than traditional sensors, which can be especially important for space applications where cost is a major consideration.

However, they may also have some limitations, such as lower sensitivity or accuracy compared to traditional microchip-based sensors.

Typically, the parameters sensed in space applications are:

- **Temperature:** Chipless temperature sensors can be used to monitor the temperature of space systems and environments, as well as to detect changes in temperature that may indicate a problem or malfunction.
- **Pressure:** Chipless pressure sensors can be used to measure the pressure of gases and liquids, as well as to monitor changes in pressure that may indicate a problem or malfunction.
- **Strain:** Chipless strain sensors can be used to measure the deformation of materials due to external loads or forces, such as the strain on structural components of a spacecraft.

Space applications often require sensors that are able to withstand harsh environmental conditions, such as extreme temperatures, high levels of radiation, and high levels of vibration and acceleration.

For instance, the temperature range that aerospace sensors need to be able to measure can vary significantly, with some applications requiring sensors that can handle temperatures as low as −150 °C and as high as 1000 °C [15, 16]. Other applications may only need sensors that can measure temperatures within a narrower range, such as from −50 to 125 °C [17].

These environmental constraints can pose significant challenges for the design and operation of sensors and other electronic devices in space applications.

Indeed, semiconductor-based sensors might suffer from extremely high/low temperatures and radiation levels [18], electrostatic charging and outgassing as shown in [19–21]. Consequently, the employment of specialized materials and design approaches, as well as careful testing and validation to ensure the reliability and performance of the sensors and electronics in space applications is extremely important.

10.3.1 Temperature Sensors

Researchers are exploring the use of high-temperature materials, such as aluminum nitride and gallium phosphate, for wireless sensor applications [22]. For instance, aluminum nitride has the potential to be used in temperature-compensating sensors that can operate at extremely high temperatures (800 °C) [23]. Other exotic materials such as Langasite, langatite, and langatate have been studied in the temperature range from −100 to 900 °C [24, 25].

Gallium phosphate has been used as a substrate in the development of a temperature sensor operating at 433 MHz which is able to operate at temperatures of 600 °C for 192 hours [26]. The sensor, based on SAW technology, employs an IDT with platinum/zirconium metallization's fabricated with direct write e-beam lithography and a lift-off process. The results shown in [26] demonstrate that the sensor's resonators exhibit no signs of oxidations at 600 °C and stability for more than 3 500 hours. The sensor proved to be suitable for applications in harsh environments such as in aerospace, power, nuclear, chemical, and petrochemical process plants.

In [27], a wireless sensing system for remote estimation of temperature aboard space platforms is presented. The system relies on SAW sensors which can measure the temperature based on the variations of the attenuation or velocity of acoustic propagation waves. The sensors are interrogated by the reader which sends a RF pulse at the sensor's resonant frequency. This pulse is converted into a SAW (piezoelectric effect) on the sensor and propagates along the reflectors on the substrate of the sensor. The reader's signal interrogates the sensors in the frequency range 428–439 MHz. In order to evaluate the performance of the proposed sensors, thermocouple was connected to each one of the four sensors: the sensors exhibit an accuracy of ±2 °C in the range −40 to 90 °C with a resolution of 0.1 °C.

An example of chipless sensors which operate in a more limited temperature range is presented in [28]. The sensor tag which is shown in Figure 10.4a, is a half-split cylindrical ceramic resonator with permittivity and resonance frequency which depend on the temperature of the external environment. The sensor uses RF backscattering to transmit the measured value. The resonance frequency of a dielectric resonator on the sensor tag is identified as a peak in its radar cross section, and this information is encoded and transmitted through the use of

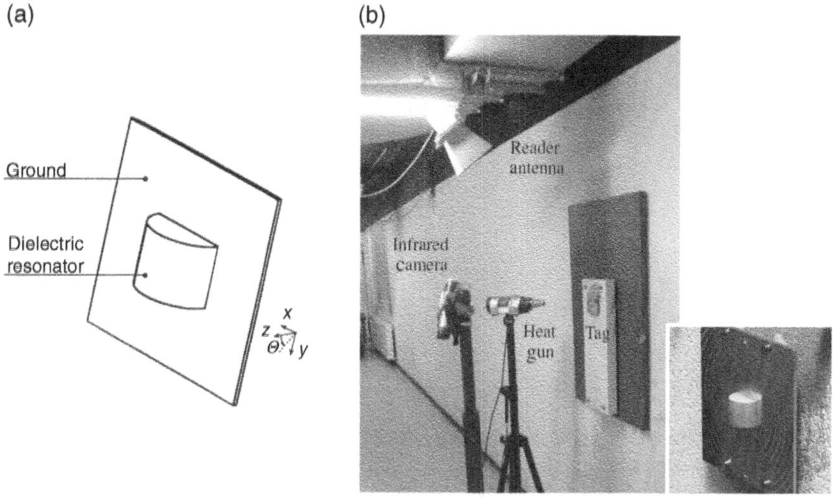

Figure 10.4 (a) Schematic representation of the temperature sensor tag; (b) measurement setup and fabricated sensor tag. *Source:* Kubina et al. [28]/IEEE.

backscattering techniques. Measurements performed with the setup shown in Figure 10.4b in the range from +20 to 370 °C where the resonance frequency of the tag lies around 3 GHz with a temperature sensitivity of 307 kHz/K.

10.3.2 Vehicle Health Monitoring Systems

A vehicle health monitoring system (VHMS) is a system that is used to monitor the condition of vehicles, such as an aircraft, to ensure their safety. VHMS is important for aerospace programs to ensure the safety of the crew and the vehicles. Passive wireless sensor networks can be used to facilitate VHMS on aerospace vehicles, which allows for the monitoring of the vehicles' health in a pervasive and continuous manner. There are several passive wireless technologies that have been developed for use in VHMS for aerospace applications.

A VHMS system which involves the use of passive wireless sensors that are attached to a vehicle's structure for health monitoring purposes is presented in [29]. As shown in Figure 10.5, the system includes an interrogator node that is used to communicate with the passive wireless sensors. This node can wirelessly transmit data from the passive wireless sensors to conventional wireless sensor networks in a heterogeneous manner. Researchers at NASA have employed SAW devices in the presented VHMS system because these devices have several desirable characteristics. They are small, low power, and able to withstand harsh environments and extreme temperatures ranging from −200 to 910 °C.

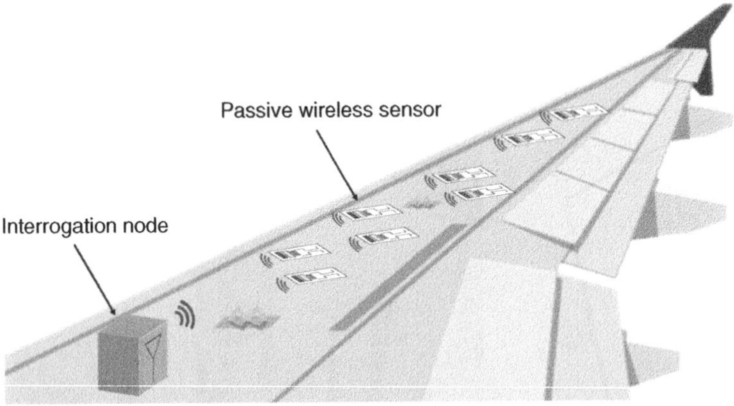

Figure 10.5 Concept for pervasive passive wireless sensors for structural health monitoring for aerospace vehicles.

A SAW device was fabricated on a Langasite ($La_3Ga_5SiO_{14}$) substrate specifically for use in VHMS applications. The sensor exhibits four Orthogonal Frequency Coded (OFC) reflector gratings (300.05, 303.04, 306.10, and 309.28 MHz), which are used to spread the device's response across multiple frequencies using OFC reflectors. The gratings are divided into two sets or banks. To prevent interference, these banks are positioned on either side of a device called an interdigitated transducer and are spaced apart in such a way that the reflections they produce do not overlap in time. A schematic representation of the SAW tag is shown in Figure 10.6.

The operating frequency of the SAW device will decrease if the device expands due to tensile strain or an increase in temperature, and it will increase if the device contracts due to compressive strain or a decrease in temperature. In order to validate the strain measurement capabilities, the device was connected to a helically

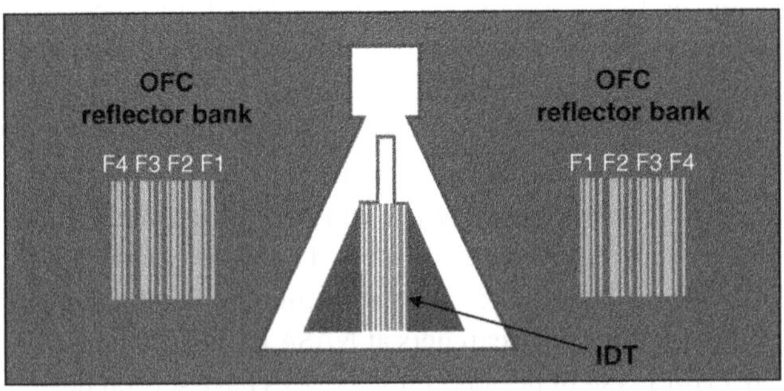

Figure 10.6 Schematic representation of the SAW tag for VHMS systems proposed in [29]. *Source:* Wilson and Juarez [29]/with permission of ELSEVIER.

wound, 315 MHz, quarter-wave antenna, and a matching antenna was connected to a network analyzer to demonstrate its ability to wirelessly measure strain. The device was placed on a panel and measurements were taken as the load on the panel was increased from 0 to 4.0 kg in 1 kg increments at room temperature. The results of these measurements demonstrate that the proposed sensor is suitable for accurate strain measurements.

10.4 Oil and Gas

One of the main problems in the oil and gas industry is the corrosion of infrastructure and equipment which leads to significant economic costs and safety risks [30]. This is particularly true in the exploration, production, processing, and transportation stages of the industry, where corrosion can have serious consequences for the operation and integrity of pipelines, drilling rigs, tanks, and other assets. Therefore, it is extremely important the adoption of sensors which are able to operate in harsh environments in order to monitor the corrosion level of infrastructure and equipment. For this reason, chipless RFID technology is a valid option for the development of sensors for oil and gas industry.

Chipless corrosion sensors are sensors that are used to detect and monitor corrosion in metal structures without the need for physical probes or chips. These sensors use various technologies to detect and measure corrosion, such as electromagnetic waves, acoustic waves, and temperature changes.

Researchers in [31] have studied the use of a passive RFID tag equipped with an interdigitated capacitor (IDT) and an LC resonator to monitor the condition of coatings on pipelines and detect water ingress. The schematic representation of the chipless tag/reader antenna and its equivalent circuit model are reported in Figure 10.7a and b, respectively. The tag has been developed for use on flexible substrates, such as coatings on pipelines. The tag is made using Ultralam 3850, a material with a dielectric constant of 2.9 and a thickness of 100 μm, produced by Rogers Corporation. In practical applications, it is expected that the chipless tag can be printed directly onto coatings using these materials, allowing for easy and cost-effective deployment.

SAW sensors are widely employed in the oil and gas industry [30]. These sensors are usually coated with chemical-sensitive materials that are specific to the target chemical being detected. These materials can include polymers, metal-organic frameworks (MOFs), metals, and metal oxides. In the gaseous phase, Rayleigh waves are often used in combination with SAW devices that have a coating of gas-absorbing or reactive layers. These layers serve as functional sensing layers and enable the detection of gases [32]. The SAW device generates a surface acoustic wave, which is then used to detect changes in the properties of the gas, such as its

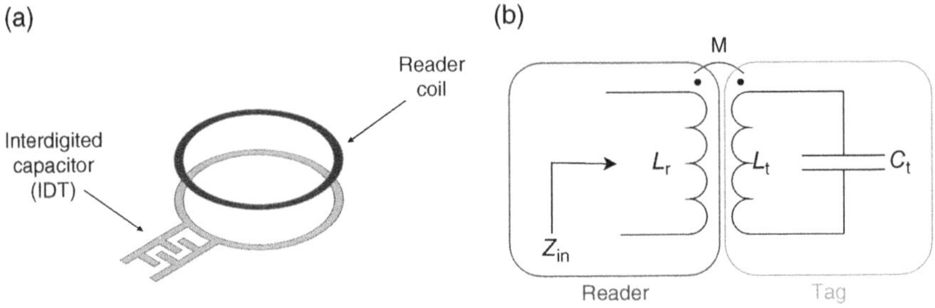

Figure 10.7 (a) Schematic representation of the chipless sensor for corrosion detection and the reader coil, (b) lumped circuit model of the sensor presented in [31]. *Source:* [31]/with permission of ELSEVIER.

concentration or composition. A pictorial representation of a chipless gas sensor based on SAW technology is presented in Figure 10.8.

It has been demonstrated that SAW sensors coated with MOF materials can be used for real-time monitoring of gases that are relevant to the oil and gas industry, such as carbon dioxide (CO_2) and methane (CH_4) [33]. These sensors have been shown to be effective in both wired and wireless operations as shown in Figure 10.9.

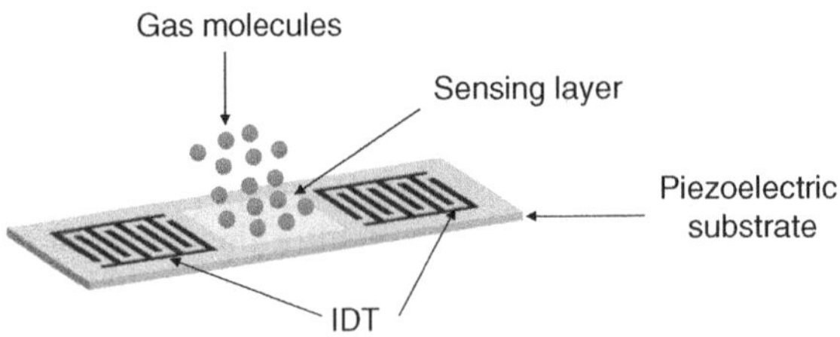

Figure 10.8 A pictorial representation of a chipless gas sensor based on SAW technology.

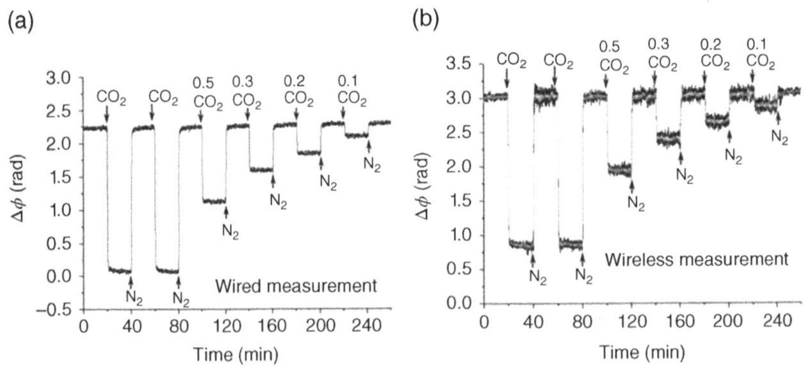

Figure 10.9 Real-time response of a ZIF-8 film-coated SAW sensor measured in wired (a) and wireless (b) mode. *Source:* Greve et al. [33]/with permission of IEEE.

10.5 Automotive

In automotive applications, several parts of the vehicle have to be monitored by using sensors. In some of the scenarios, it is useful to avoid the use of batteries since they are complicated to replace or because an active circuit can be damaged and therefore a wireless sensor is a very good candidate for this task. In this scenario, a typical applicative scenario for passive sensors is the tire monitoring and torque monitoring.

10.5.1 Tyre Monitoring

Battery-less SAW sensors have been proposed for monitoring the deformation during road contact [34, 35]. SAW devices are capable of sensing wirelessly or without physical contact, and they have long delays or high Q values that are difficult to achieve using other technologies. This allows for high-resolution wireless measurement and also helps to separate the sensor response from the environmental echo. Operating in the UHF range allows for the creation of passive transponders with small and inexpensive antennas. These devices are compact, lightweight, and low cost, and they are not sensitive to magnetic fields and tolerant of variations in the distance between the sensor and the reader antenna. In addition, SAW devices' high sensitivity to strain makes them suitable for use on standard or slightly modified automotive parts. There are two main types of wireless SAW sensors: ones based on reflective delay lines (RDL) and ones based on one-port resonators. In the 1990s, research groups at Siemens and the University of Technology in Vienna made significant contributions to demonstrating the capabilities of SAW RDL for automotive applications [36, 37]. One important advantage of RDL sensors over resonator sensors is the ability to build a multi-sensor collision-free system using TDMA, CDMA, and other orthogonal data transmission protocols [38]. On the other hand, resonators have a larger Q factor that potentially ensures a higher resolution of the wireless measurement system [34].

In Ref. [39], a single SAW die with three resonators (i.e. f_p = 434.04 MHz, f_{t1} = 433.88 MHz, f_{t2} = 433.45 MHz) connected in parallel are placed on two ledges formed on a base surface. Two resonators are used as a reference and one resonator is employed to measure the pressure. By measuring $F_p = f_p - f_{t1}$ and $F_t = f_{t1} - f_{t2}$ both pressure and temperature are measured simultaneously. In Ref. [40], a prototype of a tire pressure sensor unit with an accuracy of approximately 15 kPa within a pressure range of 100–400 kPa has been demonstrated. In Ref. [41], a SAW transponder was developed for passive wireless monitoring using a touch-mode pressure sensor. These kinds of SAW sensors have long-range readings (about 1 m) and can tolerate longer delays but deal with a weak signal. Therefore, a reader unit with a large dynamic range of over 60 dB and a very good Tx/Rx isolation of better than 100 dB is required. Tire pressure and temperature monitoring systems (TPMS) have been developed for both passenger cars and trucks in Ref. [34]. Various designs of the valve TPMS sensors are shown in Figure 10.10a together with the reader installed in a car boot Figure 10.10b [39].

Since resonant SAW sensors do not have an identification function, they must be combined with a semiconductor RFID tag that holds calibration information in order to be used in a TPMS. This combination has been implemented in a commercially available TPMS system produced by Stack Ltd. in small quantities, primarily for the racing car market. A TPMS for trucks was also developed, which used the same SAW sensing element but embedded it into a rubber patch that also contained an RFID tag. The patch was attached to the sidewall of a truck tire. Rather than using an on-board reader, the system included handheld readers for manual inspection of truck tires and a drive-by reader for automatic measurement

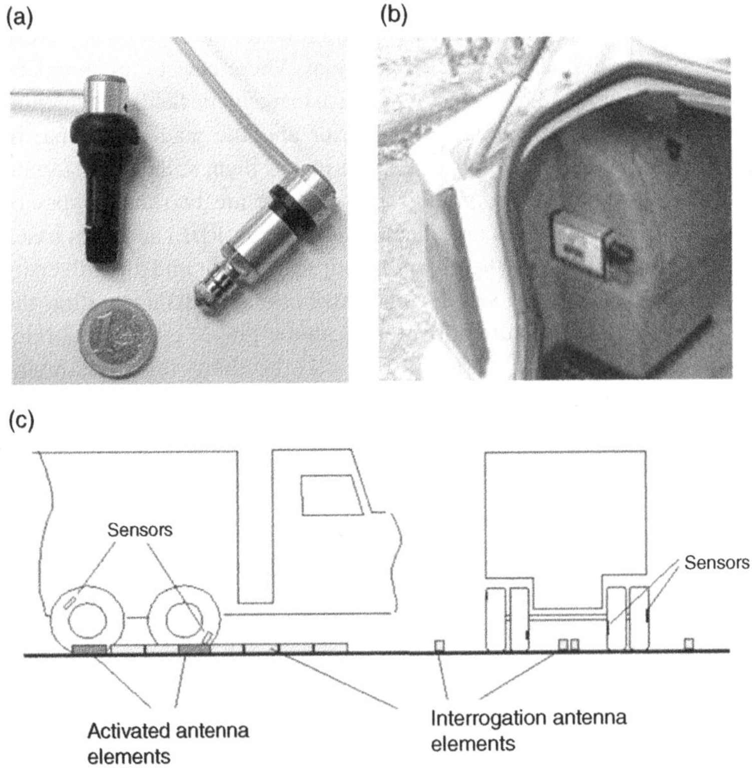

Figure 10.10 (a) Valve TPMS sensors (b) reader installed in a car boot. (c) Drive-by reader for the truck SAW TPMS. *Source:* Kalinin [34]/IEEE.

of pressure and temperature in each wheel of an 18-wheel truck, as shown in Figure 10.10c. All these sensors, however, are made of high-stiffness materials. Flexible materials or other techniques which do not require the installation of rigid components may be more robust in time.

LC resonators can be employed to monitor the state of the tyres in terms of pressure and temperature. In Ref. [42], C pressure sensitive capacitor is used in parallel with a temperature sensitive inductor and together they make a LC tank circuit. The scheme of the proposed sensor and its predicted behavior is shown in Figure 10.11. Changing the applied pressure affects the resonant frequency of the circuit while the temperature affects the bandwidth and amplitude of the impedance at this frequency. The external readout circuit is coupled with the sensor inductor and the impedance of the sensor is reflected to the external circuit by induced electromagnetic field.

Strain monitoring of tires in automobiles is also a useful parameter for improving the reliability of tires. Tire deformation or strain monitoring enables one to know the amount of friction between the tires and road surface, which can then be used for the optimization of automobile tire control systems, such as the

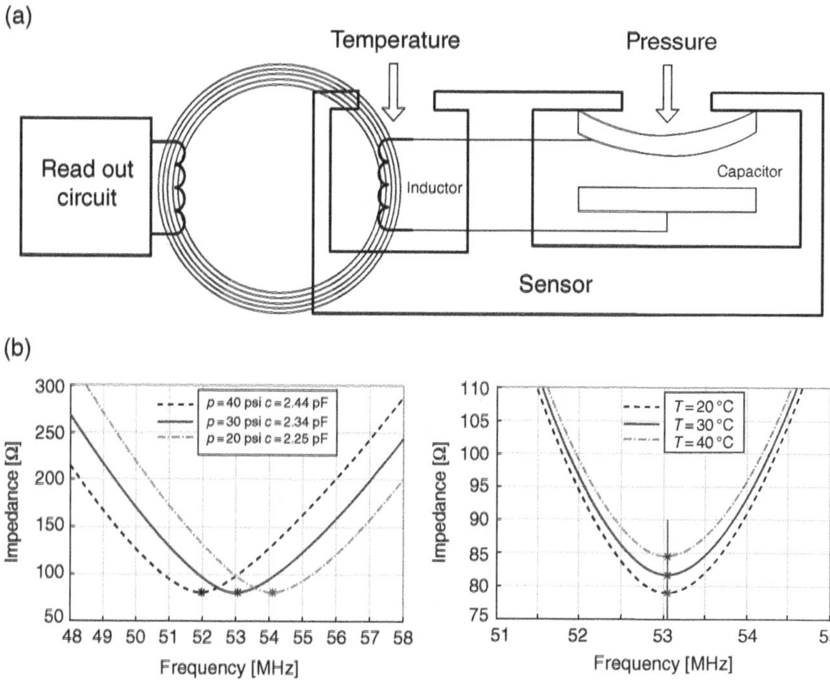

Figure 10.11 (a) The working principle of the LC pressure and temperature sensor. (b) Predicted sensor response as a function of the pressure change and temperature change. *Source:* Nabipoor and Majlis [42]/with permission of IOP Publishing.

ABS [43]. For direct strain monitoring, a strain gauge, based on polyimide film, is the best-known and most widely used method. The sensor, however, has a very high degree of stiffness and low elongation compared with tire rubber. This large difference in stiffness may disturb the deformation and stress of tires and may also cause the debonding of sensors and rubber over a long period of usage.

Tyres comprise rubber with carbon black, steel wire, and organic fiber. Steel wire belts are inserted under the tread rubber. The steel wire itself is an electrically conductive material and the rubber is a dielectric material. By exploiting these conductive wires it is possible to engineer a passive strain sensor [44]. A tyre deformation changes spacing between the steel wires, which consequently changes the capacitance. Measurement of that tyre capacitance change indicates the tyre strain or deformation. The interrogation system is based on a tuning resonant LC circuit inserted in the wheel. The transmitter emits radio waves of white noise which is picked up by the tuning circuit antenna. The tuning circuit comprising the inductance of a coil, the capacitance of a tyre specimen, and a resistance creates an LC parallel resonant circuit. The capacitance change determines a deviation from the tuning circuit resonance. The frequency of the radio wave at the receiver is calculated by means of fast Fourier transformation (FFT). Since the antenna at the external receiver can receive the signal from the sensor as well as the direct signal from the transmitter, an electromagnetic shield is needed between the white noise area and the sensor output area. Since the actual tire structure acts as a sensor, no additional sensor is required. Therefore, there is no debonding of the sensor, even during prolonged service, because there is no stiffness difference between the sensor and tire rubber.

10.5.2 Torque Monitoring

In many new vehicle designs, hydraulic power-assisted steering is being replaced by electric power steering (EPS), which requires sensors to detect the angular position of the steering wheel and the resulting torque. Rotary torque and position sensors that use sliding electrical contacts can be problematic due to vibration. An alternative solution is to use non-contact torque sensors [45]. Non-contact position and torque sensing can be implemented by using Hall-effect technology or SAW technology. Hall-effect sensors are commonly used in EPS systems to measure the angular position of the steering wheel and the resulting torque applied to the steering column. These sensors work by detecting the presence of a magnetic field and producing a corresponding output voltage. They are non-contact sensors, which means that they do not wear out or suffer from degradation due to physical contact or friction. Hall-effect sensors are durable, reliable, and accurate, and they are resistant to vibration and temperature changes [46]. In a SAW sensor, when a torque is applied, it causes the piezoelectric material to deform, producing a corresponding electrical signal.

The magnitude of the signal is proportional to the applied torque. SAW sensors are highly sensitive and can measure torque with high accuracy and resolution. They are also resistant to wear and degradation, and they can operate over a wide temperature range. A typical torque measurement range for passenger cars is ±10 Nm. The required minimum torque update rate is 1 ms, and the accuracy is 1% of the full range within a temperature range of −40 to +125 °C. This performance can be readily achieved by a SAW sensor. For short-range applications such as torque sensors, the distance between the stator and rotor can be less than 1 cm, allowing for close coupling. Short-range applications are characterized by a strong input signal, small receiver dynamic range around 20–30 dB, and ease of separation of the SAW response from the CW interrogation signal, but typically they require fast interrogation with the update period Tup < 1 ms [34]. Figure 10.12a shows a prototype of an SAW EPAS sensor that has been developed at Transense Technologies [47].

Powertrain sensors are also used in vehicles to monitor and control various aspects of the powertrain, such as engine speed, transmission gear ratio, and wheel speed. A powertrain SAW torque sensor is a flexplate torque transducer for automatic transmissions [48] (Figure 10.12b). The flexplate is a component that

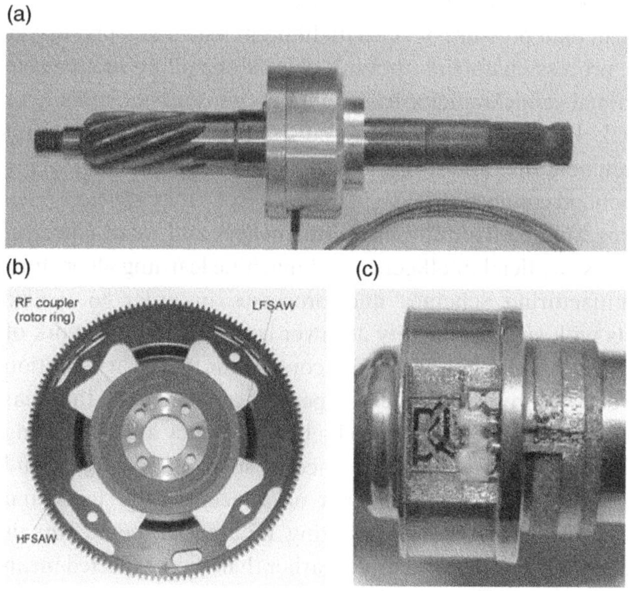

Figure 10.12 (a) SAW EPAS torque sensor installed on a steering shaft with a pinion. *Source:* Transense [47]/https://www.transense.com/sawsense/saw-technology/last accessed December 06, 2023. (b) Flexplate torque transducer for V6 engine with two SAW sensing elements, LFSAW and HFSAW. *Source:* Kalinin et al. [48]/IEEE. (c) Assembled torque sensor for F1 KERS. *Source:* Kalinin et al. [49]/IEEE.

connects the crankshaft to a torque converter, so the transducer allows for measurement of torque directly at the output of the internal combustion engine. The torque measurement range for the flexplate transducers varies from ±400 to ±1 000 Nm depending on the engine size. The flexplate transducer differs from the EPAS sensor in the required torque update rate [34]. Another example of a driveline torque sensor that operated in a very harsh environment was the sensor developed for a Formula 1 kinetic energy recovery system (KERS) [49]. A picture of the KERS sensor is shown in Figure 10.12c.

10.6 Sensors for Industrial Tools Monitoring

In the last years, the research on wireless sensor technology has undergone a tremendous attention from both researchers and industry in order to seek technical solutions capable to improve efficiency and product quality as well as reduce costs [50]. Among the different applicative scenarios, the harsh and hazardous environments represent a key technology driving the technological revolution of Industry 4.0 in the framework of smart factories [51]. Indeed, the presence of severe conditions comprising a combination of intense vibrations, high temperatures, dust, high pressure or acids in different machine tools such as continuous furnaces, process chambers, rotating objects or milling tools, make the usage of conventional semiconductor-based wireless sensors not suitable or highly inefficient [52]. In fact, electronics based on semiconductors cannot operate at such high temperatures and strong vibrations turn out to be challenging for the durability of bonded connections.

Smart manufacturing aims to fruitfully integrate cheaper and more powerful sensors, big data analysis, artificial intelligence and machine learning algorithms into traditional manufacturing schemes and processes in order to realize customizable products with superior quality at lower costs in the framework of Industrial Internet of Things (IIoT) [53, 54]. In this context, the tool wear condition monitoring (TCM) process represents a crucial aspect since, in general, the wear condition of industrial tools directly affects production efficiency and processing quality [50]. Indeed, the production of workpieces, such as aerospace and automobile parts, poses stringent requirements on the employed tools and machines. In general, with the aim of preventing failure that could impair workpieces, industrial tools are exchanged much earlier than their expected mean life. This procedure is clearly inefficient and costly and has an important weight on the final processing costs [51]. Therefore, for a more efficient usage of industrial tools throughout their lifetime as well as to ensure high-quality production, an effective predictive maintenance system during the working process is extremely desirable for achieving an early detection and diagnosis of critical components faults. For instance, a sudden temperature increase at the tip of a tool, probably due to the additional friction with the workpiece, is a common sign of tool failure that could be advantageously exploited to predict and detect tool failure.

Owing to the harsh and severe conditions inside the tools the exploitation of direct methods-based condition monitoring such as laser, optical, or contact sensors is hinder by the highly cluttered environments and reflections as well as limited line-of-sight (LOS) conditions. Therefore, wireless sensors integrated in the tools capable of providing continuous monitoring information would be advantageous.

In this framework, passive chipless wireless microwave sensors represent a key technology capable to overcome or drastically reduce the several drawbacks of electronic-based sensors within the harsh and severe environments. Indeed, although the presence of an IC in a RFID tag guarantees interesting features (in terms of large data storage and reprogrammable capability), it also presents some constraints on the item price and operating environmental conditions. The terms passive and chipless mean the absence of any batteries and integrated circuits for the sensing and identification procedure [55]. To date, the designed chipless RFID sensing systems for harsh environments have been successfully applied to temperature [28, 56–58], pressure [52, 59, 60] and strain monitoring [61, 62].

SAWs [63] and Dielectric Resonators (DRs) represent the most promising wireless passive sensing technologies for industrial tools monitoring. In general, they fruitfully combine the radar systems, for radiating the investigated environment and gathering the scattered one, and the passive tags for sensing and identification purpose.

In a microwave SAW-based sensor system, an electromagnetic wave is emitted from a radar unit and received by antenna of the SAW sensors. The sensing principle is based on the piezoelectric effect converting back the acoustic energy into an electromagnetic pulse train detected by a radar receiver. Since the speed of a SAW is much lower than radio signals, it can be easily separated from environment echo in the receiver unit. In general, the piezoelectric material losses and the interdigitated transducers degradation represent the most common causes that limit the maximum allowable temperature. In Ref. [64], a wireless SAW temperature measurement application for high-speed high-voltage motors is presented. A Frequency Modulated Continuous Wave (FMCW) radar operating at ISM band (2.4–2.5 GHz) is employed as a reader. The basic working principle of the temperature sensor is represented by the time delay as a function of the temperature of the SAW impulse response. As example, the signal level for three different temperature levels is shown in Figure 10.13. The peak time delay enhances with temperature increases. In addition to the temporal shift the amplitude signal undergoes a higher and higher degradation with the temperature.

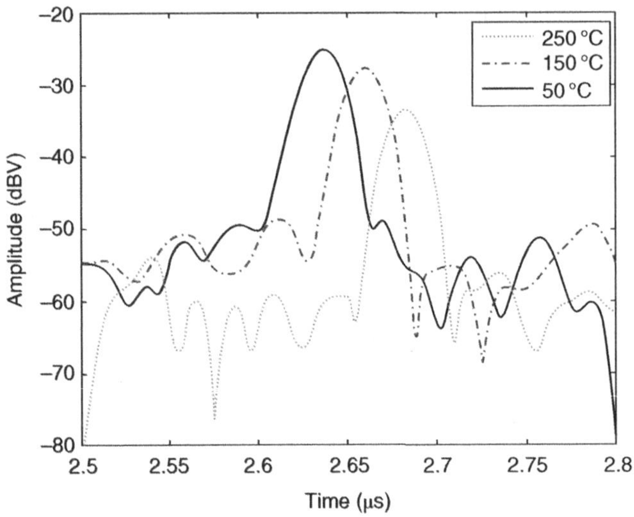

Figure 10.13 Signal level of a SAW-based temperature sensor at different temperatures. *Source:* Binder and Fachberger [64]/with permission of IEEE.

The developed sensor is characterized by a read range of 0.5 m and a temperature range from −40 to 250 °C.

In case of a DR-based sensor systems, the impinging electromagnetic wave is reflected at the resonator resonance frequency. Therefore, the DR acts as a cooperative target for a FMCW radar interrogation. A detailed overview concerning passive temperature sensors based on DRs for machine tools is reported in [51]. More in detail, the basic working principle consists of detecting the resonant frequency of the tag as a peak in the RF spectrum, corresponding to the tag's radar cross section (RCS). Then, the sensing capability is encoded through the detection of the resonance frequency shift due to the changing of some physical parameters. For instance, a wireless DR-based temperature sensor designed for the monitoring of machine tools is reported in [56] and depicted in Figure 10.14.

The DR is designed to have a resonance within the ISM band (2.4–2.5 GHz). The sensors have an outer and an inner diameter of 30 and 10 mm respectively, and a height of 15 mm. The reading procedure involves the use of a VNA connected to the horn antenna that radiates inside a kiln as depicted in Figure 10.15a. The received signal as a function of the frequency and time, with and without the time gating procedure, is shown in Figure 10.15b. It highlights that, by isolating the tag's response, the clutter is considerably reduced and so a more robust readout is achieved. Indeed, a peak around 2.68 GHz is well visible after performing the time gating procedure, namely a multiplication of the received time domain signal with a window with a certain time length.

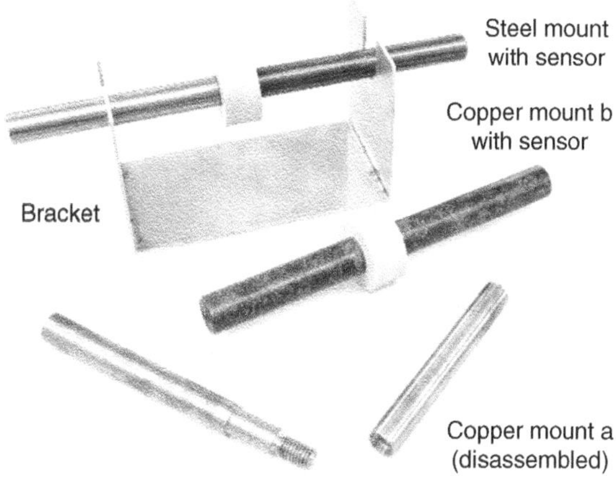

Figure 10.14 DR chipless temperature sensor mounted to copper and steel shanks for machine tools. *Source:* Mandel et al. [56]. With permission of EurAAP.

(a)

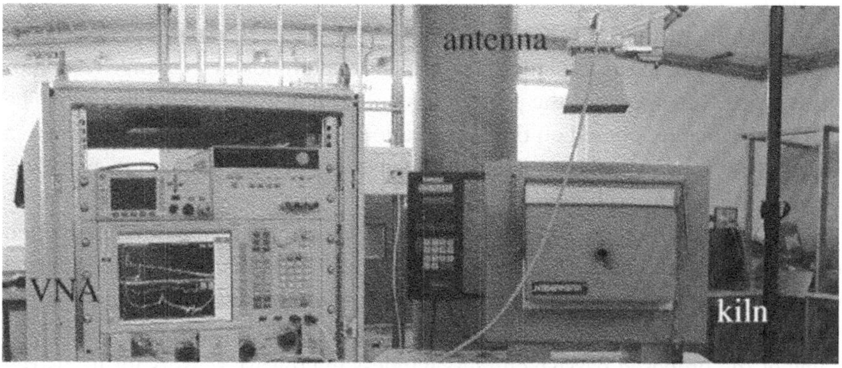

(b)

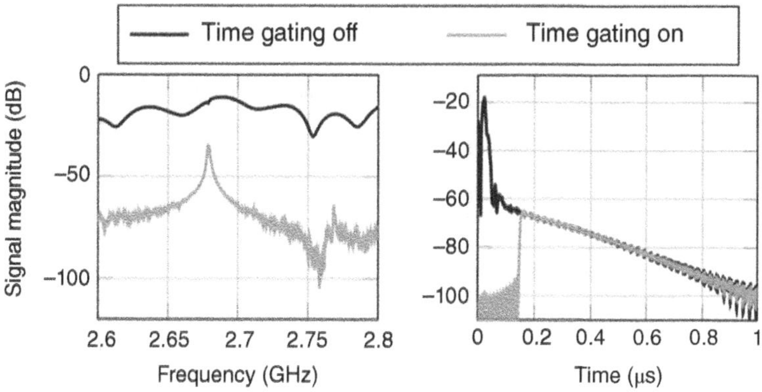

Figure 10.15 (a) Measurement setup for temperature sensor characterization and (b) frequency and time domain of the received signal. *Source:* Mandel et al. [56]. With permission of EurAAP.

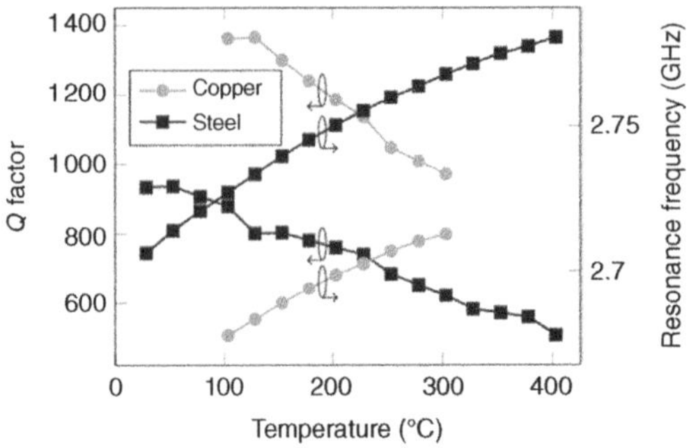

Figure 10.16 Q-factor and resonance frequency as a function of the temperature. *Source:* Mandel et al. [56]/with permission of IEEE.

In addition, Figure 10.16 emphasizes the sensor assessment in terms of both DR Q-factor and resonance frequency as a function of the temperature. As it can be drawn, the resonance frequency tends to enhance with the increasing of the temperature whereas, the Q-factor trend is the opposite.

Another interesting example concerning the DR-based sensor is reported in [59] where a pressure sensor exploiting the near-field alteration of a ceramic resonator is proposed. The sketch of the proposed sensor is illustrated in Figure 10.17. The working principle is based on the perturbation of the DR resonance mode with a metal membrane near the resonator whose deflection leads to a pressure-dependent resonance frequency.

Figure 10.18 shows the comparison between simulated and measured reflection coefficient magnitude (S_{11}) in dB as a function of the frequency by considering different distances of the metallic membrane. More in detail, the greater is the pressure the higher is the DR resonance frequency. Finally, the performance of some of the proposed passive chipless wireless microwave sensors for industrial tools monitoring is summarized in Table 10.1.

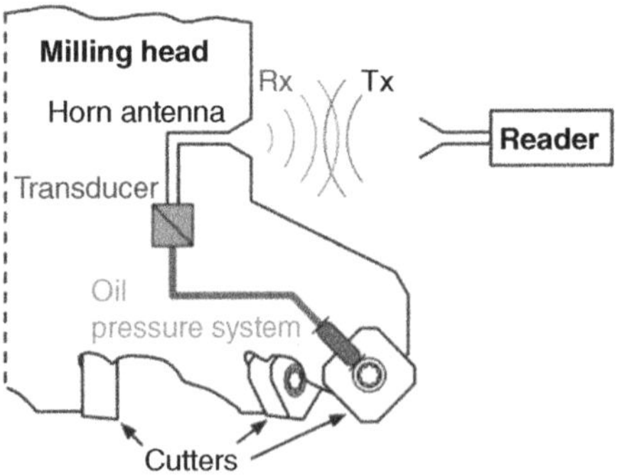

Figure 10.17 Sketch of the DR-based passive chipless pressure sensor. *Source:* Schumacher et al. [59]/with permission of Rolf Jacoby.

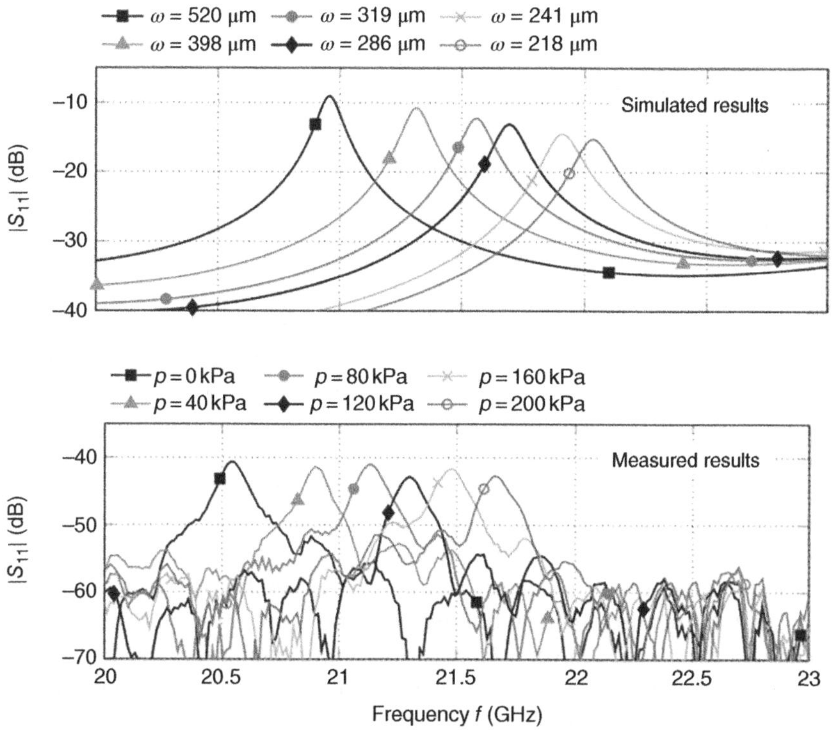

Figure 10.18 Comparison between simulated and measured magnitude of reflection coefficient (S_{11}) for different distances between DR and the metallic membrane. *Source:* Schumacher et al. [59]/with permission of Rolf Jacoby.

Table 10.1 Chipless RFID sensors for industrial tools monitoring.

Reference	Application	Technology	Sensing	Read range (m)	Frequency (GHz)	Range of sensing
Binder and Fachberger [64]	High-speed motors	SAW	Temperature	0.5	2.4	−40 to 250 °C
Fachberger and Werner [58]	Rotating rollers	SAW	Temperature	0.75	2.4	40—60 °C
Sáez et al. [57]	Machine Tools	DR	Temperature	—	2.7	100—300 °C
Mandel et al. [56]	Tool monitoring	DR	Temperature	—	2.4	Up to 400 °C
Schuster et al. [60]	Milling tools	DR	Pressure	0.5	18	Up to 200 kPa
Schumacher et al. [59]	Milling tools	DR	Pressure	1.5	20–23	Up to 200 kPa

10.7 Conclusion

In this chapter, a classification of the different paradigms exploited for chipless RFID sensors operating in harsh environment was presented. Differences with respect to IC-equipped solutions have been summarized and the advantages of a wireless sensor have been analyzed. A detailed survey of the implementation of these sensors in different application realms has been offered as well spanning from space to industrial tool monitoring. Particular attention has been devoted to individuate the materials adopted for the realization of sensors that must withstand extreme conditions related to high temperatures, corrosion hazards and mechanical stresses.

Future efforts in this domain have to be made for searching new materials that can undergo hostile environment without compromising operativity and to increase the life service. In this sense, the characterization of the behavior of materials in the microwave frequency regime can boost the realization of wireless miniaturized sensor tags. Another challenge that must be faced is the realization of readers that can collect data from larger and larger wireless sensor nodes that are required by new Internet of Things (IoT) developments and the integration with 5G and beyond 5G networks. More emphasis is also likely to be given to solutions improving autonomous guiding vehicles and the monitoring of hybrid future networks such as the Space-Air-Ground Integrated Networks (SAGINT).

References

1 Costa, F., Genovesi, S., Borgese, M. et al. (2021). A review of RFID sensors, the new frontier of internet of things. *Sensors* 21 (9): https://doi.org/10.3390/s21093138.

2 da Cunha, M.P. (2013). Wireless sensing in hostile environments. *2013 IEEE International Ultrasonics Symposium (IUS)*, 1337–1346. https://doi.org/10.1109/ULTSYM.2013.0342.

3 Fahrner, W.R., Job, R., and Werner, M. (2001). Sensors and smart electronics in harsh environment applications. *Microsystem Technologies* 7 (4): 138–144. https://doi.org/10.1007/s005420100089.

4 Pohl, A. (2000). A review of wireless SAW sensors. *IEEE Transactions on Ultrasonics, Ferroelectrics, and Frequency Control* 47 (2): 317–332. https://doi.org/10.1109/58.827416.

5 Wilson, W.C., Malocha, D.C., Kozlovski, N. et al. (2009). Orthogonal frequency coded SAW sensors for aerospace SHM applications. *IEEE Sensors Journal* 9 (11): 1546–1556. https://doi.org/10.1109/JSEN.2009.2027403.

6 Canabal, A., Davulis, P.M., Dudzik, E., and da Cunha, M.P. (2009). CDMA and FSCW surface acoustic wave temperature sensors for wireless operation at high

temperatures. *2009 IEEE International Ultrasonics Symposium*, 807–810. https://doi.org/10.1109/ULTSYM.2009.5441748.

7 Aubert, T., Kokanyan, N., and Elmazria, O. (2021). Langasite as piezoelectric substrate for sensors in harsh environments: investigation of surface degradation under high-temperature air atmosphere. *Sensors* 21 (17): https://doi.org/10.3390/s21175978.

8 Bhar, I. and Mandal, N. (2022). A review on advanced wireless passive temperature sensors. *Measurement* 187: 110255. https://doi.org/10.1016/j.measurement.2021.110255.

9 Huang, Q.-A., Dong, L., and Wang, L.-F. (2016). LC passive wireless sensors toward a wireless sensing platform: status, prospects, and challenges. *Journal of Microelectromechanical Systems* 25 (5): 822–841. https://doi.org/10.1109/JMEMS.2016.2602298.

10 Varadharajan Idhaiam, K.S., Caswell, J.A., Pozo, P.D. et al. (2022). All-ceramic passive wireless temperature sensor realized by tin-doped indium oxide (ITO) electrodes for harsh environment applications. *Sensors* 22 (6): 2165. https://doi.org/10.3390/s22062165.

11 Ghamsari, B.G., Abrahams, J., Remillard, S., and Anlage, S.M. (2013). High-temperature superconducting spiral resonator for metamaterial applications. *IEEE Transactions on Applied Superconductivity* 23 (3): 1500304–1500304. https://doi.org/10.1109/TASC.2012.2232343.

12 Ji, Y., Tan, Q., Wang, H. et al. (2019). A novel surface LC wireless passive temperature sensor applied in ultra-high temperature measurement. *IEEE Sensors Journal* 19 (1): 105–112. https://doi.org/10.1109/JSEN.2018.2872915.

13 Boccard, J., Aftab, T., Hoppe, J. et al. (2014). Far-field passive temperature sensing up to 700°C using a dielectric resonator. *2014 IEEE International Conference on Wireless for Space and Extreme Environments (WiSEE)*, October 2014. 1–5. https://doi.org/10.1109/WiSEE.2014.6973077.

14 Jimenez-Saez, A., Schumacher, P., Hauser, K. et al. (2019). Chipless wireless high temperature sensing based on a multilayer dielectric resonator. *2019 IEEE SENSORS*, Montreal, QC, Canada, October 2019, 1–4. https://doi.org/10.1109/SENSORS43011.2019.8956863.

15 Wilson, W.C., Perey, D.F., Atkinson, G.M., and Barclay, R.O. (2008). Passive wireless SAW sensors for IVHM. *2008 IEEE International Frequency Control Symposium*, May 2008, 273–277. https://doi.org/10.1109/FREQ.2008.4623003.

16 Mc Gee, K., Anandarajah, P., and Collins, D. (2022). Use of chipless RFID as a passive, printable sensor technology for aerospace strain and temperature monitoring. *Sensors* 22 (22): 8681. https://doi.org/10.3390/s22228681.

17 Thomes, Jr., W.J., Chuska, R.F., Ott, M.N. et al. (2008). Fiber optic cable thermal preparation to ensure stable operation. *Presented at the Optical Engineering +*

Applications, San Diego, California, USA, August 2008, 70700B. https://doi.org/10.1117/12.796977.

18 Rodríguez-Madrid, J.G., Iriarte, G.F., Williams, O.A., and Calle, F. (2013). High precision pressure sensors based on SAW devices in the GHz range. *Sensors and Actuators A: Physical* 189: 364–369. https://doi.org/10.1016/j.sna.2012.09.012.

19 Jiao, Z., Jiang, L., Sun, J. et al. (2019). Outgassing environment of spacecraft: an overview. *IOP Conference Series: Materials Science and Engineering* 611 (1): 012071.

20 de Groh, K.K., Banks, B.A., Miller, S.K.R., and Dever, J.A. (2018). Degradation of spacecraft materials, Chapter 28. In: *Handbook of Environmental Degradation of Materials*, 3e (ed. M. Kutz), 601–645. William Andrew Publishing https://doi.org/10.1016/B978-0-323-52472-8.00029-0.

21 De Rooij, A. (1989). The oxidation of silver by atomic oxygen. *ESA Journal* 13 (4): 363–382.

22 Wilson, W.C. and Atkinson, G.M. (2013). Wireless sensor applications in extreme aeronautical environments. *IEEE International Conference on Wireless for Space and Extreme Environments*, Baltimore, MD, USA, November 2013, 1–6. https://doi.org/10.1109/WiSEE.2013.6737554.

23 Trang, H., Patrice, R., Marie-Helene, V. et al. (2007). Temperature-compensated structure for saw pressure sensor in very high temperature. *2007 IEEE International Frequency Control Symposium Joint with the 21st European Frequency and Time Forum*, May 2007, 40–44. https://doi.org/10.1109/FREQ.2007.4319026.

24 Puccio, D., Malocha, D.C., Saldanha, N., and Pereira Da Cunha, M. (2007). SAW parameters on Y-cut langasite structured materials. *IEEE Transactions on Ultrasonics, Ferroelectrics, and Frequency Control* 54 (9): 1873–1881. https://doi.org/10.1109/TUFFC.2007.471.

25 Bardong, J., Schulz, M., Schmitt, M. et al. (2008). Precise measurements of BAW and SAW properties of Langasite in the temperature range from 25°C to 1000°C. *2008 IEEE International Frequency Control Symposium*, May 2008, 326–331. https://doi.org/10.1109/FREQ.2008.4623013.

26 Hamidon, M.N., Skarda, V., White, N.M. et al. (2006). High-temperature 434 MHz surface acoustic wave devices based on $GaPO_4$. *IEEE Transactions on Ultrasonics, Ferroelectrics, and Frequency Control* 53 (12): 2465–2470. https://doi.org/10.1109/TUFFC.2006.194.

27 Gamba, P., Goldoni, E., Savazzi, P. et al. (2013). Wireless passive sensors for remote sensing of temperature on aerospace platforms. *IEEE International Conference on Wireless for Space and Extreme Environments*, Baltimore, MD, USA, November 2013, 1–6. https://doi.org/10.1109/WiSEE.2013.6737567.

28 Kubina, B., Schüßler, M., Mandel, C. et al. (2013). Wireless high-temperature sensing with a chipless tag based on a dielectric resonator antenna. *2013 IEEE SENSORS*, November 2013, 1–4. https://doi.org/10.1109/ICSENS.2013.6688181.

29 Wilson, W.C. and Juarez, P.D. (2014). Emerging needs for pervasive passive wireless sensor networks on aerospace vehicles. *Procedia Computer Science* 37: 101–108. https://doi.org/10.1016/j.procs.2014.08.018.

30 Wright, R.F., Lu, P., Devkota, J. et al. (2019). Corrosion sensors for structural health monitoring of oil and natural gas infrastructure: a review. *Sensors* 19 (18): 3964. https://doi.org/10.3390/s19183964.

31 M. H. Zarifi, S. Deif, and M. Daneshmand, 'Wireless passive RFID sensor for pipeline integrity monitoring', *Sensors and Actuators A: Physical*, vol. 261, pp. 24–29, Jul. 2017, doi: https://doi.org/10.1016/j.sna.2017.04.006.

32 Devkota, J., Ohodnicki, P.R., and Greve, D.W. (2017). SAW sensors for chemical vapors and gases. *Sensors* 17 (4): https://doi.org/10.3390/s17040801.

33 Greve, D.W., Devkota, J., and Ohodnicki, P. (2018). Wireless CO_2 SAW sensors with a nanoporous ZIF-8 sensing layer. *2018 IEEE International Ultrasonics Symposium (IUS)*, October 2018, 1–4. https://doi.org/10.1109/ULTSYM.2018.8580119.

34 Kalinin, V. (2011). Wireless physical SAW sensors for automotive applications. *2011 IEEE International Ultrasonics Symposium*, October 2011, 212–221. https://doi.org/10.1109/ULTSYM.2011.0053.

35 Pohl, A., Steindl, R., and Reindl, L. (1999). The "intelligent tire" utilizing passive SAW sensors measurement of tire friction. *IEEE Transactions on Instrumentation and Measurement* 48 (6): 1041–1046. https://doi.org/10.1109/19.816111.

36 Reindl, L., Scholl, G., Ostertag, T. et al. (1998). Theory and application of passive SAW radio transponders as sensors. *IEEE Transactions on Ultrasonics, Ferroelectrics, and Frequency Control* 45 (5): 1281–1292. https://doi.org/10.1109/58.726455.

37 Pohl, A. and Seifert, F. (1997). Wirelessly interrogable surface acoustic wave sensors for vehicular applications. *IEEE Transactions on Instrumentation and Measurement* 46 (4): 1031–1038. https://doi.org/10.1109/19.650822.

38 Kuypers, J.H. (2006). Passive 2.45 GHz TDMA based multi-sensor wireless temperature monitoring system: results and considerations. *Proceedings of IEEE Ultrasonics Symposium, Vancouver, 2006.*

39 Dixon, B., Kalinin, V., Beckley, J., and Lohr, R. (2006). A second generation in-car tire pressure monitoring system based on wireless passive SAW sensors. *2006 IEEE International Frequency Control Symposium and Exposition*, June 2006. 374–380. https://doi.org/10.1109/FREQ.2006.275414.

40 Schimetta, G., Dollinger, F., and Weigel, R. (2000). A wireless pressure-measurement system using a SAW hybrid sensor. *IEEE Transactions on Microwave Theory and Techniques* 48 (12): 2730–2735. https://doi.org/10.1109/22.899037.

41 Oh, J.-G., Choi, B., and Lee, S.-Y. (2008). SAW based passive sensor with passive signal conditioning using MEMS A/D converter. *Sensors and Actuators A: Physical* 141 (2): 631–639. https://doi.org/10.1016/j.sna.2007.09.012.

42 Nabipoor, M. and Majlis, B.Y. (2006). A new passive telemetry LC pressure and temperature sensor optimized for TPMS. *Journal of Physics Conference Series* 34 (1): 770. https://doi.org/10.1088/1742-6596/34/1/127.

43 Matsuzaki, R. and Todoroki, A. (2008). Wireless monitoring of automobile tires for intelligent tires. *Sensors* 8 (12): https://doi.org/10.3390/s8128123.

44 Matsuzaki, R. and Todoroki, A. (2005). Passive wireless strain monitoring of tyres using capacitance and tuning frequency changes. *Smart Materials and Structures* 14 (4): 561. https://doi.org/10.1088/0964-1726/14/4/014.

45 Swee, T.H. (2017). Contactless torque and position sensing paves the road to success for electric power steering. *Fierce Electronics*. https://www.fierceelectronics.com/components/contactless-torque-and-position-sensing-paves-road-to-success-for-electric-power (accessed 30 December 2022).

46 Nazars, A. (2002). Hall effect torque sensor. US20020079156A1, June 27, 2002. https://patents.google.com/patent/US20020079156A1/en (accessed 30 December 2022).

47 V. Kalinin, 'Wireless physical SAW sensors for automotive applications', in 2011 IEEE International Ultrasonics Symposium, Orlando, FL, USA: IEEE, Oct. 2011, pp. 212–221. doi: https://doi.org/10.1109/ULTSYM.2011.0053.

48 Kalinin, V., Lohr, R., Leigh, A., and Bown, G. (2007). Application of passive SAW resonant sensors to contactless measurement of the output engine torque in passenger cars. *2007 IEEE International Frequency Control Symposium Joint with the 21st European Frequency and Time Forum*, May 2007, 499–504. https://doi.org/10.1109/FREQ.2007.4319124.

49 Kalinin, V., Lohr, R., Leigh, A. et al. (2010). High-speed high dynamic range resonant SAW torque sensor for kinetic energy recovery system. *EFTF-2010 24th European Frequency and Time Forum*, April 2010, 1–8. https://doi.org/10.1109/EFTF.2010.6533643.

50 Zhou, C., Guo, K., and Sun, J. (2021). An integrated wireless vibration sensing tool holder for milling tool condition monitoring with singularity analysis. *Measurement* 174: 109038. https://doi.org/10.1016/j.measurement.2021.109038.

51 Jiménez-Sáez, A. (2022). *Towards THz Chipless High-Q Cooperative Radar Targets for Identification, Sensing, and Ranging*. Cham: Springer International Publishing https://doi.org/10.1007/978-3-031-04976-7.

52 Fricke, S., Friedberger, A., Seidel, H., and Schmid, U. (2012). A robust pressure sensor for harsh environmental applications. *Sensors and Actuators A: Physical* 184: 16–21. https://doi.org/10.1016/j.sna.2012.05.043.

53 Mehrpouya, M., Dehghanghadikolaei, A., Fotovvati, B. et al. (2019). The potential of additive manufacturing in the smart factory industrial 4.0: a review. *Applied Sciences* 9 (18): 3865. https://doi.org/10.3390/app9183865.

54 Wu, D., Jennings, C., Terpenny, J. et al. (2017). A comparative study on machine learning algorithms for smart manufacturing: tool wear prediction using random forests. *Journal of Manufacturing Science and Engineering* 139 (7): 071018. https://doi.org/10.1115/1.4036350.

55 Genovesi, S., Costa, F., Borgese, M. et al. (2016). Enhanced chipless RFID tags for sensors design. *2016 IEEE International Symposium on Antennas and Propagation (APSURSI)*, June 2016, 1275–1276. https://doi.org/10.1109/APS.2016.7696345.

56 Mandel, C., Jiménez-Sáez, A., Polat, E. et al. (2017). Dielectric ring resonators as chipless temperature sensors for wireless machine tool monitoring. *2017 11th European Conference on Antennas and Propagation (EUCAP)*, Paris, France, March 2017, 3912–3916. https://doi.org/10.23919/EuCAP.2017.7928618.

57 Sáez, A.J., Polat, E., Mandel, C. et al. (2016). Chipless wireless temperature sensor for machine tools based on a dielectric ring resonator. *Procedia Engineering* 168: 1231–1236. https://doi.org/10.1016/j.proeng.2016.11.428.

58 Fachberger, R. and Werner, C. (2013). Wireless temperature monitoring in an electrolytic galvanizing plant. *2013 IEEE International Ultrasonics Symposium (IUS)*, Prague, Czech Republic, July 2013, 251–254. https://doi.org/10.1109/ULTSYM.2013.0065.

59 Schumacher, P., Schuster, C., Jiménez-Sáez, A. et al. (2018). Passive chipless wireless pressure sensor for Harsh and reflective environments. *2018 11th German Microwave Conference (GeMiC)*, March 2018, 227–230. https://doi.org/10.23919/GEMIC.2018.8335071.

60 Schuster, C., Schumacher, P., Schüßler, M. et al. (2017). Passive chipless wireless pressure sensor based on dielectric resonators. *2017 IEEE SENSORS*, October 2017, 1–3. https://doi.org/10.1109/ICSENS.2017.8234269.

61 Rodini, S., Genovesi, S., Manara, G., and Costa, F. (2022). Contactless waveguide characterization of piezoresistive materials for wireless strain sensors. *Sensors* 22 (11): 4085. https://doi.org/10.3390/s22114085.

62 Thai, T.T., Aubert, H., Pons, P. et al. (2011). A newly developed radio frequency wireless passive highly sensitive strain transducer. *2011 IEEE SENSORS*, 211–214. 10.1109/ICSENS.2011.6127239.

63 Scholl, G., Schmidt, F., Ostertag, T. et al. (1998). Wireless passive SAW sensor systems for industrial and domestic applications. *Proceedings of the 1998 IEEE International Frequency Control Symposium (Cat. No.98CH36165)*, Pasadena, CA, USA, 595–601. https://doi.org/10.1109/FREQ.1998.717961.

64 Binder, A. and Fachberger, R. (2011). Wireless SAW temperature sensor system for high-speed high-voltage motors. *IEEE Sensors Journal* 11 (4): 966–970. https://doi.org/10.1109/JSEN.2010.2076803.

Section 3

Systems

11

Energy-Autonomous Wireless Architectures for Predictive Maintenance in Harsh Closed Applications

Alessandra Costanzo, Diego Masotti, Francesca Benassi, and Giacomo Paolini

DEI "Guglielmo Marconi", University of Bologna, Bologna, Italy

11.1 Introduction and State of the Art

In modern architectures, wireless sensor network (WSN) technologies have been applied with multiple goals, among which the predictive maintenance has played an important role in minimizing machinery failures, regardless of the application scenarios [1, 2].

These spaces are generally made of metal parts, creating an environment that is hostile to the electromagnetic fields, challenging researchers to find suitable solutions to be able to safely operate even when the hosting environment is considered "harsh."

Such technologies, implemented within industrial internet of things (IIoT) applications, enable seamless connectivity between components that need constant monitoring, acquiring real-time data on parameters such as position, speed, temperature, or pressure [3].

In this scenario, energy autonomy is indeed one of the key points to be addressed to perform an accurate design of device that boasts miniaturization and lightness, thanks to the possibility to remove bulky batteries, extending this way the device's lifetime and cycles and avoiding periodic battery replacement, especially when the device to be monitored is located in areas that are not easy to be accessed [4–7]. To achieve this purpose, the power needs to be provided to the sensor node wirelessly either directly through wireless power transfer (WPT) techniques, or it must be collected by the surrounding environment, namely exploiting the so-called energy

Wireless Identification and Sensing Systems for Harsh and Severe Environments,
First Edition. Edited by Smail Tedjini and Valentina Palazzi.

Published 2025 by John Wiley & Sons, Inc.

harvesting (EH); in that sense, wide interest has been shared on the exploitation of radiofrequency (RF) waves, which can be a valid option to avoid the need of batteries and cables.

When implementing a system whose energy autonomy relies on EH [8–10], the choice of the operative frequency is a mandatory and crucial design step, also related to the main constrain that electromagnetic (EM) EH experiences, namely high availability but low-power level to be exploited [11]. Near-field solutions, operating mainly in the low MHz range, allow to transfer higher power at the expense of having the transmitter and receiver in close proximity to each other and in line-of-sight (LOS) conditions [12], whereas far-field EH solutions, operating in the GHz range, suffer from lower received power levels but enable longer transmitting distance, even in non-LOS conditions, when operating in electromagnetically harsh environments [13, 14].

To ensure robustness with respect to the environment, a detailed design of the sensor must be achieved, starting from the antenna that needs to be electrically decoupled from the surrounding metal parts while guaranteeing at the same time high radiation performance and stability. Nevertheless, the metalized surrounding environment can also be exploited in favor of an enhancement of the received power due to multiple reflections that the EM waves are subjected to when traveling in such spaces. Given the variability of those reflections, EH and WPT solutions to be adopted in metalized environments must be designed accurately by means of full-wave simulations of the entire structure to be able to predict how the scattering effects influence the power received by each sensor node [15].

Furthermore, both academia and enterprises have raised interest on the concepts of wireless information and power transfer (WIPT) [16], and simultaneous wireless information and power transfer (SWIPT) [17], paradigms that can find applications both in the industrial or consumer sectors. This approach results to be a robust and suitable solution to be applied for predictive maintenance purposes, when sensor nodes are attached to machinery critical parts and they are able to be charged wirelessly while, or before, sending all the information related to the monitored components.

Figure 11.1 shows a representation of the entirely unique system created by WPT and WSN for a particular automobile application example. The ability to combine electromagnetic far-field WPT capabilities with low-power sensing and communication components that are designed to operate in the same frequency band is a key selling point. With the ability to store a certain amount of energy, this enables the simultaneous wireless transfer of power and information, allowing for the full-cycle operation of a microcontroller unit, a real-time clock, a three-axis accelerometer, but most importantly bidirectional communication with a Long Range (LoRa) transceiver working at 2.4 GHz.

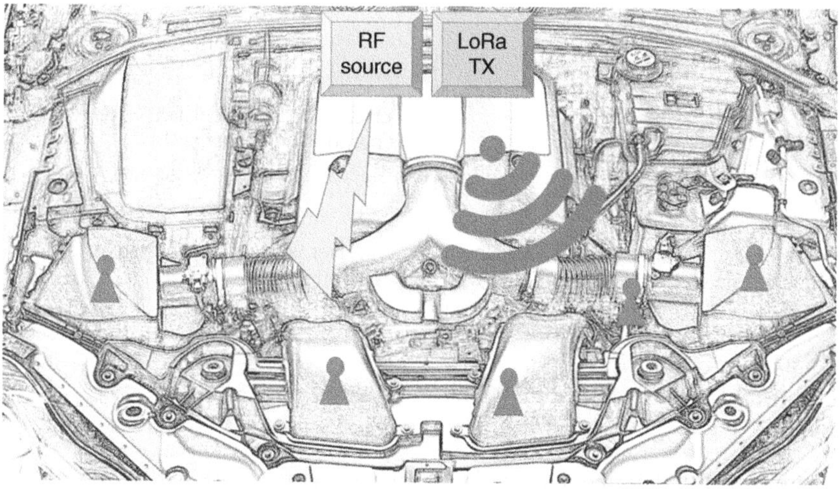

Figure 11.1 Illustration of the ideal scenario for a WSN system powered by WPT technology in a challenging automotive environment. The power supply (RF source), the communication hub (LoRa TX), and the sensor nodes are all depicted in one of the available locations. *Source:* Giacomo Paolini et al. [18]/MDPI/CC BY 4.0.

This chapter describes the operative steps for the design of wireless systems able to operate in electromagnetically harsh indoor spaces, giving an overview of the constraints and challenges that need to be tackled to achieve thorough solutions that can be applied in different branches of the IIoT sector.

11.2 Transmitter/Receiver Link Analysis and Illuminators Best Positioning Simulations

Regarding the link analysis between transmitter and receiver, the simplest solution consists of a wireless device equipped with a unique antenna that receives one interfering monochromatic RF signal (at frequency f_{RF}) generated by an uninformed source of volume V_i and current density J_i, placed in the far-field zone: the incoming field (E_i) can be assumed as a uniform plane wave, around the wireless receiving device.

The receiving antenna can be considered a linear active system and can be represented through the Norton equivalent circuit (see Figure 11.2): the single-port antenna is described as a dispersive admittance (Y_A), as a result of the corresponding EM analysis. Thus, the whole receiving antenna can be described by the complex equivalent current generator $J_{eq}(f_{RF})$; the following formula reveals the equivalent current generator thanks to the reciprocity theorem:

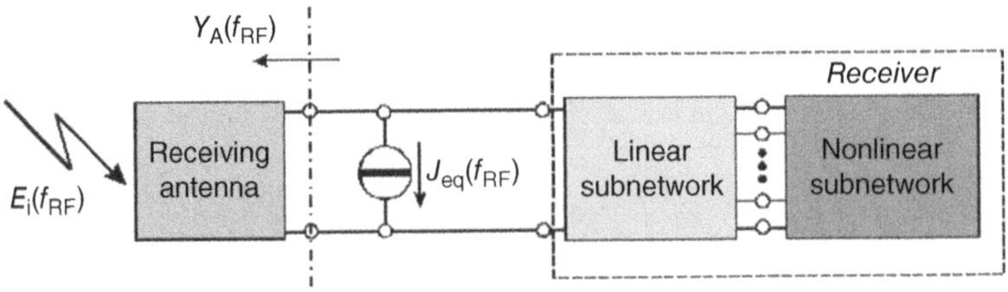

Figure 11.2 Block diagram of a single-antenna receiver with incidence field at frequency f_{RF}. *Source:* Monti et al. [19]/with permission from IEEE.

$$J_{eq}\left(f_{RF}\right)=\frac{1+R_0Y_A\left(f_{RF}\right)}{U}\int_{V_i} J_i\left(P_i\right)\bullet E_R\left(P_i\right)dV_i \tag{11.1}$$

where E_R is the field radiated by the receiving antenna working in transmitting mode and fed by a source characterized by sinusoidal voltage at frequency f_{RF}, an electro-motive force U, and internal impedance R_0. Despite E_R can be easily evinced by means of a full-wave EM simulation, the use of (11.1) results quite demanding, mainly due to the fact that the current density of the RF source is hard to estimate. From the EM theory, we can simplify (11.1) with the following formulation, as long as the hypothesis of uniform plane wave still remains valid [20]:

$$J_{eq}\left(f_{RF}\right)\approx j\left[1+R_0Y_A\left(f_{RF}\right)\right]\frac{2\lambda_{RF}re^{j\beta r}}{\eta}\frac{E_i\bullet E_R\left(\theta_L,\varphi_L\right)}{U} \tag{11.2}$$

where E_R (θ_L, φ_L) represents the field radiated by the receiving antenna in the link direction (θ_L, φ_L) estimated in the far-field zone at a distance r from the phase center of the RF source antenna, and η and β are the free-space wave impedance and the phase constant, respectively.

The mutual locations of the transmitting and the receiving devices are of great importance in terms of the evaluation of phase and amplitude errors, which could be significant in case of two antennas or coils that are not in the far-field region of each other, or the link direction does not correspond to the direction of maximum radiation (e.g. in wearable devices for body area networks [BAN] or EH applications). These are some cases where the formulation of Norton equivalent generator is absolutely needed, but at the same time cannot be calculated by means of (11.2), especially in cases of low-power RF energy levels, e.g. electromagnetic compatibility (EMC)/electromagnetic interference (EMI) problems [19], or EH from environmental sources [21]. Therefore, the current density J_{eq} can be alternatively achieved through (11.3) with the aid of the field equivalence theorem replacing the 3D current density J_i on V_i with the equivalent surface electric (J_{surf}) and magnetic currents (M_{surf}) due to the electromagnetic field (E_i, H_i) generated by J_i, and evaluated on the surface S of an infinite plane placed in between the two (transmitting and receiving) antennas [22]:

$$\begin{aligned} J_{eq}\left(f_{RF}\right) &= \frac{1+R_0Y_A\left(f_{RF}\right)}{U}\int_S \left(J_{surf}\left(P_S\right)\bullet E_R\left(P_S\right) - M_{surf}\left(P_S\right)\bullet H_R\left(P_S\right)\right)dS \\ &= \frac{1+R_0Y_A\left(f_{RF}\right)}{U}\hat{n}\bullet\int_S \left[E_i\left(P_S\right)\times H_R\left(P_S\right) - E_R\left(P_S\right)\times H_i\left(P_S\right)\right]dS \end{aligned} \tag{11.3}$$

where the symbols are the same of (11.1) (also for the magnetic field), P_S is a generic point and $\hat{n}$ is the unit vector orthogonal to the surface S (Figure 11.3).

The rigorous formulation of J_{eq} and its evaluation can be achieved thanks to two full-wave simulations, one for each antenna: in this way, it is possible to estimate the EM fields of (11.3) on a surface S placed in among the two antennas. The amplitude of J_{eq} given by (11.2) and (11.3) can differ also by 50% in worst cases, especially when low RF power are involved and situations where distance and/or orientation of the two antennas do not allow to consider valid the plane wave approximation [22], thus altering drastically the RF-to-dc efficiency.

In case of multi-port antenna at the receiver side (here, R number of ports), the situation turns out to be more complicated; also, in this case, we are talking about RF EH systems with low-power density of energy involved, or unwanted signals coming from other nearby devices (EMI and EMC cases) [19, 23]. The system represented in Figure 11.4 describes these possible situations from a schematic point of view: Y_A is now an $R\times R$ matrix that considers all the electromagnetic couplings between the different ports of the multi-port receiving antenna.

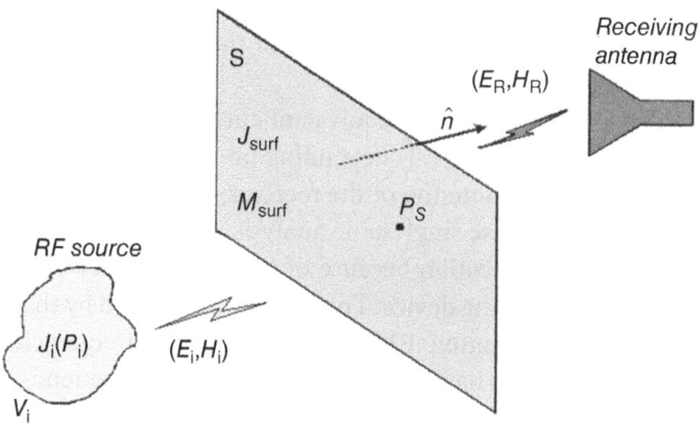

Figure 11.3 Schematic for the rigorous calculation of (11.3) for both far-field and near-field scenarios. *Source:* Monti et al. [19]/with permission from IEEE.

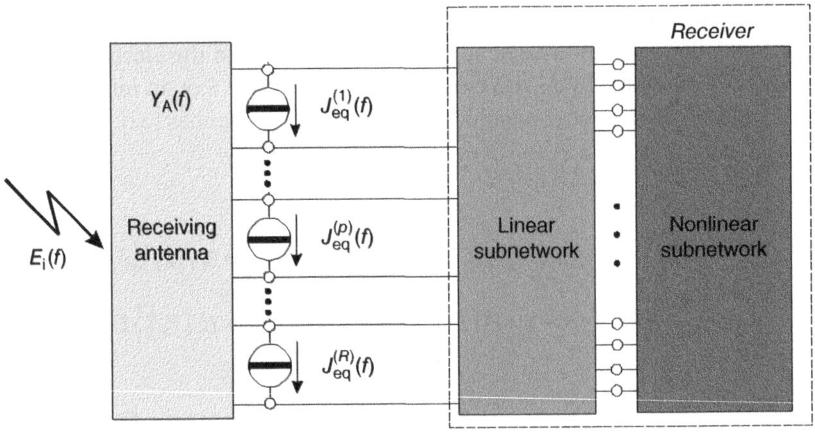

Figure 11.4 Scheme of a multi-port antenna receiver with incidence field at a frequency *f*. *Source:* Monti et al. [19]/with permission from IEEE.

For this specific case, referring to the *p*th port of the multi-port antenna, the expression of the equivalent current generator turns out to be:

$$J_{\mathrm{eq}}^{(p)}(f) = \frac{1 + R_0 Y_{pp}(f)}{U} \int_{V_i} J_i(P_i) \bullet E_R^{(p)}(P_i)\, dV_i \tag{11.4}$$

where Y_{pp} is the admittance and $E_{\mathrm{R}}^{(p)}$ the radiated field of the considered *p*th port in substitution of the admittance (Y_{A}) and the radiated field (E_{R}) of the single-port antenna of (11.1). In this situation, the effects of the trans-admittances Y_{pm} $(m \neq p)$ on the antenna's *p*th port (automatically included in the Y_{pp} value) have to be taken into account: in the particular cases of close placement ($<\lambda/10$) of the antennas in the multi-port solution, this led to important consequences (e.g. multi-rectenna with differently polarized close antennas, or EMI situations where the ports of each nonlinear device of the receiving system can be coupled with ports of other devices [20]).

The formulation of the equivalent current generator (11.4) can be obtained by means of (11.2) or (11.3) depending on the reciprocal position between the RF source and each antenna of the receiver; the harmonic balance (HB) simulations are also in this case single-tone analysis, but they could need not negligibly times and memory allocation because of the high number of nonlinear devices of the multi-port receiving device. This is also emphasized by the simultaneous presence of multiple incoming EM fields at different frequencies ($f_{\mathrm{RF},k}$): for this reason, (11.2)–(11.4) have to be calculated for each frequency and the corresponding HB analysis becomes a multitone one.

With the aim of selecting the positions where to place the RF power sources inside the engine compartment, a simplified version of a car engine environment under evaluation has been electromagnetically simulated by means of the abovementioned method that exploits the reciprocity theorem in order to estimate the power received by the nodes.

This is performed thanks to the integral equation (IE) solver [24] of the EM simulation software Computer Simulation Technology (CST) Studio Suite, allowing to decrease the simulation problem size, thus leading to faster simulations; but at the same time, accurate results in more complex scenarios.

A schematic picture of the electromagnetically harsh environment to be computed is represented in Figure 11.5, i.e. the 1248 cc engine compartment of a generic hatchback (dimensions: $62 \times 54 \times 60\,\text{cm}^3$) inside a metallic closure (in light red) measuring $76 \times 143 \times 95\,\text{cm}^3$. Different engine parts, such as the battery, air filter, and metal holders for the motor body, are also shown in Figure 11.5.

The custom-method based on IE solver that is used to evaluate the received power at the nodes exploiting the reciprocity theorem but without considering the hypothesis of far-field, which is not verified in this case. First, the positions of the illuminators and the sensor nodes are decided; then, a virtual surface Σ (for this case, the surface is placed horizontally between the RF sources and the engine parts represented in Figure 11.5) is generated and placed between the transmitting and receiving elements. From the numerical evaluation of the EM fields generated by the transmitting illuminator (E_I, H_I) and by the receiving tag (E_T, H_T) on a point $P \in \Sigma$, the expression of the Norton equivalent current at the load of the receiving tag load (R_0) can be calculated as:

$$J_{eq} = \frac{1 + R_0 Y_T(\omega)}{U} \hat{n} \bullet \iint_\Sigma \left[E_I(P) \times H_T(P) - E_T(P) \times H_I(P) \right] d\Sigma \tag{11.5}$$

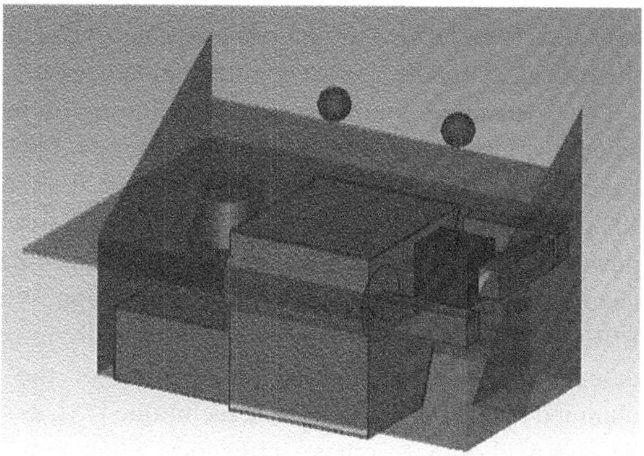

Figure 11.5 Representation of the simulated electromagnetically harsh scenario under evaluation, i.e. the engine compartment of a car. RF power sources are represented as spheres, and the surface Σ is place horizontally between the RF sources and the engine compartment's items. *Source:* Paolini et al. [15]/ with permission from IEEE.

where Y_T is the receiving tag antenna input admittance, $\hat{n}$ is the unit vector orthogonal to the Σ surface and pointing at the tags side, and U is a sinusoidal voltage source whose internal impedance is R_0.

This method brings to the evaluation of the current produced in each receiving node, without taking care of the polarization and the incoming direction of the fields; given the current evaluated with (11.5), the available power at each receiver can be straightforwardly calculated as:

$$P_T(\omega) = \frac{|J_{eq}|^2}{8\mathrm{Re}(Y_T(\omega))} \tag{11.6}$$

11.3 Design and Realization of the 2.45 GHz RF Power Source

The device dedicated to the RF powering of the wireless sensor nodes is designed merging both off-the-shelf components and customized radiating elements. To guarantee an output power of −3 dBm in the 2.4 GHz industrial, scientific, and medical (ISM) band, a voltage-controlled oscillator (VCO), here Maxim MAX2750, has been used also due to the chosen operating frequency band that allows designing reduced-dimension antennas both on the transmitter and receiver side of the system, while maintaining high efficiency when transferring an effective isotropic radiated power (EIRP) of maximum 27 dBm (maximum allowed band: 8 MHz). Using a VCO allows for dynamically selecting the illuminator's operating frequency to perform frequency division between the powering and the communication operations. The VCO tuning voltage input is defined by a voltage control element (Linear Technology LT6650), that is set to have the oscillator signal at 2.45 GHz and allows a voltage sweep between 0.8 and 1.4 V. This operative expedient is performed to guarantee a sufficient distance in the spectrum between the EH signal and the LoRa communication one (at 2.401 GHz).

The other main components of the illuminator are the RF amplifiers, dedicated to an uncoherent feeding of the antenna. This operative choice allows to have three separated energy sources instead of an antenna array. The selected RF amplifiers are the Skyworks SE2598L which provides a 23 dBm transmitted power to each circularly polarized (CP) transmitting antenna, whose gain has been measured to be 4 dBi in broadside direction. With these values, the overall EIRP turns out to be 27 dBm as expected, fully in compliance with the up-to-date regulations. A circular polarization of the transmitting antennas is obtained by the corner-trimming technique [25], as shown in Figure 11.6a, and has been chosen to ensure efficiency when powering sensor nodes whose

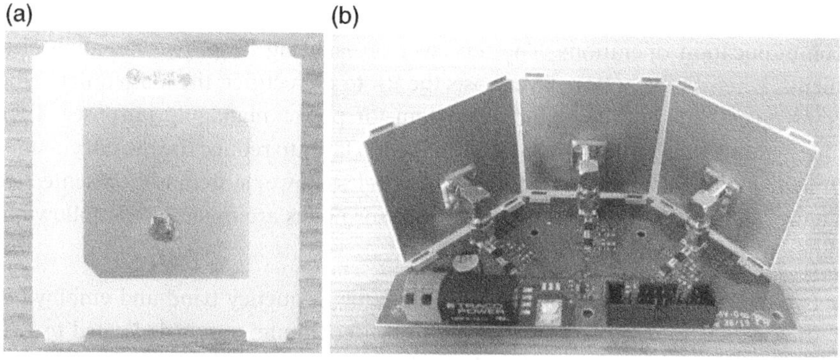

Figure 11.6 (a) Picture of one of the transmitting patch antennas with trimmed corners for CP; (b) photo of the illuminator with the three CP patch antennas and the board hosting the circuitry. *Source:* Paolini et al. [15]/IEEE.

location, and thus the orientation with respect to the illuminator, which is not known a priori.

The patch antenna represented in Figure 11.6a is realized on a Rogers substrate, the RO4350B (thickness: 1.524 mm, $\varepsilon_r = 3.48$, $\tan(\delta) = 0.0037$), whose characteristics are exploited to enhance the antenna radiation performance. Each antenna of the illuminator is coaxially fed using a right-angle connector and hosted on a printed circuit board (PCB) derived on a different Rogers substrate, the RO4360G2 (thickness: 0.610 mm, with $\varepsilon_r = 6.4$, $\tan(\delta) = 0.0038$), denser and highly efficient in terms of loss reduction. This operative choice allows not only to diminish unwanted wasted power-on the circuitry path, but also to reduce the overall dimensions of the microstrip lines inside the PCB. The final illuminator, hosting the three patch antennas located in a semi-circular arrangement, is shown in Figure 11.6b. The designed illuminator has total dimensions of $13 \times 6 \times 5.54\,\text{cm}^3$ which easily allows its installation in locations where not such extended spaces are present, such as car compartment or industrial machinery.

11.4 Design of Low-Power Wireless Battery-Less Sensor Nodes

The full-wave design and characterization of the whole tag have been conceived with the twofold goal of miniaturizing the tag encumbrance while performing electrical decoupling from the surrounding metallic environment. Furthermore, the design expedient and optimization are led by the ambition of obtaining energy autonomy while performing communication operation at ultra-low-power levels.

The proposed tags embed two antennas, one dedicated to WPT and the other to communication operations. The top layer of each tag hosts two radiating elements, whereas the bottom layer hosts the RF-to-dc rectifier, the radio unit (LoRa transceiver), and the baseband subsystem for power managing purposes. Two energy storage capacitors are located on the top layer to reduce the overall encumbrance. A block diagram of the presented wireless power system is represented in Figure 11.7a; in particular, the wireless sensor nodes are made by the following different subsystems:

- two coplanar antennas operating in the same frequency band and employing the previously discussed frequency-diverse technique; one is dedicated to EH (the custom-designed linearly polarized patch) and the other to communication (the Johanson Technology 2450AT42E0100), see Figure 11.7b.
- A voltage-doubler rectifier, with the presence of two Skyworks SMS7630-079LF Schottky diodes, to perform signal transduction from RF dc. The output of the rectifier is then connected to a buck-boost dc/dc converter (Texas Instruments bq25570) designed specifically for harvesting energy from low-voltage and low-power sources.
- A power management section, which includes the bq25570, dedicated to the efficient harvesting of the rectifier out power and to the management of the correct power supply to the sensor node.
- A low-power microcontroller unit (MCU), STMicroelectronics STM32L476RE, to control the node peripherals.
- A 2.4 GHz Semtech SX1280 transceiver supporting LoRa communication protocol, feeding an off-the-shelf Johanson Technology 2450AT42E0100 surface mount technology (SMT) mini-chip-antenna specific for on-metal applications.
- An ultra-low-power three-axis accelerometer (STMicroelectronics IIS2DLPC) to perform inertial sensing.
- An ultra-low-power real-time clock (RTC), Abracon AB1805, for duty-cycling the node operation, which involves turning on the node only at specific time intervals.

Figure 11.7b shows the top layer of the proposed wireless sensor node, where a patch antenna dedicated to WPT, and a Johanson mini-chip antenna dedicated to communication are present.

The patch antenna has been specifically designed for the sensor tag, with the main goal of achieving high efficiency while reducing the overall encumbrance at the operating frequency. The size reduction has been performed by means of the shorting pin techniques, involving seven pins equally spaced on the top edge of the radiating part. This operation allows for reducing the overall size of about a 55% of the theoretical one, guaranteeing that the sensor's antenna overall dimensions of $2.50 \times 3.00 \times 0.32\,\text{cm}^3$ are in full agreement with the given project

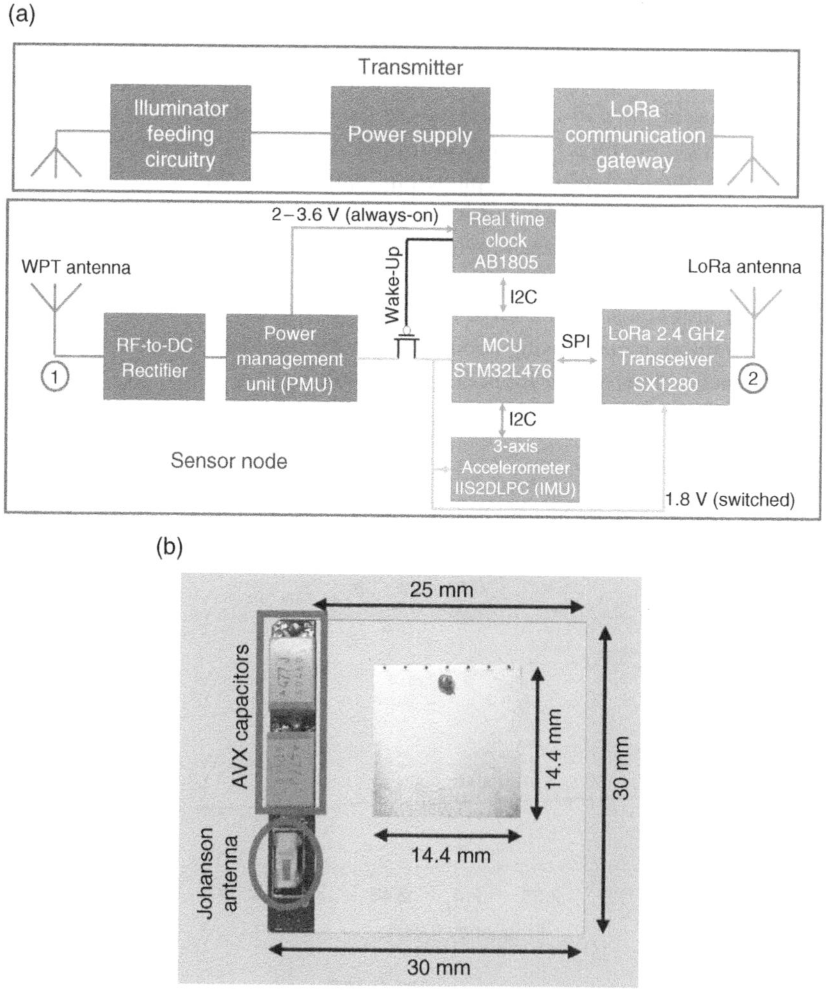

Figure 11.7 (a) Block diagram of the whole system: transmitter subsystem and receiving wireless sensor node; (b) pictorial representation of the top layer of the tag, highlighting the antenna for WPT, the metal-mounted LoRa antenna, and the two storage capacitors. *Source:* Giacomo Paolini et al. [18]/MDPI/CC BY 4.0.

goals. The abovementioned thickness includes two laminates of RO4350B (thickness: 1.524 mm, $\varepsilon_r = 3.48$, $\tan(\delta) = 0.0037$) that are attached together by means of a prepreg FR-4 film (thickness: 0.120 mm, $\varepsilon_r = 4.5$). Despite the achieved reduced dimensions, the antenna boasts a realized gain of 1.84 dBi, a radiation efficiency of 74% and a bandwidth of 35 MHz. Having the two coplanar antennas on the top layer implies that also their mutual position must be optimized by means of full-wave simulations. The obtained results have shown that the location described in Figure 11.7b allows for obtaining the lowest EM coupling possible between the two radiating elements, also because the chip antenna is position parallel to one of the non-radiative sides of the patch antenna. The decoupling between the two antennas has been confirmed by means of experimental validations and the obtained results are plotted in Figure 11.8a, where the decoupling can be efficiently quantified by mean of the transmitting coefficient (S_{21}) of the two-port antennas (see the stack-up of Figure 11.8b), obtaining a value of around −20 dBm.

The WPT antenna is connected to the circuitry hosted on the node's bottom of the energy-autonomous sensing node by means of a plated through-hole to connect the patch to the rectifier input (Figure 11.8b).

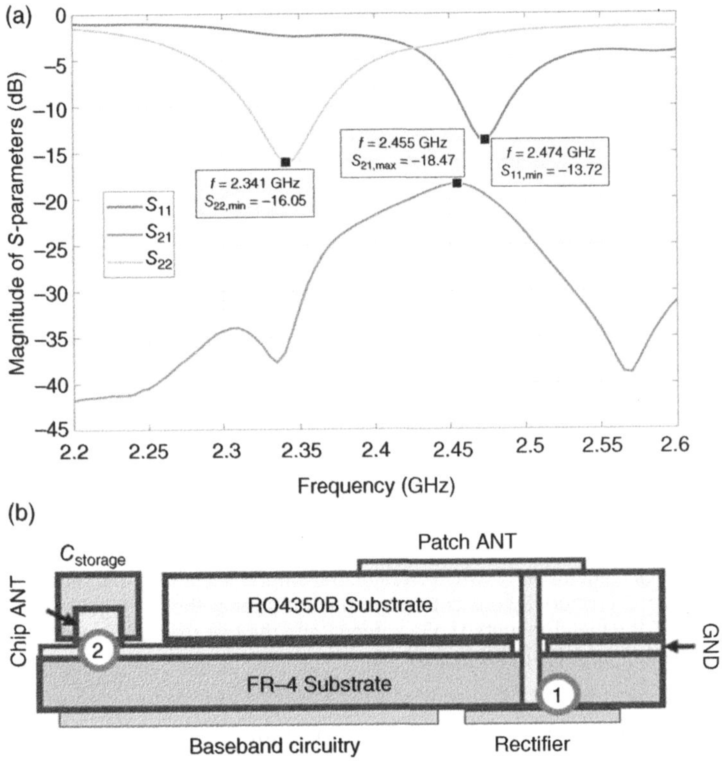

Figure 11.8 (a) Measured scattering parameters of the two-port antenna located on the top layer of the wireless sensor node; (b) stack-up of the sensor node with ports 1 and 2 corresponding to the patch and the metal-mounted chip antennas, respectively. *Source:* Giacomo Paolini et al. [18]/MDPI/CC BY 4.0.

11.5 EH and Power Management: Rectification and WPT Performance Characterization

The operative context in which this system is designed, together with the technological base operative principles of energy-autonomous sensor nodes, implies having a harvested power in the order of microwatts. For this reason, the power management subsystem must be extremely efficient to limit the amount of power that can be wasted in favor of the maximization of the collected energy. To begin, the storage elements must operate with extremely low-leakage currents while storing enough energy to support node operation. Two 470 μF ultra-low-leakage tantalum capacitors (AVX TMJE477K006RCQYA) are selected and added to the node, showing the capability of limiting leakage currents to less than 2 μA, while boasting a package with reduced dimensions ($7.3\times4.4\,mm^2$ footprint; thickness: 4.1 mm). As shown in Figure 11.7b, in order to not exceed extremely the sensor node thickness, the capacitors are allocated coplanar to the antenna dedicated to the WPT.

The harvesting circuitry, a voltage-doubler rectifier is implemented using two Skyworks SMS7630-079LF Schottky diodes, and the linear subnetwork, dedicated to impedance matching between the antenna and the rectifier, is realized by means of one 0.3 pF capacitor and a short-circuited stub (width: 0.51 mm; length: 5.70 mm), used for minimizing losses with respect to lumped components. The overall RF-to-dc power conversion efficiency (PCE) [26] is measured, resulting in a value of 25% for an input power level of −10 dBm and of 34% when the input power level is −5 dBm.

As aforementioned, the rectifier output is connected to a buck-boost dc/dc converter (Texas Instruments bq25570), which has been properly modeled to be able to harvest energy efficiently when low-voltage and low-power sources, such as 100 mV and 15 μW respectively. It is known that the best RF-to-dc performance of a rectifier can be achieved when the voltage on the load reaches ~50% of its open-circuit voltage V_{OPEN}, i.e. the voltage at the output of the rectifier in an open-circuit condition [21]. To achieve this, the maximum power point tracker (MPPT), which is integrated inside the bq25570 power management integrated circuit (IC), is programmed to measure the open-circuit input voltage every 16 seconds, to subsequently transfer the maximum available power to the output.

A buck-boost dc/dc converter, the Analog Devices LTC3106, is used to provide power to the sensor node. Its input is directly connected to the storage capacitors through a p-type metal-oxide-semiconductor field-effect transistor (p-MOS) switch controlled by the RTC circuit. This IC allows it to operate with a minimum of 700 mV input voltage, allowing for an available energy of 4.8 mJ provided by the $C_{storage}$ operating between 0.7 and 3.3 V.

After characterizing the component elements separately, as previously described, the entire sensor nodes have been tested to evaluate the charging times and the harvested power levels when a static LOS condition is considered. Real-case environmental conditions have been recreated inside a laboratory context, accounting for the simultaneous presence of multiple RF sources, such as Bluetooth and Wi-Fi, operating at the same working frequency of the WPT system (2.4 GHz band). The effectiveness and robustness in using circular polarization on the transmitting have been tested considering different orientations of the sensor node with respect to the maximum link direction when only air is interposed. Results have confirmed the expectations; a safe wireless power transmission is correctly performed, regardless the mutual orientation between the illuminator and the sensor node.

Figure 11.9 shows the measured results of the charging time obtained for increasing illuminator-node distance from 30 up to 100 cm (in free-space conditions), which, for the frequency of operation, is equivalent to work from almost near-field location up to the far-field condition. In the plots, the measured charging time values are reported and separated to highlight each cycle: power management unit (PMU) cold-start, the charge of the storage capacitance $C_{storage}$ up to 2 V (first power-on), and the complete charge of the capacitor up to 3.3 V.

Furthermore, it is worth noting that, for these sets of measurements conducted in free space, the maximum achievable illuminator-node distance is less than 120 cm; further distances lead to retrieve voltage levels at the bq25570 input that are lower than 600 mV, which is the minimum voltage level needed to start the cold-start operations.

Subsequently, a second measurement campaign has been performed to delineate the MPPT performance and the operating conditions of the bq25570, to also confirm the nonlinear optimization carried out to maximize the rectification efficiency when the optimum loading condition is considered. For this analysis, the RF input power is set to a fixed value equal to −10 dBm.

To conduct this characterization, the open voltage at the rectifier's output is firstly measured; subsequently, the rectifier dc output is connected to a potentiometer to obtain one-half of the open voltage V_{OPEN}. This procedure is needed to simulate the actual operations carried out by the MPPT of the PMU and the load values measured in correspondence of $V_{OPEN}/2$ are reported in Table 11.1.

It is interesting to notice that, for an illuminator-node distance of 80 cm, which corresponds to an input power of −10 dBm, the open voltage is 0.92 V, and the value of the matched rectifier load (R_{LOAD}) is 9.40 kΩ. This computed load value is close to the optimum one (6.44 kΩ) estimated by the optimizer of the circuit simulation software Keysight Advanced Design System (ADS) and measured (8.85 kΩ) for an input power of −10 dBm.

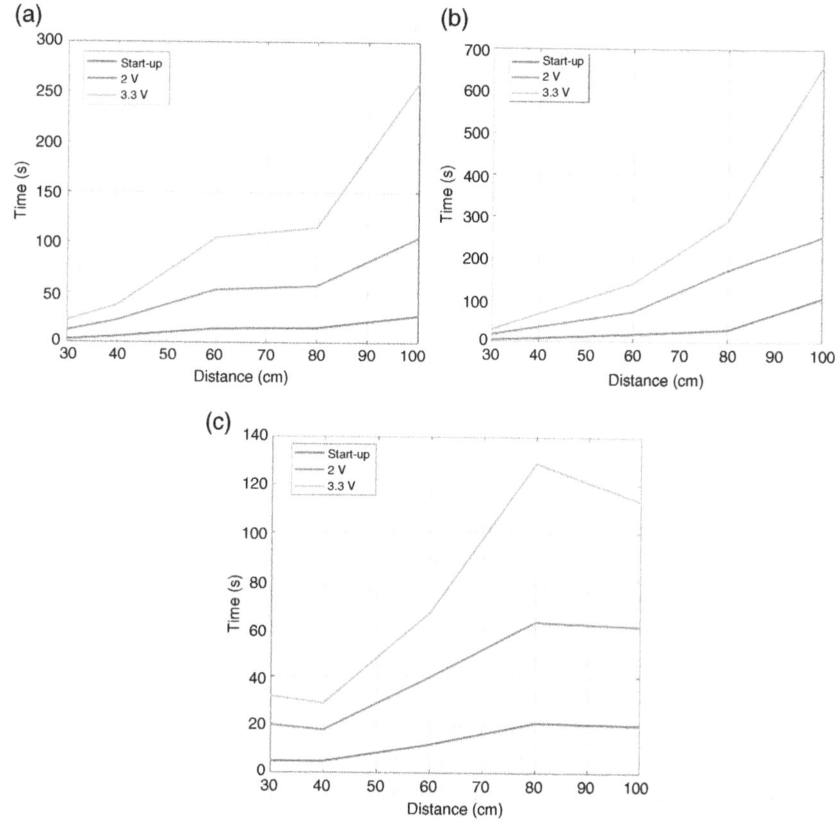

Figure 11.9 Time required for the nodes to exit cold-start and charge the storage capacitance at 2 and 3.3 V. Different separations and angles from the illuminator have been taken into consideration: frontal node with (a) 0°, (b) 30°, and (c) 45° inclination with respect to the vertical axis. *Source:* Giacomo Paolini et al. [18]/MDPI/CC BY 4.0.

Table 11.1 Measurements of the values of the load emulating the bq25570 MPPT operations.

Illuminator-node distance (cm)	V_{OPEN} (V)	R_{LOAD}@$V_{OPEN}/2$ (kΩ)
30	3.24	7.58
40	2.12	10.43
60	1.58	12.38
80	0.92	9.40

Source: Giacomo Paolini et al. [18]/MDPI/CC BY 4.0.

11.6 Case Study: Measurement Campaign in the Engine Compartment of a Car

As first characterization, the energy consumption for the different block elements inside the node has been quantified, in particular taking into account two sampling frequencies of the inertial measurement unit (IMU) data, i.e. 400 and 1 600 samples per second (SPS). The energy required by the node, when operating in active mode, ranges between 1.45 and 1.65 mJ, depending on the accelerometer's configuration, also considering the energy requirements of the other components, in particular the MCU, the LoRa transceiver, and the RTC.

Finally, the entire system has undergone an extensive measurement campaign carried out in a hatchback engine compartment, that resembles quite realistically an electromagnetically harsh environment due to the presence of metal parts and interposed elements that lead to multiple and unpredictable reflections of the EM waves. Figure 11.10 show the measurements setups: the first one (Setup 1), Figure 11.10a, hosts the illuminator upfront whereas the second configuration (Setup 2), Figure 11.10b, has the illuminator located on the left side. For both configurations, six sensor nodes are considered and located throughout the engine compartment in non-LOS positions.

The measurement campaign is conducted in a car engine compartment, as shown in Figure 11.10, where different distance between the illuminator and the sensor nodes are considered, with a maximum achieved one of about 112 cm.

It is worth noticing that all the sensor nodes are surrounded by a quasi-total metallic environment, which generates interfering reflections and implies non-LOS propagation conditions. Furthermore, all the sensors are located in close proximity

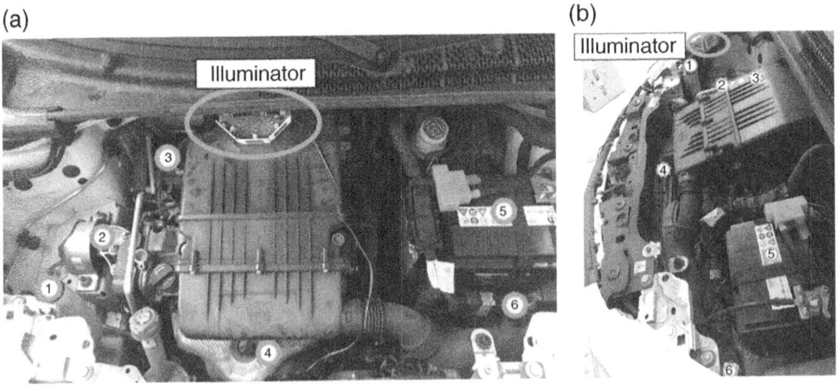

Figure 11.10 Picture of the measurement setups with the illuminator located (a) front and (b) right side. *Source:* Giacomo Paolini et al. [18]/MDPI/CC BY 4.0.

or even in contact with metallic parts of the car engine, with the main purpose of directly monitor some parameters of the compartment itself. For these measurements, the bonnet is kept closed, and the instrumentation is located outside the car to make the measurement setup as close as possible to a real-case scenario.

The charging time of each sensor node has been computed and shown in Figure 11.11 for the two abovementioned configuration setup illustrated in Figure 11.10. Three noteworthy charging times are highlighted, namely the end of the cold-start period, first power-on, and full charge of the two capacitors. Furthermore, the minimum sleep time between two subsequent turn-on periods of the node (minimum sleep interval) has also been measured and reported in Figure 11.11.

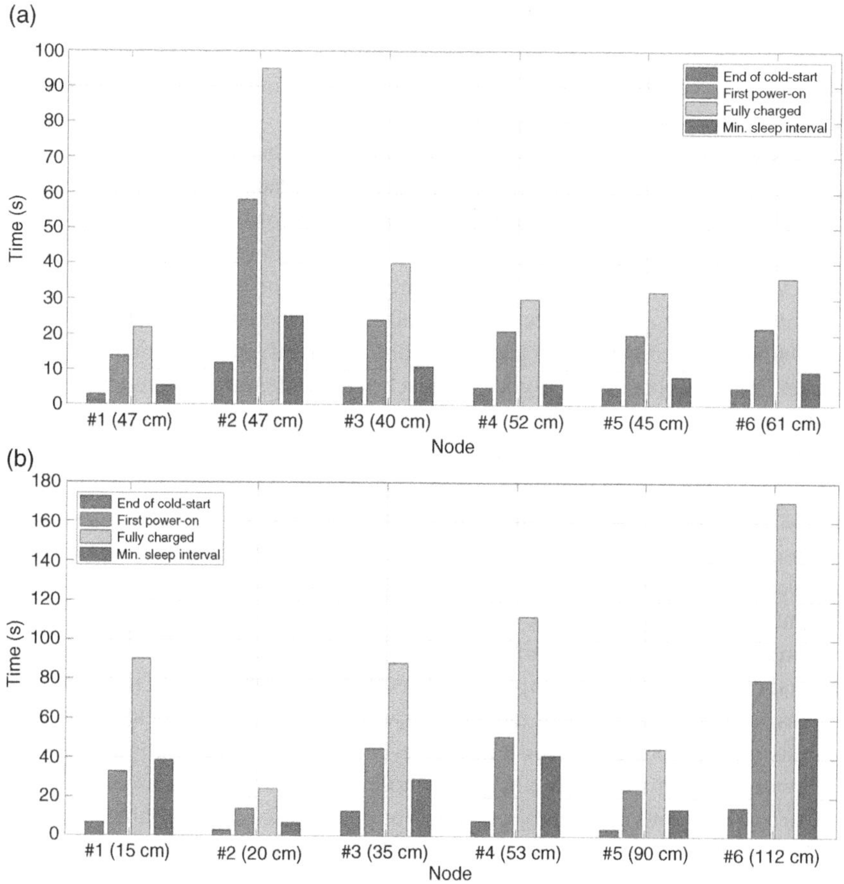

Figure 11.11 Charging times computed for the six different nodes for (a) Setup 1 and (b) Setup 2. *Source:* Giacomo Paolini et al. [18]/MDPI/CC BY 4.0.

11.7 Conclusions

In this chapter, a full end-to-end design of an energy-autonomous wireless node to be applied in IIoT applications is described and extensive measurements have been conducted, starting from the preliminary design of the RF source described in [15]. The main goal has been to guarantee insensitivity from the environment, which is considered electromagnetically harsh, and efficiency and robustness of the technological implementation in terms of EH capabilities. A conformal illuminator, made by three incoherently fed CP antennas, is designed and realized to provide wireless power to linearly polarized sensor nodes. The operative choice of selecting a patch antenna layout has been justified by the presence of a background plane that allows for electrical decoupling of the antenna from the surrounding environment, that for the presented scenario is made by metal. Furthermore, having a complete metal environment can be positively exploited by the antenna itself, which experiences an increased radiation efficiency due to the virtually enlarged background plane.

Most of the works implementing EH applications in industrial environments exploit light, RF, thermal, and kinetic energy, but the lack of specific applications related to predictive maintenance of mechanical parts allows the presented system to boast novelty. In the context of far-field WPT [27], high-value RF-to-dc PCEs are achieved for input power levels equal or higher than 0 dBm, whereas in the presented scenario the main goal has been the optimization of the entire sensor node when expecting lower received power level, in the order of about −10 dBm. For this received power level, the obtained RF-to-dc PCE reaches 25%, whereas if the received power is increased to −5 dBm the obtained PCE is 34%, meaning that, although power levels are low, both dc power (25 μW) and voltage (466 mV) at the output of the rectifier (thus at the input of the PMU), are sufficient to complete the cold-start phase of the bq25570 (whose minimum required values are 15 μW and 330 mV).

Based on literature reviews, devices dedicated to the wireless charging of sensor networks have been designed to exploit more commercially available solutions operating in a lower frequency range with respect to this work, as for example around 915 MHz RFID ultrahigh frequency (UHF) band [28]. The resulting performance is comparable for input power level of −10 dBm but with the drawback of having a receiver side with more extensive dimensions. In particular, a PCE of 45% for an input power of −10 dBm is achieved in [29] when operating at 900 MHz, but with a bigger monopole antenna, whose dimensions would exceed the one given by constraints of this project. The solution proposed in this chapter accounts for an antenna whose dimensions of $2.50 \times 3.00 \times 0.32\,\text{cm}^3$ allow it to be placed in small machinery cavities, such as a car engine compartment. Furthermore, the presented solution makes it possible to perform sensing operations on the node itself.

For what concerns the power budget of the low-power sensor node, the overall energy required to perform an active cycle is between 1.45 and 1.65 mJ, with a current consumption that is below 150 nA for sleep mode configuration. This implies a reduction of the capacity of the storage element up to a 1 mF, which is intrinsically related to the leakage currents and size, which are reduced too.

This research has demonstrated not only the feasibility but also the experimental characterization of a system that, to the authors' best knowledge, represents a novel implementation of a fully self-sustainable, battery-free multi-node sensor system designed and validated for electromagnetically harsh environments. The system performs communications and wireless power transmission between one or more RF sources, the illuminators, and the receiver sensor nodes located in any possible position and orientation inside the environment to be monitored.

The system is developed around specialized energy sources called illuminators that were specifically made to withstand severe environments and support the operation of battery-free nodes powered by low-power technologies like LoRa or Bluetooth Low Energy (BLE). The system has been methodically defined to show remote node IDs and parameters monitoring communications without requiring batteries to power up the nodes. With extremely effective WPT capabilities and robustness with regard to the environment, the design choices for the illuminator and the nodes have been meticulously crafted with the goal of providing wireless monitoring in difficult contexts. The nodes' full functionality has been proven with charging times appropriate for engine monitoring purposes used in real world applications.

References

1 Markiewicz, M., Wielgosz, M., Bocheński, M. et al. (2019). Predictive maintenance of induction motors using ultra-low power wireless sensors and compressed recurrent neural networks. *IEEE Access* 7: 178891–178902. https://doi.org/10.1109/ACCESS.2019.2953019.

2 Magno, M., Boyle, D., Brunelli, D. et al. (2014). Extended wireless monitoring through intelligent hybrid energy supply. *IEEE Transactions on Industrial Electronics* 61 (4): 1871–1881. https://doi.org/10.1109/TIE.2013.2267694.

3 Fleming, W.J. (2008). New automotive sensors – a review. *IEEE Sensors Journal* 8 (11): 1900–1921. https://doi.org/10.1109/JSEN.2008.2006452.

4 Hayashi, H. and Ueda, T. (2012). Requirements and considerations of energy harvesting for industrial wireless transmitter. *2012 Proceedings of SICE Annual Conference (SICE)*, Akita, Japan (20–23 August 2012). IEEE. 1414–1415.

5 Viehweger, C., Keutel, T., and Kanoun, O. (2014). Energy harvesting for wireless sensor nodes in factory environments. *2014 IEEE 11th International*

Multi-Conference on Systems, Signals & Devices (SSD14), Barcelona, Spain (11–14 February 2014). IEEE. 1–4. https://doi.org/10.1109/SSD.2014.6808913.

6 Kim, S., Vyas, R., Bito, J. et al. (2014). Ambient RF energy-harvesting technologies for self-sustainable standalone wireless sensor platforms. *Proceedings of the IEEE* 102 (11): 1649–1666. https://doi.org/10.1109/JPROC.2014.2357031.

7 Berger, A., Hörmann, L.B., Leitner, C. et al. (2015). Sustainable energy harvesting for robust wireless sensor networks in industrial applications. *2015 IEEE Sensors Applications Symposium (SAS)*, Zadar, Croatia (13–15 April 2015). IEEE. 1–6. https://doi.org/10.1109/SAS.2015.7133585.

8 Pizzotti, M., Perilli, L., Del Prete, M. et al. (2017). A long-distance RF-powered sensor node with adaptive power management for IoT applications. *Sensors* 17 (8): 1732. https://doi.org/10.3390/s17081732.

9 Okba, A., Takacs, A., and Aubert, H. (2019). Compact rectennas for ultra-low-power wireless transmission applications. *IEEE Transactions on Microwave Theory and Techniques* 67 (5): 1697–1707. https://doi.org/10.1109/TMTT.2019.2902552.

10 Palazzi, V., Hester, J., Bito, J. et al. (2018). A novel ultra-lightweight multiband rectenna on paper for RF energy harvesting in the next generation LTE bands. *IEEE Transactions on Microwave Theory and Techniques* 66 (1): 366–379. https://doi.org/10.1109/TMTT.2017.2721399.

11 Piñuela, M., Mitcheson, P.D., and Lucyszyn, S. (2013). Ambient RF energy harvesting in urban and semi-urban environments. *IEEE Transactions on Microwave Theory and Techniques* 61 (7): 2715–2726. https://doi.org/10.1109/TMTT.2013.2262687.

12 Pacini, A., Costanzo, A., Aldhaher, S., and Mitcheson, P.D. (2017). Design of a position-independent end-to-end inductive WPT link for industrial dynamic systems. *2017 IEEE MTT-S International Microwave Symposium (IMS)*, Honolulu, HI, USA (4–9 June 2017). IEEE. 1053–1056. https://doi.org/10.1109/MWSYM.2017.8058774.

13 Zannas, K., Matbouly, H.E., Duroc, Y., and Tedjini, S. (2019). Flipping a coin, heads or tails. Flipping an RFID tag on metal, ETSI or FCC bands. *2019 IEEE MTT-S International Microwave Symposium (IMS)*, Boston, MA, USA (2–7 June 2019). IEEE. 283–285. https://doi.org/10.1109/MWSYM.2019.8700808.

14 Popovic, Z. (2013). Far-field wireless power delivery and power management for low-power sensors. *2013 IEEE Wireless Power Transfer (WPT)*, Perugia, Italy, 1–4. https://doi.org/10.1109/WPT.2013.6556867.

15 Paolini, G., Shanawani, M., Costanzo, A., et al. (2020). RF energy on-demand for automotive applications. *2020 IEEE/MTT-S International Microwave Symposium (IMS)*, Los Angeles, CA, USA (4–6 August 2020). IEEE. 1191–1194. https://doi.org/10.1109/IMS30576.2020.9224078.

16 Clerckx, B., Zhang, R., Schober, R. et al. (2019). Fundamentals of wireless information and power transfer: from RF energy harvester models to signal and system designs. *IEEE Journal on Selected Areas in Communications* 37 (1): 4–33. https://doi.org/10.1109/JSAC.2018.2872615.

17 Ji, L., Wang, L., Liao, C., and Li, S. (2019). Simultaneous wireless power and bidirectional information transmission with a single-coil, dual-resonant structure. *IEEE Transactions on Industrial Electronics* 66 (5): 4013–4022. https://doi.org/10.1109/TIE.2018.2831196.

18 Paolini, G., Guermandi, M., Masotti, D. et al. (2021). RF-powered low-energy sensor nodes for predictive maintenance in electromagnetically harsh industrial environments. *Sensors* 21 (2): 386. https://doi.org/10.3390/s21020386.

19 Monti, G., Masotti, D., Paolini, G. et al. (2018). EMC and EMI issues of WPT systems for wearable and implantable devices. *IEEE Electromagnetic Compatibility Magazine* 7 (1): 67–77. https://doi.org/10.1109/MEMC.0.8339550.

20 Rizzoli, V., Costanzo, A., and Monti, G. (2004). General electromagnetic compatibility analysis for nonlinear microwave integrated circuits. *2004 IEEE MTT-S International Microwave Symposium Digest (IEEE Cat. No.04CH37535)*, Fort Worth, TX, USA (6–11 June 2004), vol. 2. IEEE. 953–956. https://doi.org/10.1109/MWSYM.2004.1339135.

21 Masotti, D., Costanzo, A., Francia, P. et al. (2014). A load-modulated rectifier for RF micropower harvesting with start-up strategies. *IEEE Transactions on Microwave Theory and Techniques* 62 (4): 994–1004. https://doi.org/10.1109/TMTT.2014.2304703.

22 Costanzo, A., Romani, A., Masotti, D. et al. (2012). RF/baseband co-design of switching receivers for multiband microwave energy harvesting. *Sensors and Actuators A: Physical* 179: 158–168.

23 Popović, Z., Korhummel, S., Dunbar, S. et al. (2014). Scalable RF energy harvesting. *IEEE Transactions on Microwave Theory and Techniques* 62 (4): 1046–1056. https://doi.org/10.1109/TMTT.2014.2300840.

24 Bleszynski, E., Bleszynski, M., and Jaroszewicz, T. (1994). A fast integral-equation solver for electromagnetic scattering problems. *Proceedings of IEEE Antennas and Propagation Society International Symposium and URSI National Radio Science Meeting*, Seattle, WA, USA, vol. 1, 416–419. https://doi.org/10.1109/APS.1994.407725.

25 Sharma, P. and Gupta, K. (1983). Analysis and optimized design of single feed circularly polarized microstrip antennas. *IEEE Transactions on Antennas and Propagation* 31 (6): 949–955. https://doi.org/10.1109/TAP.1983.1143162.

26 Del Prete, M., Costanzo, A., Georgiadis, A. et al. (2015). A 2.45-GHz energy-autonomous wireless power relay node. *IEEE Transactions on Microwave Theory and Techniques* 63 (12): 4511–4520. https://doi.org/10.1109/TMTT.2015.2494003.

27 Popović, Z., Falkenstein, E.A., Costinett, D., and Zane, R. (2013). Low-power far-field wireless powering for wireless sensors. *Proceedings of the IEEE* 101 (6): 1397–1409. https://doi.org/10.1109/JPROC.2013.2244053.

28 Sung, G.-M., Chung, C.-K., Lai, Y.-J., and Syu, J.-Y. (2019). Small-area radiofrequency-energy-harvesting integrated circuits for powering wireless sensor networks. *Sensors* 19 (8): 1754. https://doi.org/10.3390/s19081754.

29 La Rosa, R., Livreri, P., Trigona, C. et al. (2019). Strategies and techniques for powering wireless sensor nodes through energy harvesting and wireless power transfer. *Sensors* 19 (12): 2660. https://doi.org/10.3390/s19122660.

12

Implanted NFC Tags: Study of Energy Harvesting and Reading by Means of Smartphones

Antonio Lázaro, Martí Boada, Ramón Villarino, and David Girbau

Department of Electronic, Electric and Automatic Engineering, Universitat Rovira i Virgili, Tarragona, Spain

12.1 Introduction

Nowadays, the implantation of micrometric devices is increasingly accepted due to the large number and diversity of possible applications. Some of these examples include implantable glucose and oxygen sensors for diabetics [1, 2], implantable chemical devices [3], neural implants [4], cochlear implants [5], ocular implants [6, 7], and cardiovascular devices [8]. However, aspects such as the one related to the energy supply of the implanted devices remain still to be solved [5]. There are transcutaneous battery recharging options using wireless telemetry or direct replacement using surgical techniques. However, in support of greater biocompatibility, there is a growing trend to implant passive devices [9]. On the other hand, regarding communications in these applications, there are technologies such as ultrasonic [10, 11] or RF, the latter being the most widely used. However, the disadvantage of the presence of cellular tissues surrounding the implant must be considered, since they present dielectric loss and consequently attenuate electromagnetic signals, degrading communications. Therefore, the read range, the minimum space occupancy, and the conditions inside the body will be decisive when putting the system into operation, so that it must be considered a severe environment. In addition, when implanted devices are to be powered from an external device, wireless power transfer (WPT) is also a challenge. Among the WPT techniques, the inductive, optical [12], and ultrasonic [13] stand out, as well as the use of radiofrequency electromagnetic waves [14–16].

Wireless Identification and Sensing Systems for Harsh and Severe Environments,
First Edition. Edited by Smail Tedjini and Valentina Palazzi.

Published 2025 by John Wiley & Sons, Inc.

Near Field Communication (NFC) is increasingly used as a wireless communication technology to transfer data in many types of applications, being body implants one of them. NFC technology derives from radiofrequency identification (RFID) used in short-distance applications, and although it has existed for more long time [17, 18], it did not reach its greatest popularity until its massive incorporation into payment systems. Most modern smartphones incorporate NFC technology because it enables simple and safe communication between electronic devices, allowing consumers to perform contactless transactions, access digital content and connect electronic devices with a single tap. NFC technology is also relevant in IoT and Industry 4.0 applications [19]. This technology is also useful for developing low-cost, low-range sensors since it provides a fast and easy way to obtain data from them simply by bringing the reader close to the sensor tag without needing to pair the devices. In addition, there are a variety of commercial NFC integrated circuits (IC) that include the capability of energy harvesting, with enough capacity to power various sensors and microcontrollers [20–23]. The energy transfer between the reader and the tag through NFC reaches the maximum efficiency in the case of using coil antennas of similar size [18, 20]. When implanted coils are considered, they must be as small as possible, for obvious reasons.

The aim of this chapter is to review and study the possibility of powering and communicating with implanted NFC tags, which are implemented with conventional energy harvesting NFC IC available on the market (so a specific ASIC does not need to be designed) [20], using the smartphone as a reader. To achieve it, a system based on three resonant coils will be used to increase the reading range and compared to a conventional system based on two coils. In [15, 24], work related to this type of system can be found. In short, two of the objectives related to the implant will be to power it and to create a communication link in order to send and receive data from it, at a rate that can be relatively high (e.g. 26 kbps under ISO 15693-compliant IC). The number of applications associated with this technology is limited by its short range; however, most applications of body-implanted devices with NFC communication might fall within this range and may be useful for future implanted sensors, such as glucose sensors, chemical sensors, and neuroprosthetics.

In consequence, this chapter verifies the potential of battery-less sensors based on NFC technology for novel implanted biomedical and wearable applications using smartphones as readers. The operation conditions are not simple, and several factors must be considered such as the size limitation, the presence of biological tissues surrounding the tag as well as the metallic parts of the reader near the antenna. All these issues make the system work in very harsh conditions, which limits the communication. To study and overcome these difficulties, a study regarding the adequate depth at which the implant can be placed is performed, to activate the energy harvesting output of the NFC IC using the energy transferred from the smartphone and thus power the sensors that are part of the implant. For this reason, the two- and three-coil systems are studied, proposing a new model, and focusing on the maximum range of each of them. Previous works are available [15], based on a linear analysis that includes the effect of coupling between the coils and the frequency, consisting in using a simplified model of the antennas, but it does not consider the nonlinear behavior of the tag. On the contrary, here a commercial circuit simulator is used to create a model that includes the NFC IC in order to obtain more realistic efficiency measures. The problem arises because these measurements are not made with a truth reader and the load is usually modeled too simply with a constant value, which is not true because this value depends on the received power. In addition, the detuning effect on the coils due to the metal casing of the smartphone must be considered. This chapter presents experimental measurements made with two or three coils, to determine the depth at which the implant can be placed. One of the objectives is to measure the average magnetic field received at the tag, which depends on the number and shapes of coils used, the coupling between them and the power transmitted. The read range is determined when the average magnetic field is higher than a minimum magnetic field, which is a function of the NFC IC and the coil used in the tag. Therefore, its value can be interpreted as a figure of merit of the implanted tag.

12.2 General Considerations on the Proposed Systems

One of the essential aspects that must be considered when working with externally-powered implanted tags is obviously related to energy. The power source is the reader itself (here a mobile phone), which transfers the energy through magnetic coupling between its coil (the one used in the NFC transceiver) and the coil in the implant, a process known as inductive coupling. Consequently, it is essential to analyze which coil (shape, dimensions, coating...) is the most effective both to power the implanted sensor and to extract information from it.

There are several configurations to implement a communication link based on inductive coupling [24]. The basic configuration consists of two coils, one of which transfers the energy and the other receives it. In body applications, one of the advantages is that there is no difference between the magnetic permeabilities outside and inside the body. However, since the magnetic field in the near field region is inversely proportional to $1/d^3$, where d is the distance between coils, it is crucial to carry out a detailed study of alignment and distance between coils, to optimize the energy transfer. In addition, the inductive coupling is smaller the greater the ratio between d and the size of the implant is, which unfortunately is the most common case.

Several works [25, 26] provide solutions to the mentioned problems. In Figure 12.1, three different electrical diagrams of inductive coupling can be observed, depending on whether they are: nonresonant (a) or resonant (b and c), where in the latter case the systems consist of two (b) or three (c) coils, respectively. This chapter will insist on the system that incorporates a third coil located between the reader and the implant (addressed as relay coil or relay antenna), demonstrating its benefits when compared to the system with two coils.

Figure 12.2 shows the schematic of the complete system proposed in this chapter. The magnetic field that provides the energy to power the implant's circuits comes from a smartphone with NFC capability, which is also used to obtain data from the implanted device and transfer them to the cloud. It can also be observed that the communication link is made up of three coils: one in the NFC transceiver of the mobile phone, a relay coil placed on the skin surface and the last one integrated into the implanted device. The implanted device integrates the NFC IC with energy harvesting capability, in addition to a microcontroller and a sensor.

Firstly, it is necessary to carry out a power transfer efficiency (PTE) study considering the reader-implant separation and the misalignment between coils. There are several studies on the WPT generated from a source (R_S) that will be delivered to the load (R_L) [15]. Efficiency is defined as the ratio between the power delivered to the load (P_L) and the power generated by the source (P_{in}). The analytical

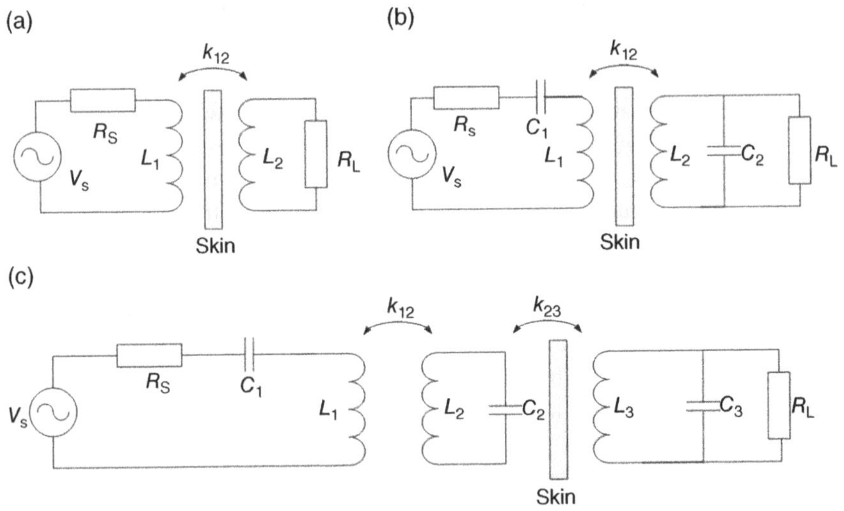

Figure 12.1 (a) Nonresonant inductive link, (b) resonant coupling with two coils, and (c) resonant coupling with three coils.

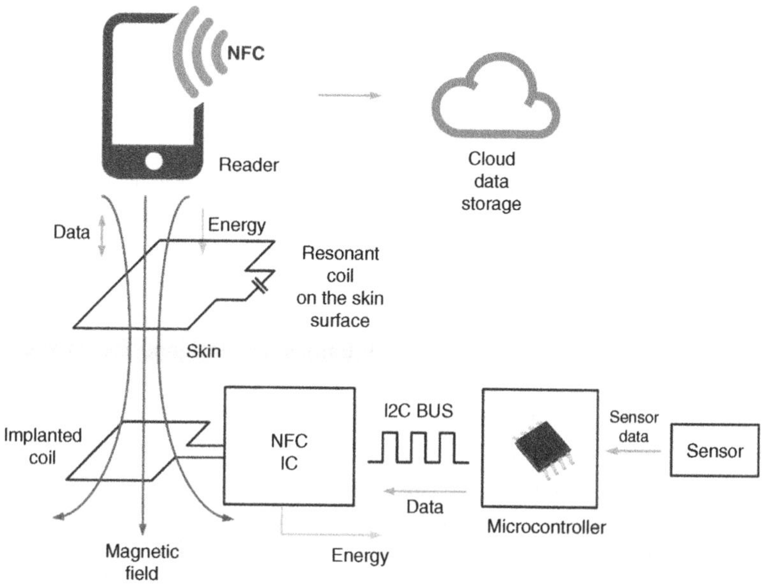

Figure 12.2 Schema of the proposed system.

expressions of efficiency for resonant circuits depending on whether they have two or three coils are given in [15]. Unfortunately, it is shown that the optimal efficiency value does not correspond to the maximum power delivered to the load. Therefore, in the system proposed in this chapter, a design-oriented to transfer the maximum power to the load is considered, since the energy needed to power the electronic circuits that make up the implant depends exclusively on the smartphone. This means that it is essential to take into account the impedance of the IC, which affects the values of both the load impedance and the quality factor. Different works [27–29] demonstrate this nonlinear behavior of the IC impedance that must be considered both when designing and simulating the system. Therefore, it will no longer be enough to use linear simulation models, but it will be necessary to bet on more complex ones, which can be implemented in harmonic balance simulations.

12.3 Description of the Two Systems

In order to study the behavior of the systems based on two and three coils in terms of WPT, a series of simulations will be carried out. Firstly, the models of the reader and the tag are described to then study the effect produced by the materials placed around the body both in the relay and implanted antennas. The coupling between the reader and the tag will also be analyzed to determine the maximum depth the tag can be implanted.

12.3.1 Reader Modeling

The reader used in this chapter is a smartphone with NFC technology, which will be demonstrated capable of reading tags implanted under the skin. Consequently, the reader loop antenna used will be that of the chosen smartphone. Many modern smartphones with plastic casing have their antenna attached to a ferrite layer above the battery [30, 31]. However, in others with a metal case, the antenna is located around the camera [32–34]. These antennas are designed for payment applications on payment terminals whose dimensions are around 85.6×54 mm. The dimension of the coil in the reader, which is normally rectangular in shape, ranges between 3 and 5 cm. Since the implant antenna needs to be much smaller, the coupling between both antennas is expected to be very weak. Consequently, it is necessary to carry out a detailed study.

One essential parameter for communications based on NFC technology is related to the quality factor Q of the reader antenna. Its maximum value will be the one that allows for recovering the information from the implant generated by the modulation of the tag (working as backscatter). Therefore, the bandwidth at -3 dB should be considered, defined as f_c/Q, where f_c is the carrier frequency of 13.56 MHz. This bandwidth coincides with that of the sidebands that modulate the carrier f_c [20]. The maximum loaded Q of the transmitter is given by $f_c \times T_p$, where T_p is the pause time on the carrier modulation. According to NFC forum standard ISO 14443, the quality factor is limited to 40 (35 if design tolerances are considered) at 106 kbps bit-rate transfers [18]. For applications that use ISO 15693, whatever the bit rate, the shortest time present in the uplink communication protocol is a pause lasting $T_p = 9.44\,\mu s$. In this case, $Q_{1max} = 128$, which can generally be reduced to a usable value of 100, assuming design tolerances. In any case, the final value of Q will depend on the type of standard supported by the smartphone used as a reader and, according to some of the publications [19, 35] its final value is around 35. One way to reduce the quality factor is by using a series resistor connected to the antenna.

Figure 12.3 shows the electronic circuit that models the reader in which the series resistance R_Q used to adjust the quality factor is observed. The circuit also includes both the matching network and the electromagnetic interference (EMI) filter. One of the drawbacks of these readers is their difficult operation at short distances, due to the load effect that the tag produces on the reader [36, 37] regardless of the quality of the received signal. Due to the proximity of the reader antenna both to the metallic parts (which cause a variation in inductance) and to the presence of materials with high permittivity (typical in applications related to

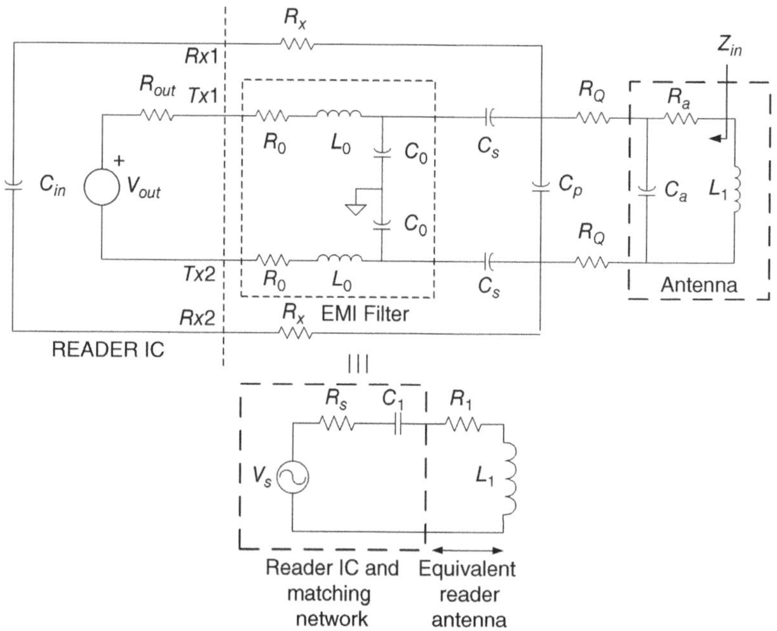

Figure 12.3 Model of the reader, including the matching network (top) and EMI filter, and the simplified model (bottom).

body implants) that introduce a parasitic capacity, detuning of the resonant frequency occurs. The use of coils covered by ferrite foils reduces the variation of the inductance but not completely the mismatch in the reader due to the coupling. This phenomenon, besides excess coupling, limits the maximum distance between the reader and the relay antennas.

In practice, the reader model used to perform the simulations is simplified using a Thevenin equivalent circuit consisting of a voltage source (V_s) in series with a resistance (R_s) and a series capacitance (C_1), which has the purpose of making the circuit resonating at the working frequency. The reader antenna is modeled with a series L–R network (L_1, R_1), which allows losses to be included in order to achieve the desired quality factor. To obtain a correct match at the resonance frequency ($R_s = R_1$), the reader model includes the matching network that also considers the parasitic capacity of the antenna incorporated in the Thevenin equivalent model.

The power delivered by the reader is an essential parameter in the calculation of the link budget, in which the smartphone model used must be considered. The typical transceiver IC for NFC in commercial smartphones can achieve transmit powers in the range of 20–23 dBm (for example, NXP PN7120 or TI TRF7970A Transceiver ICs).

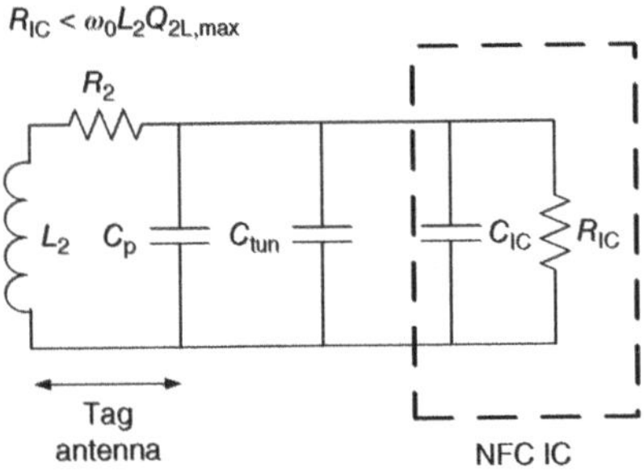

Figure 12.4 Simplified model of the tag including the tag antenna, tuning capacitor and NFC IC.

12.3.2 NFC IC Tag Modeling

The simplified electrical circuit that models the tag is presented in Figure 12.4 and consists of a resistor (R_{IC}) connected in parallel with a capacitor (C_{IC}). It should be noted that the impedance of the IC has a nonlinear behavior and depends on the received power. To tune the circuit to the working frequency, a tuning capacitor (C_{tun}) connected in parallel is included. Besides parameters such as the quality factor of the coils, the coupling between antennas and the load resistance of the tag, which are decisive in the wireless power transferred, a series of aspects must also be considered when using NFC technology, since WPT in NFC systems imposes certain restrictions on the receiver [20]. One of them is the quality factor that is once again limited by the signal that modulates the subcarrier. In this way, the work [38] estimates that the maximum value of the quality factor for a topology that works with two coils is approximately $Q_{2Lmax} = 8\pi = 25$. In the case of exceeding this limit, the time constant for the transition would be increased and the load modulation would decrease. The total quality factor of the tag is obtained as a result of the hyperbolic average of the quality factors of the tag antenna and that associated with the impedance of the IC [15]. To send the information from the tag to the reader with backscatter modulation, the R_{IC} value should be less than 1 500 Ω, considering a value of the implanted inductance L_2 around 0.7 μH.

$$R_{IC} < \omega_0 L_2 Q_{2L,max} \tag{12.1}$$

Figure 12.5 shows the block diagram of the NFC IC which is connected to the tag antenna, which consists of two parts. The first part (upper part, inside dashed line) is the WPT unit and is made of three blocks, whose detailed electrical circuits are represented in Figure 12.6. Its function is to power the IC. The first block is a limiter, which has the function of avoiding damaging the internal circuitry in the event of any unwanted high signal that may appear at the input of the device, since the reader operates in the vicinity of the tag. The second block is a high-efficiency rectifier to perform RF to DC conversion. Finally, the last block is a shunt regulator. The second part (lower part, inside dashed line) of the IC is the communication unit.

The modeling of the limiter is based on the circuit of Figure 12.6a [39]. It consists of two large-area NMOS shunt transistors that turn on when the input power is large enough to produce a current through the diodes that is larger than the

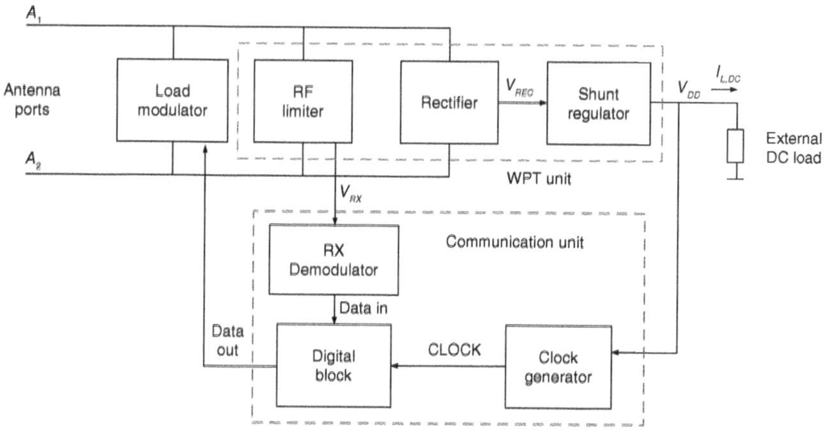

Figure 12.5 Block diagram of the NFC tag IC.

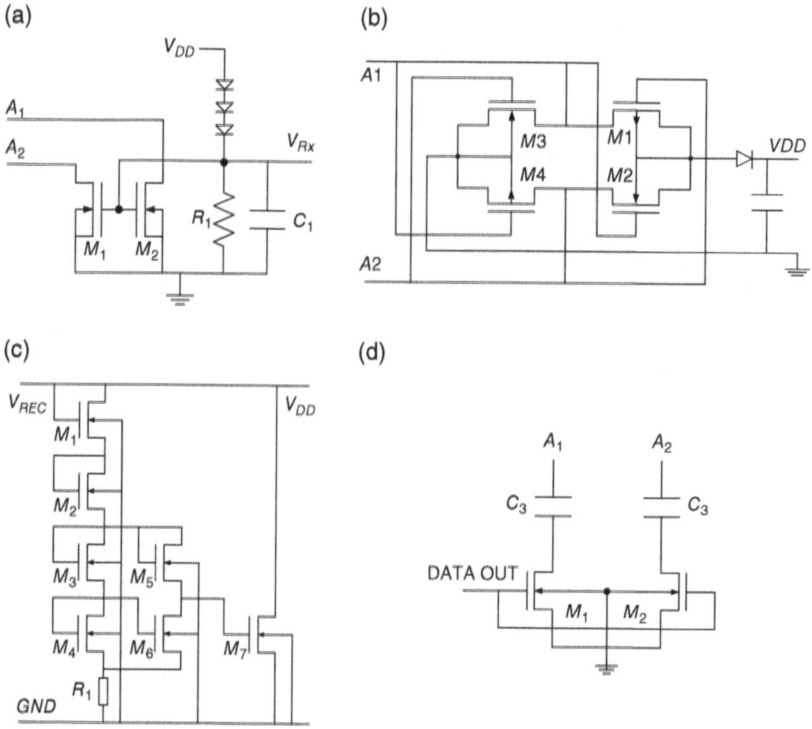

Figure 12.6 Circuits of the WPT unit of the NFC IC: (a) RF limiter, (b) full-wave rectifier, (c) shunt regulator, and (d) load modulator.

threshold voltage needed to turn on the transistors. The limiter is also used to implement an envelope detector using the diodes and the R_C filter (R_1 and C_1) for ASK demodulation. The output of this circuit V_{RX} is used as the input of a slicer circuit for data demodulation.

Although the typical full-wave signal rectifier circuit is based on the diode bridge, its efficiency is diminished by the threshold voltage of the diodes themselves. Schottky diodes could be used to solve this problem, but unfortunately, these are not compatible with CMOS technology. However, there are various alternatives to CMOS rectifiers to alleviate the problem. The most used are the NMOS bridge and the CMOS gate cross-coupled (or cross-connected) rectifier [40]. The rectifier used in the simulations is based on the CMOS gate cross-coupled rectifier (Figure 12.6b). This topology presents a lower turn-on level than an NMOS bridge.

The regulator circuit can be implemented with a Zener diode, which is compatible with CMOS processes [41]. However, another possibility is to implement the circuit with transistors [36] (Figure 12.6c). In this circuit, the shunt impedance to reduce the voltage is controlled by comparing a band-gap-based reference voltage and the output voltage.

Concerning the load modulator (Figure 12.6d), it is composed of two NMOS transistors that act as switches, each of which connects a capacitor from the RF input to GND [42, 43]. The information that is transmitted to the reader is governed by the data that acts on the gate of the transistors (turning them on or off). Consequently, the impedance of the antenna changes according to the data, which results in a modulation of the backscatter. The modulator is modeled as a parasitic shunt capacitance of the IC. The nonlinear model is implemented in Keysight ADS using the TSMC 0.18 μm CMOS design kit from MOSIS [44].

12.3.3 Effect of the Body on the Implant

The communication between the relay antenna and the implanted antenna depends on the body tissues that are between them, influencing the measurements made. Figure 12.7 illustrates the composition of these tissues. Starting from the top, the first layer is the epidermis, which is about 100 μm thick and has the function of acting as a waterproof barrier. Just below is the dermis, it is a resistant tissue with a thickness between 1 and 2 mm, varying according to the age of the person [45, 46]. It is made up of pilose follicles and sweat glands. The immediately inferior layer that precedes the muscle is the hypodermis, it consists of fatty matter and connected tissues. Its thickness varies between 1 and 15 mm. The electrical parameters (permittivity and conductivity) used [47–49] to verify the effect of these tissues using the electromagnetic simulator (Keysight Momentum) are illustrated in Table 12.1.

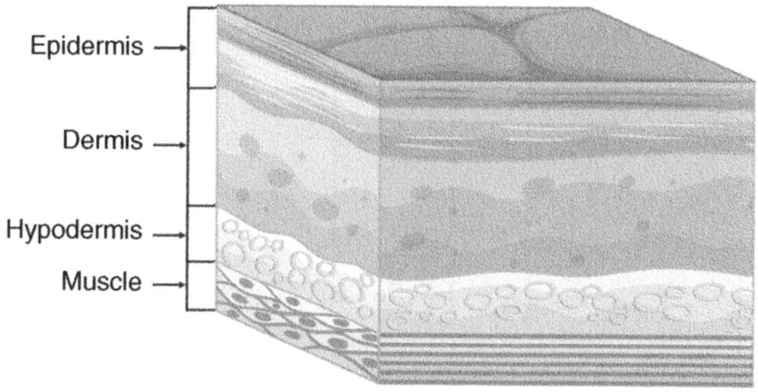

Figure 12.7 Skin cross-section.

Table 12.1 Dielectric layers used in the simulation of the body at 37 °C.

Layer	Relative permittivity	Conductivity (S/m)	Thickness (mm)
Skin (epidermis + dermis)	120	0.25	1.5
Fat (hypodermis)	38	0.21	4
Muscle	138	0.62	25

The value of the inductance of the loops at the frequency of 13.56 MHz, considering the restrictions set by the WPT, is within the range from 0.5 to 1 μH. The antennas used in applications such as implants can be of the following types: based on coiled wires, based on multi-stranded wires, or printed [50]. In this work, the latter is chosen because they are simple to manufacture and are easily connectable with the rest of the electronic circuits. Furthermore, they can adopt different shapes and their inductive value can be approximately calculated from rules that are documented [51, 52]. However, if more rigorous calculations considering the effect of parasitic elements of the antenna or via holes are required, more sophisticated tools such as electromagnetic simulators are needed.

12.3.3.1 Study of the Implanted Antenna

Figure 12.8a shows the typical circuit model of an antenna suspended in the air. The value of the inductance L_a is not very different from that of the implanted antenna since the dielectric permittivity of the tissues does not substantially modify its value. However, the parasitic capacity is affected by the high value of the dielectric constant, which increases the resonance frequency of the tag. On the other hand, the presence of the body increases the dielectric loss, degrading the quality factor of the antenna. These losses are simulated with the resistance R_p.

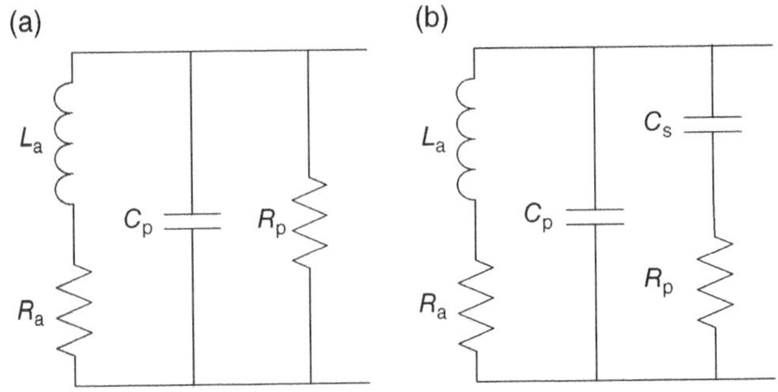

Figure 12.8 (a) Circuit model for a loop antenna in the air, and (b) improved circuit model proposed for an implanted loop antenna.

However, in an implanted antenna, other effects must also be considered. In the first place, it is to be expected that the implant will not only be formed by the antenna but will also include other electronic and mechanical elements that must be isolated to guarantee their correct operation. It has been shown in [53] that silicone compounds used as implant coatings are chemically inert and can embedded in the body for long periods of time, since they do not cause infections when in contact with body tissues, thus being biocompatible for use in humans. However, the electrical properties of the antenna are modified when covered with this type of material. In this way, a gap appears caused by the coating that isolates the antenna from losses introduced by the body. The immediate consequence is that two phenomena occur with respect to the case of the uncovered implant: the detuning of the tag is not as pronounced, and the quality factor is higher compared to the case when it is not protected. The extended electrical model that includes the effect of the coating is the one represented in Figure 12.8b. The silicon coating parameters used for the simulation are $\varepsilon_r = 3$, $\sigma = 2.26\ 10^{-6}$ S/m and $\tan\delta = 0.001$.

Thanks to the values of R_a and R_p, it is possible to adjust the quality factor at a specific frequency, since R_a produces a displacement of the response toward higher frequencies, while R_p does so toward lower frequencies. The equivalent capacitance C_s considers the distributed capacitances of the coating layers. The series antenna resistance R_a is frequency-dependent mainly due to the skin effect. Because the conductivity of the body tissues is much lower than that of the antenna, the eddy currents generated in the tissues are negligible. R_a can be computed from [54]:

$$R_a = R_{dc} \frac{t}{\delta\left(1 - e^{-t/\delta}\right)} \cdot \frac{1}{1 + \frac{t}{w}} \tag{12.2}$$

Where the DC resistance R_{dc} is given by:

$$R_{dc} = \frac{L}{\sigma W t} \tag{12.3}$$

And the skin depth δ is:

$$\delta = \frac{1}{\sqrt{\sigma \pi f \mu_0}} \tag{12.4}$$

In (12.2)–(12.4), L is the length of the strip, t is the thickness of the strip, W is the width of the strip, μ_0 is the vacuum magnetic permeability constant ($4\pi \cdot 10^{-7}$ H/m), and σ is the conductor conductivity.

The manufactured antenna is a rectangular loop, chemically etched on a 0.8 mm thick substrate (FR4), whose both surfaces are plated with copper (35 μm). The loop has a total of three turns on each side electrically connected, with a strip width and a gap between them of 0.5 mm. The area of the loop is $15 \times 15\,\text{mm}^2$ and the measured value of inductance L_a is 0.68 μH, in the order of the necessary to achieve the specifications defined in expression (12.1).

Figure 12.9a shows the simulated and measured results of the quality factor of the antenna, both uncoated and coated with silicone (with different thicknesses) and surrounded by muscle tissue.

The quality factor Q_a is obtained from the real and imaginary parts of the input impedance (12.5) using the electrical parameters and thicknesses of the different layers that make up the body's tissues the ones in Table 12.1.

$$Q_a = \frac{Im(Z)}{\text{Re}(Z)} \tag{12.5}$$

The values of the model are calculated as follows. The value of R_{dc} is optimized from the value of the length L of the antenna, R_a is obtained from the expressions (12.2) to (12.4). The initial values of C_p and C_s are set to half the value of C_p of the antenna in the air, obtained from its resonance frequency. The initial value of the inductance L_a is obtained from the value of the imaginary part of the input impedance at low frequency. Table 12.2 summarizes the model parameters, the quality factor and the resonance frequency of the antenna depending on whether or not it is coated with the silicone layer. In addition, this table includes the value of the tuning capacitor (C_{tun}) necessary to adjust the resonance at the frequency of 13.56 MHz. This capacitance is calculated (12.6) from the imaginary part of the impedance and includes the capacitance of the IC, which is $C_{IC} = 27.5$ pF, obtained from the manufacturer's specifications M24LR04E.

$$C_{tun} = \frac{1}{(2\,f_0)Im(Z(f_0))} - C_{IC} \tag{12.6}$$

The measurements were made by covering the antenna with layers of polylactic acid (PLA) of different thicknesses, manufactured with a 3D printer, surrounded by a pork sirloin to emulate the muscle tissue. The permittivity of PLA is very close to

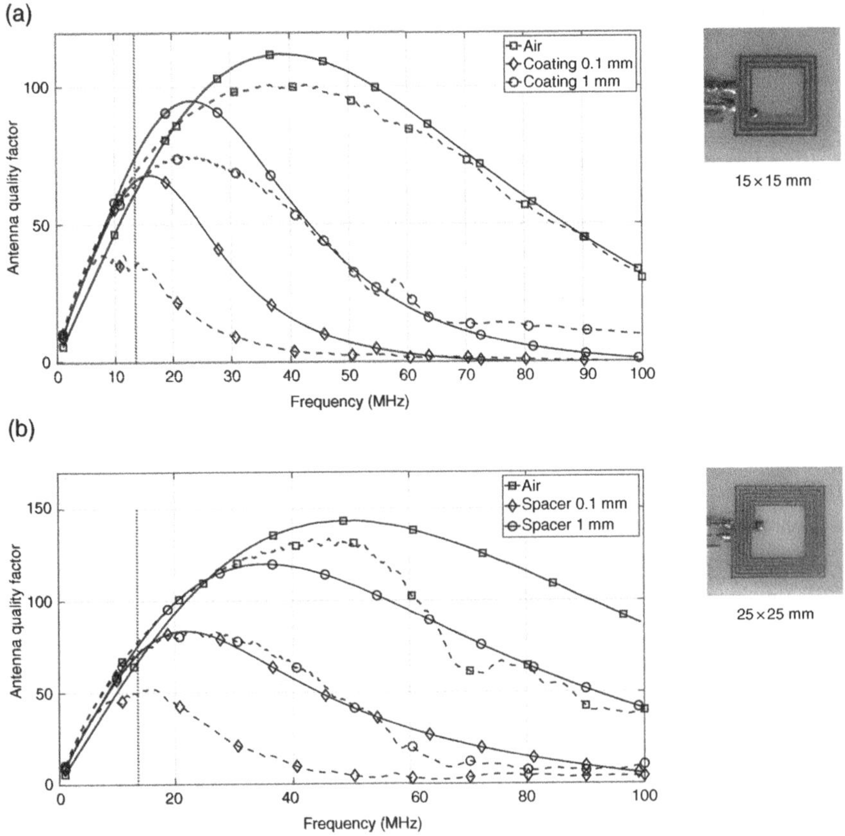

Figure 12.9 Quality factor as a function of frequency for (a) antenna in the air and implanted with coatings of different thicknesses and (b) relay antenna in the air and on the skin separated with spacers of different thicknesses. Continuous line: electromagnetic simulations, dashed line: measurements.

that of silicone [55]. The results show quality factor values lower than those obtained in the simulations and dependent on the thickness of the PLA layer, obtaining values of 40 and 63 for coating thicknesses of 0.1 and 1 mm, respectively. One of the reasons for the difference in the responses obtained in the measurements and the simulations is the high-water content of the pork fillet, especially on its surface, which increases its conductivity compared to that of the simulated tissue.

Finally, as can be seen in Figure 12.9a, the increase in the thickness of the PLA layer contributes to the reduction of the resonance frequency since the effective electrical permittivity decreases. The electromagnetic fields are concentrated between the electrodes and the material that covers them and, consequently, the

Table 12.2 Parameters of the circuit model of the implanted antenna obtained by optimization from electromagnetic simulations.

Parameter	On air	Implanted antenna with different thicknesses of silicone coating		
Thickness of coating (mm)	0	0.1	0.5	1.0
R_{dc} (Ω)	0.38	0.38	0.38	0.38
R_a (Ω)	0.95	0.89	0.78	0.78
f_r (MHz)	135.8	75.1	104.6	120.9
C_p (pF)	2.02	0.93	0.94	0.90
L_a (μH)	0.68	0.68	0.68	0.68
R_p (Ω)	—	78.5	72.56	192.01
C_s (pF)	—	7.20	3.45	2.78
Q_a[a] (13.56 MHz)	61.4 (61.8)	65.9 (35.0)	74.2 (60.5)	73.2 (64.5)
C_{tun} (pF)	201.04	200.18	203.8	143.05

[a] Quality factors obtained from simulation (the values in parenthesis are obtained from measurements).

losses decrease. The value of the thickness of the covering layer that maximizes the quality factor is 0.5 mm. Furthermore, even though this value is lower than that obtained with the same antenna in the air, the resonance occurs at a value closer to the working frequency.

12.3.3.2 Considerations of the Relay Antenna in a Three-Coil System

The relay antenna analysis procedure is done like that of the implanted antenna (see Section 12.3.3.1). Now the antenna is located on top of the epidermis (see Figure 12.7). Once again, the electrical parameters of the tissues given in Table 12.1 are used to carry out the electromagnetic simulations using Keysight Momentum. However, the dielectric properties of tissues and skin, which differ according to liquid content and vary from person to person, result in a large dispersion in resonant frequency measurements. One way to minimize this dispersion is to use a barrier layer of biocompatible material with the skin that has both low permittivity and low losses. In addition, it is useful when protecting the electronics connected to the antenna. This layer is again made of silicone, although other breathable materials with dielectric constant values in the range between 2 and 3 can be used.

The designed relay antenna is a four-turn square loop with an area of $25 \times 25\,mm^2$, and a width and a separation between strips of 0.5 mm, manufactured on the same substrate as the implanted antenna. Figure 12.9b represents the quality factor obtained from the electromagnetic simulations and the measurements, as a function of the thickness of the material that covers the antenna (spacer). The value of the quality factor approaches that of the antenna in free space as the thickness of the covering layer increases. Its value is greater than for the case of the implanted antenna and it is around 75.

Agreement is good between the simulated model and the improved model in Figure 12.9b, which models the quality factor. The values of the model components have been optimized using the Keysight ADS optimization tool and are shown in Table 12.3. The quality factor measurements will be made first with the relay antenna suspended in the air and then placing several layers of PLA of different thicknesses between the antenna and a pork sirloin used as a phantom. As in the case of the implanted antenna, the excess water in the fillet degrades the expected response. The difference in values of this response 40 and 73 is relevant, depending on the thickness of the PLA layer 0.1 and 2 mm respectively.

12.3.4 Coupling Coefficient in Systems Based on Two- and Three-Coils

In order to estimate how deep the device can be implanted, starting from a known distance between the reader and the body, it will be necessary to obtain the coupling coefficient k between the antennas. This parameter depends on the distance

Table 12.3 Parameters of the circuit model of the relay antenna obtained by optimization from electromagnetic simulations.

Parameter	On air	Implanted antenna with different thicknesses of silicone coating		
Thickness of coating (mm)	0	0.1	0.5	1.0
R_{dc} (Ω)	0.43	0.42	0.42	0.42
R_a (Ω)	0.98	0.98	0.98	0.98
f_r (MHz)	173.3	113.5	140.3	150.0
C_p (pF)	2.0	0.07	0.07	0.07
L_a (μH)	0.70	0.70	0.70	0.70
R_p (Ω)	—	790.7	685.9	727.3
C_s (pF)	—	1.0	0.5	0.4
Q_a[a] (13.56 MHz)	61.7 (77.8)	69.9 (49.4)	74.2 (58.5)	74.0 (62.1)
C_{tun} (pF)	194.8	196.7	196.7	196.7

[a] Quality factors obtained from simulation (the values in parenthesis are obtained from measurements).

and alignment between them as well as the type of smartphone antenna and the restrictions related to the maximum quality factor when using smartphones as reader. One way to obtain the value of k between two coupled coils is through the Z parameters (12.7), which are related to the S parameters.

$$k = \frac{\sqrt{Im(Z_{12}) \cdot Im(Z_{21})}}{\sqrt{Im(Z_{11}) \cdot Im(Z_{22})}} \tag{12.7}$$

To estimate the coupling coefficient between the reader antenna and the implanted one as a function of the separation between them, various prototypes are manufactured. The measurements are carried out by placing the reader antenna on a mobile support controlled by a stepper motor and the relay and implanted antennas on a fixed support, both mounted on a rail. All antennas have square surface printed loops with different lengths, designed to resonate at 13.56 MHz. The dimensions of each of them, in addition to the values of L_a and R_a are summarized in Table 12.4. The implanted reference loop has dimensions of 15×15 mm and a total of six turns. Two types of relay antennas of 25×25 mm of four and six turns will be proposed, in addition to three antennas that will be used for the reader unit, two of which have dimensions like those found in commercial smartphones that have a metal cover and surround the camera [22]. The third antenna that acts on the reader unit has dimensions like those of smartphones

Table 12.4 Sizes of antennas used in the coupling study.

Antenna name	Dimensions	Number of turns	L_a (nH)	R_a (Ω)
Implanted	15×15 mm	6	666 684*	0.83 0.91*
Relay 1	25×25 mm	4	697 716*	0.98 0.98*
Relay 2	25×25 mm	6	1185 1172*	1.12 0.99*
Reader 1	25×25 mm	4	697 716*	0.98 0.98*
Reader 2	25×25 mm	6	1185 1172*	1.12 0.99*
Reader 3	50×50 mm	4	2965 2893*	2.39 2.89*

Simulated and measured (*) inductance (L_a) and resistance (R_a) at 13.56 MHz.

protected with a plastic cover. On the other hand, the presence of ferrite sheets, whose objective is to electrically isolate the loop from the effects of the battery or other metallic parts of the device, hardly affects the measurement of the coupling coefficient [22].

Figure 12.10a shows the simulated and measured results of the coupling coefficient as a function of the separation between coils, considering the expression (12.7) and performed with a vector network analyzer (VNA). In this figure, the

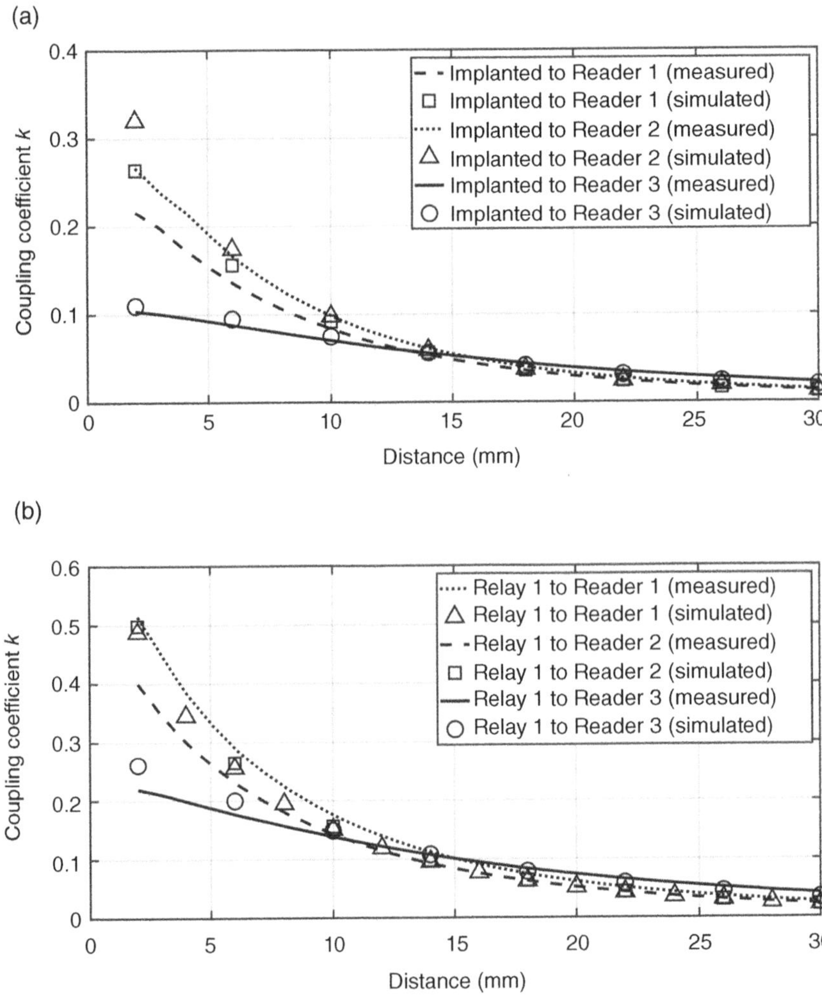

Figure 12.10 Simulated and measured coupling coefficients between (a) the implanted antenna and three reader antennas, (b) the relay antenna (Relay 1) and three different reader antennas, and (c) the implanted antenna and two different relay antennas (See Table 12.4).

(c)

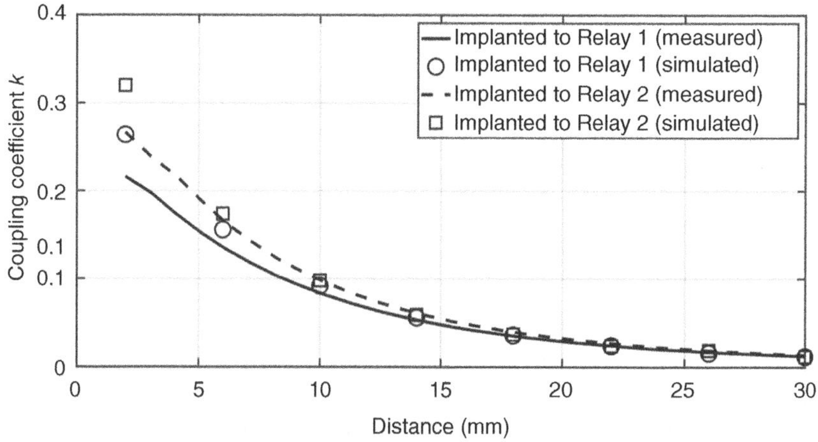

Figure 12.10 (Continued)

case of the system formed by two coils corresponding to the reader and the implant (no relay used in this case) is considered.

Figure 12.10b and c show the coupling coefficients between the reader and relay antennas and the relay and implanted antennas, respectively. The results show a good agreement between the simulations and the measurements. The differences are probably due to the high sensitivity of the system based on three coils, to the misalignment or small separation error between antennas. In any case, it can be clearly seen that loops with a similar area present a greater coupling the closer they are to each other and, on the other hand, large-area antennas maintain a relatively high coupling value at long distances.

12.3.5 Simulation of the Coupling Coefficient and the Power Delivered to the Implant

The aim of this section is to carry out several simulations including the nonlinear model of the tag for both the two-coil and the three-coil systems, to evaluate the possibility of powering them with smartphones. For this, the Harmonic Balance (HB) simulation tool of Keysight ADS is used. The circuit model of both systems is shown in Figure 12.11. Note that the part of the relay antenna must be excluded from the electrical model when the two-coil system is simulated. The part of the circuit associated with the NFC IC is replaced by a block whose circuit elements that compose it have already been described in Section 12.3.2. HB simulation results show the output voltage spectrum; consequently, the power

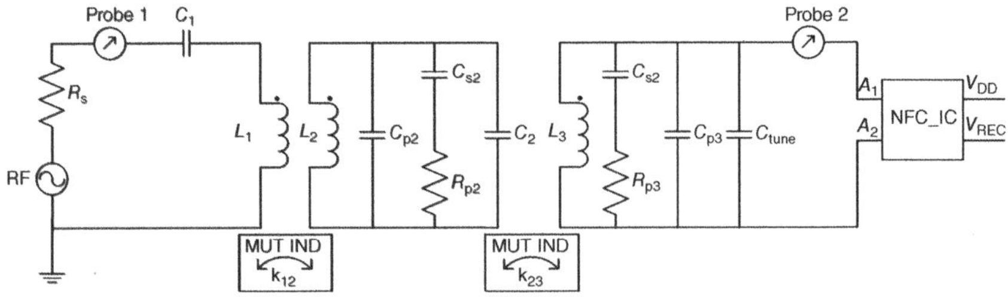

Figure 12.11 Schematic used for the Harmonic Balance simulation of the three-coil system.

received by the IC can be measured by connecting a power probe to its input. On the other hand, the simulator can calculate the AC and DC voltage components after the rectifier (V_{REC}) or regulator (V_{DD}) stage. Efficiency can be calculated from the ratio between the received and input power at the fundamental component (Probe 2 and Probe 1, respectively, in Figure 12.11). Input impedance can be obtained using a current probe at the input of the IC.

Table 12.5 shows the essential parameters to simulate both efficiency and power transfer. In these simulations, the values of the inductance and the quality factor obtained from the simulation of both the implanted antenna and the relay antenna are used. The reader is modeled from the circuit in Figure 12.3 and includes the matching network. To perform the simulations, a low-quality factor, $Q = 35$, is considered, so a resistor R_Q is connected in series with the reader antenna. In this way, a correct transmission in the overall frequency band is guaranteed. Consequently, the source resistance used in the circuit (Figure 12.11) is

Table 12.5 Parameters used in the system simulations.

Parameter	Value	Parameter	Value
Source power (dBm)	20	Relay parasitic capacitance C_{p2} (pF)	0.07
Source resistance R_s (Ω)	4.86	Relay parasitic capacitance Cs_2 (pF)	0.4
Source voltage V_s (V)	0.98	Capacitance C_2 (pF)	196.7
Source capacitance C_1 (pF)	68.8	Tag antenna inductance L_3 (μH)	0.7
Reader antenna induct L_1 (μH)	2	Tag antenna quality factor Q_3	60
Reader quality factor Q_1	35	Tag parasitic resistance R_{p3} (Ω)	192
Relay inductance L_2 (μH)	0.7	Tag parasitic capacitance C_{p3} (pF)	0.9
Relay quality factor Q_2	70	Relay parasitic capacity Cs_3 (pF)	2.8
Relay parasitic resistance R_{p2} (Ω)	700	Tuning Capacitance C_{tun} (pF)	175.9

the one calculated from the quality factor mentioned through the expression $R_s = 2\pi f_0/Q_1$. On the other hand, the signal level of the source is obtained from the transmitted power using the expression $V_s = (2 \cdot P_{TX} \cdot R_S)^{1/2}$. The value of the capacitance C_1 is chosen so that the system resonates at the frequency $f_0 = 13.56$ MHz ($C_1 = 1/(L_1\ \omega_0^2)$). Finally, the capacitors C_2 and C_{tun} are chosen so that both the relay and the tag antennas resonate at f_0.

The first simulation shows the dependence of both the resistance and the input capacitance of the IC on the coupling coefficient, considering both the two-coil and three-coil systems, being the value of k_{12} 0.05 in this last case (Figure 12.12). The nonlinear behavior of the NFC IC [27–29] makes the R_{IC} resistance dependent on the received power. In this way, when the reader interrogates the tag at the beginning, the R_{IC} value is high (several kΩ). However, the resistance value decreases at the moment the NFC IC wakes up to go into operating mode. Therefore, the resistance value decreases when either the AC voltage or the power received by the coil antenna increases. On the other hand, the values of this resistance are lower than the limit established by the maximum quality factor, to guarantee sufficient bandwidth to be able to establish communication by backscattering (12.1). During IC mode of operation, the quality factor of the tag is limited by the quality factor Q_L, which is well below the maximum value needed.

One of the problems to consider refers to the measurement of the IC impedance using standard laboratory instruments (50 Ω) since these two impedances are very

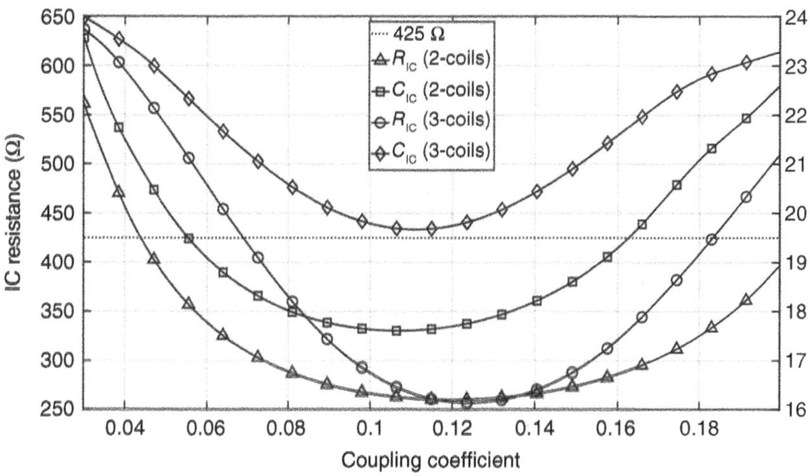

Figure 12.12 Simulated equivalent IC resistance (triangles and circles) and IC capacitance (squares and diamonds) as a function of the coupling coefficient in the two-coil and three-coil ($k_{12} = 0.05$) systems. The threshold resistance at the threshold power is shown (dotted line).

different. Consequently, it is difficult to activate the IC with a commercial VNA, as its maximum power levels are typically limited to 20 dBm referenced at 50 Ω. Therefore, it is necessary to modify the VNA setup including an amplifier and a reflectometer [27–29] to characterize the IC and thus achieve power conditions like those of a reader. Experimental results obtained in the above studies for different ICs show the power dependence of the impedance obtained here in the simulations.

Two important parameters that depend on the coupling coefficient are the AC voltage received in the implanted antenna and the rectified voltage obtained at the output of the shunt regulator. Figure 12.13a shows this dependency considering both systems. For the system to work correctly, $V_{DC} = 2.9\,V$ must be guaranteed at the energy harvesting output, since most commercial ICs have this output disabled below the mentioned voltage. Consequently, below this threshold, both the microcontroller and the sensors will not be biased and therefore it will not be possible to perform any type of measurement. In commercial ICs, the power level required is higher than the minimum required to read the data previously stored in the NFC EEPROM. For the three-coil system, the rectified AC and DC voltages depend both on k_{12} and the implant-relay antenna coupling k_{23}. In this way, for example, to achieve a sufficient DC conversion, it will be necessary for the value of the coupling k_{23} to be equal to 0.05 or 0.08, for a coupling k_{12} of 0.05 or 0.1, respectively.

On the other hand, it will be necessary to protect the circuit if the AC voltage signal level on the antenna terminals exceeds 4.8 V so that for higher signal levels, the RF limiter protects the rectifier transistors. The power conversion efficiency (PCE) of the rectifier also depends on the power received at its input and the coupling coefficient. Figure 12.13b shows this parameter, which is calculated by loading the energy harvesting output with a 3 mA sink current source, in order to simulate the consumption of the sensors. Efficiency values in the order of 44% for received powers around 14 dBm are obtained. To consider the separation between the implant and the reader, Figure 12.14a shows the power received and the efficiency as a function of the coupling coefficient, which has been obtained from the data of Figure 12.10a, considering a two-coil system, being the reading antenna the number 1 (see Table 12.4). For a coupling around 0.04, which corresponds to a reader-implant distance of about 18 mm, the power received by the tag is higher than the threshold (14 dBm). Therefore, a smartphone located at a distance range between 5 and 10 mm should supply enough energy to an implant at a depth between 8 and 13 mm. The harvesting output remains stable for power levels above the threshold and should be able to withstand DC consumptions of around 3 mA. In the event that the coupling is excessive (> 0.12), the power received is reduced by activating a protection limiter that increases the resistance of the IC.

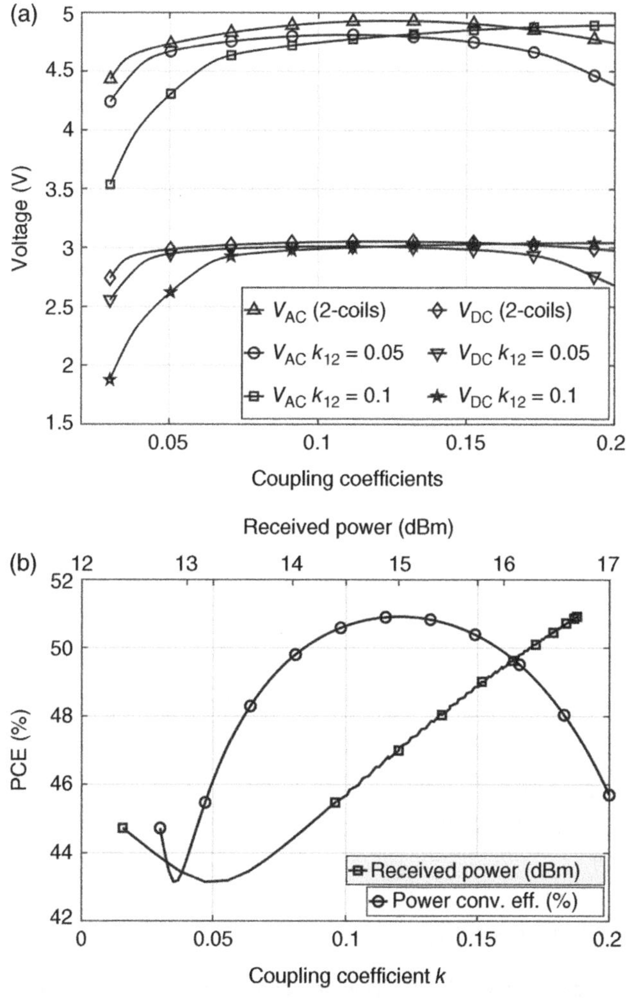

Figure 12.13 (a) Simulated AC voltage (triangles, circles, and squares) at the receiver coil and DC voltage at the output of the shunt regulator (diamonds, downward-pointing triangle, pentagram) as a function of the coupling coefficient in the two-coil and three-coil (k_{12} = 0.05 or 0.1) systems, and (b) simulated power conversion efficiency (PCE) as a function of k (circles) and received power (squares) for the two-coil system.

In the simulations carried out on the three-coil system, the relay antenna is included, in which case the coupling factors between the reader and the relay antennas (k_{12}) and the relay and the implanted antennas (k_{23}) must be considered. The results obtained for both the power received and the efficiency as a function of the coupling k_{23} are shown in Figure 12.14b, as a function of several k_{12} (or

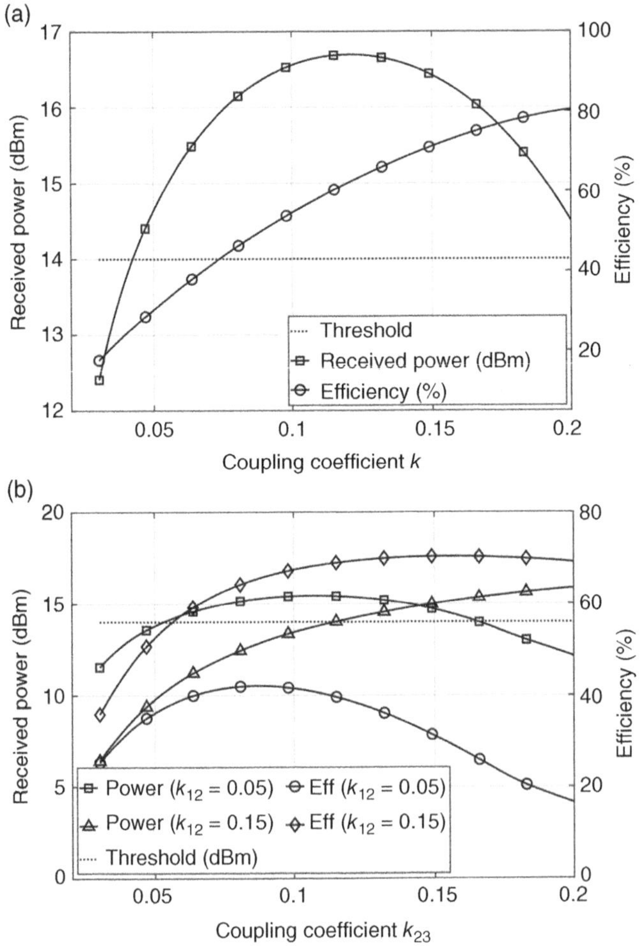

Figure 12.14 Simulated received power in dBm (squares) and efficiency (circles) as a function of the coupling coefficient and the distance between (a) the reader and the implanted antennas in the two-coil system and (b) the relay and implanted antennas in the three-coil system. The dotted line shows the power threshold.

equivalently, for several distances between the reader and relay antenna). The efficiency has an optimal coupling k_{23} that depends on the coupling k_{12} [15]. Both the received power and the efficiency depend on the distance between the implant and the relay antenna, obtained from Figure 12.10b and c, where the reader and relay antenna number 1 have been considered (see Table 12.4). For distances of the order of 1 cm between reader and relay antennas, the maximum depth of the implant ranges between values of 18 and 21 mm. Although the depth of the implant is similar to that of the two-coil system, the main advantage of the three-coil system is its tolerance to antenna misalignment. In this sense, the relay antenna can be used to locate the implant and although slight misalignments may occur, these can be accepted as long as the coupling between the reader antenna and the relay antenna is strong enough, which usually happens if both antennas are separated in a range of 3–10 mm. Consequently, in cases where these misalignments in the horizontal plane are excessive (greater than 10 mm), a significant decrease in the coupling coefficient may occur [56], especially for readers using small-area antennas (very popular in modern smartphones that are protected with metal cases). Therefore, the three-coil system allows information to be obtained from the NFC tags implanted at a depth of around 16 mm, provided that the reader reads the information at a distance from the surface of the skin that varies between 5 and 10 mm.

12.4 Experimental Measurements of Implants Using a Commercial Smartphone with NFC

To demonstrate the correct operation of both systems (based on two- or three-coils), several measurements are made, observing the agreement with the results of the simulations. Measurements are done on an implant inside a phantom consisting of pork sirloin covered with a 100-micron plastic envelope (PET) to protect it, with dielectric properties like those of silicone ($\varepsilon_r = 2.5$). In the experiments, the relay antenna number 1 is used, see Table 12.4. The antenna is a four-turn coil with an area of $25 \times 25\,\text{mm}^2$ that is manufactured on a flexible substrate Ultralam 3000 (Rogers Inc.) with a thickness of 100 μm and $\varepsilon_r = 3.14$. This is the most critical case because the relay antenna is very close to the skin. To adjust the resonance frequency to 13.56 MHz and avoid detuning due to the high permittivity of the phantom, a capacitor connected in parallel with the loop is used. The behavior of the flexible relay antenna is very similar to the original made on a rigid substrate (FR4) with a thickness of 0.8 mm.

12.4.1 Efficiency Characterization

The purpose of the following experiments is to compute both the efficiency and the power delivered to the tag from the measured S parameters. To solve the problem of different impedances between the one of the VNA (50 Ω) and the input impedance of the NFC IC and the source impedance of the reader, the ADS Keysight simulator is used. The procedure consists in loading the measurements using the two-port S parameter data item tool, connecting the tuning capacitors and the appropriate loads in cascade according to the scheme of Figure 12.1.

Considering the impedance simulations carried out in the previous sections, an R_{IC} of 525 Ω and a source resistance that allows obtaining the highest quality factor, $Q = 35$, are chosen. At the input and output, two capacitors with values of 220 and 190 pF, are connected in series and parallel, so that the system is tuned to operate at a frequency of 13.56 MHz. The implanted antenna is manufactured on a 0.8 mm thick FR4 substrate, with a surface area of $15 \times 15\,mm^2$ and a total number of turns of 6. The coil reader 1 described in Table 12.4 ($25 \times 25\,mm^2$ with four turns on FR4 substrate) is used to emulate the reader.

In order to perform the measurements, a guide is used on which the implanted phantom (with or without the relay antenna depending on the system used) and the smartphone pointing to the tag is mounted, so that it can be moved linearly along the guide controlled by a stepper motor. In this way, it is possible to obtain the measurements of the efficiency and the power received by the tag as a function of the distance. Figure 12.15 shows these results depending on whether the system has two- or three-coils, considering different implant depths between 6 and 18 mm. The thickness of the muscle (also simulated with a phantom) located behind the implant is 2 cm thick. To determine the influence of the misalignment between the reader antenna and the implanted or relay antenna (depending on the system), measurements of the efficiency and power received are made, modifying the position of the plane of the reader antenna with respect to the other antennas by means of an electronic control. For these measurements, a distance between the reading antenna and the surface of the phantom of 1 cm and an implant depth of 6 mm is considered. The results shown in Figure 12.16 refer to the measures mentioned depending on whether the system has two or three coils and demonstrate the rapid drop in both efficiency and power received due to misalignment of the antennas, much more pronounced in systems of two than in those of three coils.

12.4.2 Wireless Power Transfer Using a Smartphone as Reader

One of the most important goals of the work is to make measurements of the implanted devices using a smartphone as a reader unit and NFC ICs integrated into the implanted tag. The first challenge is to solve the problems of nonaccessibility to the mobile antenna, to consider the nonlinear behavior of the tag and to study the influence of the metal casing of the mobile that can detune the coils [20]. The solution to these problems is to conduct measurements at different distances between the mobile and the implant of the average magnetic field received by the tag.

To carry out these measurements, Figure 12.17 shows a scheme of the measurement setup and a photograph of the tag. The reader is a commercial smartphone that interrogates the tag and reads the content of the information stored in the

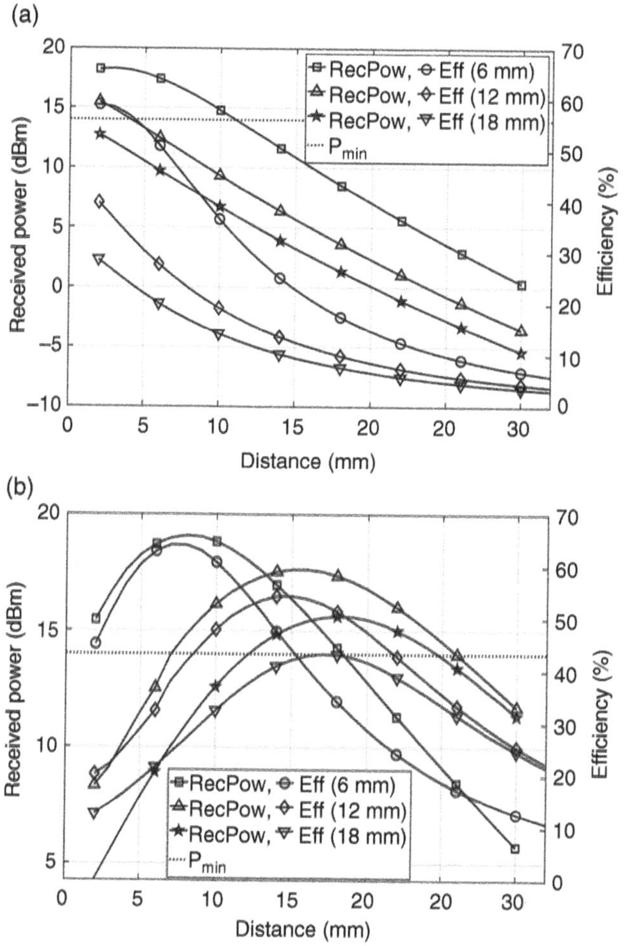

Figure 12.15 Measured received power (squares, triangles, and pentagrams) and efficiency (circles, diamonds, and downward-pointing triangles) as a function of the separation between the antenna reader and the skin surface for several depths of the implant for the (a) two-coil and (b) three-coil systems.

memory of the NFC IC through the development of a specific application. To perform the different measurements, the tag antenna is connected to both a spectrum analyzer (SA, Rohde & Schwarz model FSP3) and a vector network analyzer (VNA, Agilent model E5062A) through a microwave switch. Once the VNA has been calibrated to make electrical measurements of the antenna, the switch is periodically switched between both instruments to measure either the received power (SA) of the carrier frequency (activating the maximum hold trace) or the

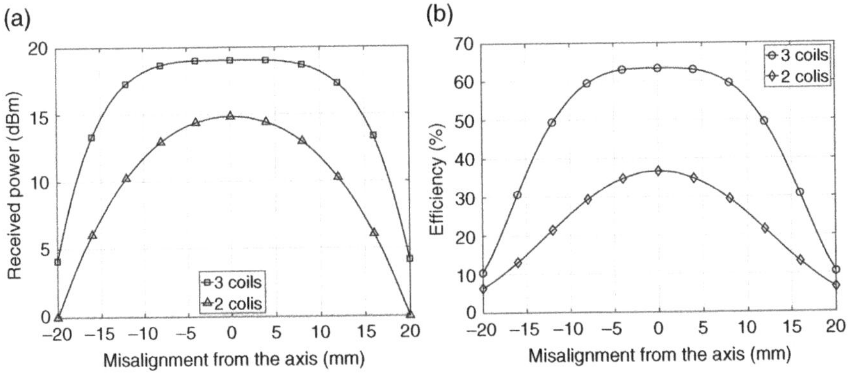

Figure 12.16 Measured (a) received power efficiency and (b) efficiency as a function of the misalignment distance from the axis for the two-coil and three-coil systems. The depth of the implanted antenna is 6 mm and the distance between the reader antenna and the skin surface is 10 mm.

reflection coefficient (S_{11}) of the antenna (VNA) to extract the impedance. Thanks to these measurements, it is possible to obtain the mean magnetic field from expression (12.8):

$$H_{av}\left(\frac{A_{RMS}}{m}\right) = V_{RMS}\cdot|AF| \tag{12.8}$$

Where V_{RMS} is extracted from the power measured with the spectrum analyzer and AF is the antenna factor obtained from the loop impedance [20, 57], which depends on its input impedance Z_{in}, and Z_0 is the reference impedance (50 Ω).

$$AF = \frac{Z_0 + Z_{in}}{j2\pi f\,\mu_0 Z_0 A\cdot N} \tag{12.9}$$

All these measurements are made for different distances between the smartphone and the implant antenna using a step-by-step controlled linear motor. Both the reader and the tag antennas are attached to the guide using PLA supports manufactured with a 3D printer.

Once the measurements have been made, the tag antenna is replaced by the tag itself, placing it on the guide and facing toward the reader (Figure 12.17a). To verify the correct bias of both the microcontroller and the rest of the tag electronics, it is necessary to activate the tag harvesting output, which is measured with an oscilloscope. The activation of the harvesting output is determined by the minimum magnetic field threshold, below which the tag is unbiased and therefore no measurements can be performed. The implanted loop has a total surface area of

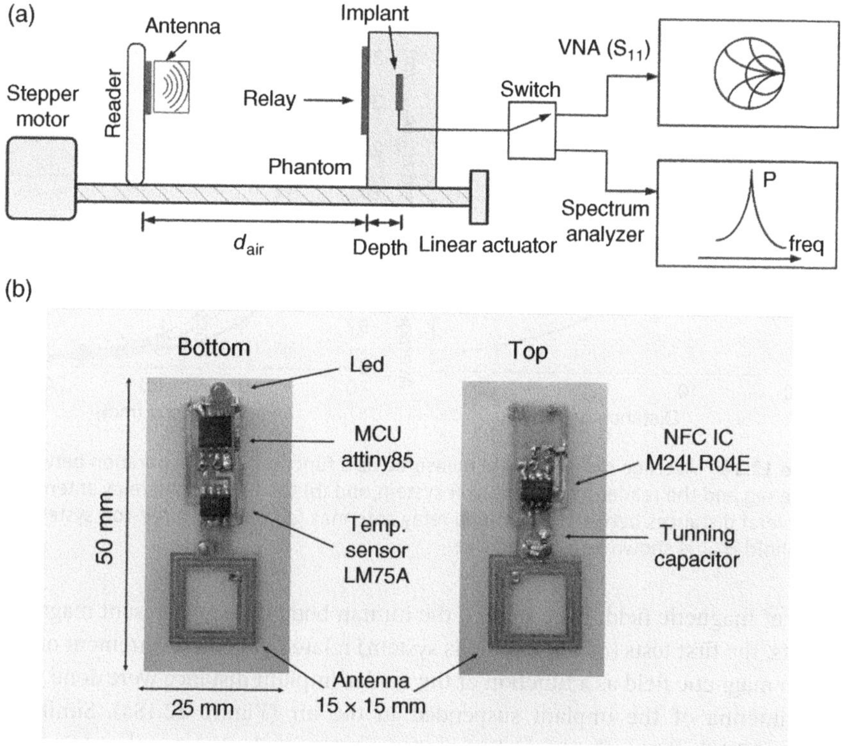

Figure 12.17 (a) Setup used to measure the average magnetic field, (b) photograph of the tag manufactured for measurements: bottom view (left) and top view (right).

$15\times15\,mm^2$ and a total of six turns (three on each side). It is made on 0.8 mm thick FR4 substrate (Figure 12.17b). The chip used for this tag is the ST M24LR04E NFC IC that has the energy harvesting capability. The microcontroller connected to the sensors is the ATtiny85 from Atmel and its clock frequency is 1 MHz. Its function is reading the information from the sensors and storing it in the internal EEPROM of the NFC IC via the I2C bus in NDEF format, so that the data can be read by the smartphone. In general, the maximum separation between the reader and the implant is conditioned by the energy necessary to power the implant circuitry. Functional tests are carried out by reading the information from a temperature sensor installed in the tag, which integrates a led connected to the energy harvesting output to instantly check whether this output is activated or not. The intensity level of the LED is adjusted until reaching the maximum consumption level using a bias resistor (Figure 12.17b). The tag current consumption in the operating mode is 3 mA at 3 V. The NFC IC is configured considering the worst case from the point of view of consumption, which implies generating a sufficient

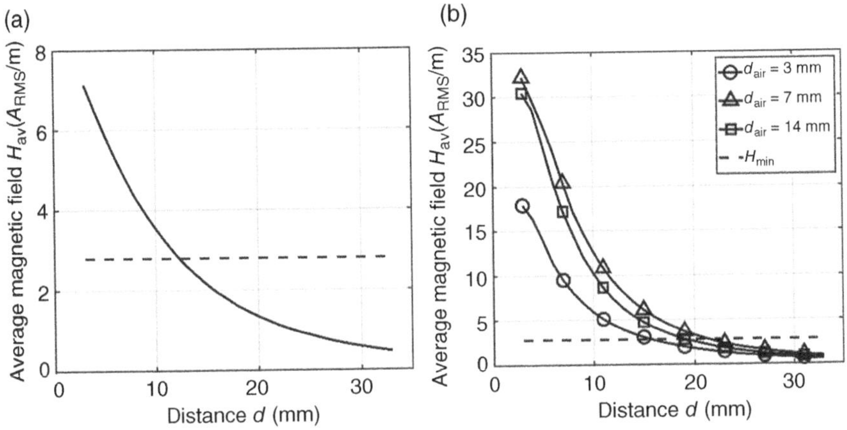

Figure 12.18 Average magnetic field measured as a function of the separation between (a) the tag and the reader for the two-coil system, and (b) the tag and the relay antenna for several distances between reader and relay antennas (d_{air}) for the three-coil system. Threshold H_{min} is shown as a dashed line.

level of magnetic field. Even though the human body does not present magnetic effects, the first tests (on the two-coils system) related to the measurement of the mean magnetic field as a function of the reader-implant distance were done with the antenna of the implant suspended in the air (Figure 12.18a). Similarly, Figure 12.18b shows the equivalent measurement for the three-coil system. The results that are obtained in both cases show a minimum threshold magnetic field H_{min} of 2.8 A_{RMS}/m, which correspond to a read range of 12 mm for the case of the two-coil system or a range of distances that varies between 15 and 20 mm for three-coil system, depending on the separation between the relay antenna and the implant.

Alternatively, the minimum magnetic field H_{min} can be calculated analytically from the expression:

$$H_{min} \approx \frac{\sqrt{\left[1-\left(\frac{f}{f_r}\right)^2\right]^2+\frac{1}{Q_{2L}^2}}}{2\pi f \mu_0 A \cdot N} \cdot U_{min} \tag{12.10}$$

where f_r is the resonance frequency of the tag, A is the loop area, N is the number of turns and U_{min} is the minimum voltage necessary for the operation of the tag and that depends on the IC impedance. Figure 12.19 shows the result of the H_{min} simulation applying the model defined by (12.10) with the following values:

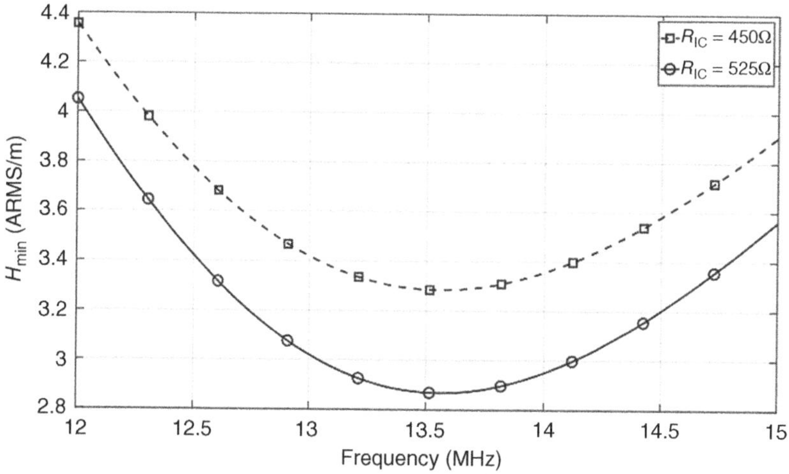

Figure 12.19 Simulated threshold magnetic field H_{min} as a function of the frequency.

$R_{IC} = 525\,\Omega$, inductance of the tag's antenna $L = 0.7\,\mu H$, $Q = 70$, $A = 15 \times 15\,mm^2$, $N = 6$ and $U_{min} = 4.8\,V$ (obtained from the measurement with a probe of low capacity at a specific distance). The experimental measurements of H_{min} agree with those obtained in the analytical model. For example, the minimum power required to activate the energy harvesting output is 13.2 dBm, a value very close to the 14 dBm obtained in the simulations. On the other hand, the estimation of the power conversion efficiency is 42.82%, considering a load current of 3 mA and a current consumption in DC of 400 μA (according to the specifications), which is close to the values obtained in the simulations (Figure 12.13b).

The values of the average magnetic field depending on the separation between the reading antenna and the skin surface, for different depths of the implant and according to the two- or three-coil system, are represented in Figure 12.20a and b, respectively. The benefits of the three-coil system are clearly visible, with which implants can be placed to depths of up to 16 mm. On the contrary, as the reader approaches the relay antenna, a detuning occurs due to the metal parts of the smartphone. Therefore, there is an optimal separation between the reader and the skin of about 15 mm that generates a maximum of the magnetic field. It should be noted that this maximum magnetic field can be modified by changing the resonance frequency of the relay antenna, connecting the appropriate capacitance. The reader used during the experiment was a Xiaomi Note 2. Its antenna is located around the camera and has dimensions around $25 \times 25\,mm^2$, that is, like the first two reader antennas in Table 12.4.

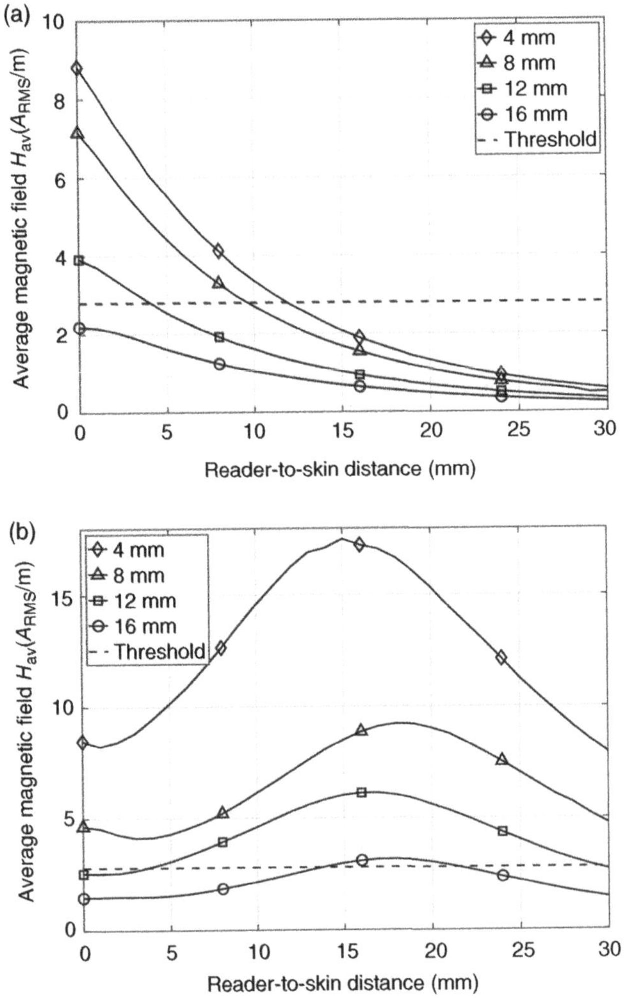

Figure 12.20 Measured average magnetic field as a function of the distance between reader and skin (d_{air}) for several depths of the implanted tag in the phantom for (a) the two-coil and (b) the three-coil systems. Threshold H_{min} is shown as a dashed line.

References

1 Anabtawi, N., Freeman, S., and Ferzli, R. (April 2016). A fully implantable, NFC enabled, continuous interstitial glucose monitor. *2016 IEEE-EMBS International Conference on Biomedical and Health Informatics (BHI)*, Las Vegas, NV, USA, 612–615.

2 A. DeHennis, S. Getzlaff, D. Grice, and M. Mailand, "An NFC-enabled CMOS IC for a wireless fully implantable glucose sensor," *IEEE Journal of Biomedical and Health Informatics*, vol. 20, no. 1, pp. 18–28, Jan. 2016.

3 R. Farra, Sheppard Jr, N.F., McCabe, L. et al., "First-in-human testing of a wirelessly controlled drug delivery microchip," *Science Translational Medicine*, vol. 4, no. 122, 122ra21, Feb. 2012.

4 G. Yilmaz and C. Dehollain, "Single frequency wireless power transfer and full-duplex communication system for intracranial epilepsy monitoring," *Microelectronics Journal*, vol. 45, no. 12, pp. 1595–1602, Dec. 2014.

5 K. Bazaka and M. Jacob, "Implantable devices: issues and challenges," *Electronics*, vol. 2, no. 4, pp. 1–34, Dec. 2012.

6 Chiou, J.-C., Hsu, S.H., Liao, Y.T. et al. (2016). Toward a wirelessly powered on-lens intraocular pressure monitoring system. *IEEE Journal of Biomedical and Health Informatics* 20 (5): 1216–1224.

7 Chitnis, G., Maleki, T., Samuels, B. et al. (2013). A minimally invasive implantable wireless pressure sensor for continuous IOP monitoring. *IEEE Transactions on Biomedical Engineering* 60 (1): 250–256.

8 P. D. Costa, P. P. Rodrigues, A. H. Reis, and A. Costa-Pereira, "A review on remote monitoring technology applied to implantable electronic cardiovascular devices," *Telemedicine and e-Health*, vol. 16, no. 10, pp. 1042–1050, Dec. 2010.

9 Hannan, M.A., Mutashar, S., Samad, S.A., and Hussain, A. (2014). Energy harvesting for the implantable biomedical devices: issues and challenges. *BioMedical Engineering Online* 13 (1): 79.

10 Seo, D., Neely, R.M., Shen, K. et al. (2016). Wireless recording in the peripheral nervous system with ultrasonic neural dust. *Neuron* 91 (3): 529–539.

11 G. E. Santagati and T. Melodia, "Experimental evaluation of impulsive ultrasonic intra-body communications for implantable biomedical devices," *IEEE Transactions on Mobile Computing*, vol. 16, no. 2, pp. 367–380, Feb. 2017.

12 K. Goto, T. Nakagawa, O. Nakamura, and S. Kawata, "An implantable power supply with an optically rechargeable lithium battery," *IEEE Transactions on Biomedical Engineering*, vol. 48, no. 7, pp. 830–833, Jul. 2001.

13 Phillips, W.B., Towe, B.C., and Larson, P.J. (Sept. 2003). An ultrasonically-driven piezoelectric neural stimulator. *Proceedings of the 25th Annual International Conference of the IEEE Engineering in Medicine and Biology Society (IEEE Cat. No.03CH37439)*, Cancun, Mexico, 1983–1986.

14 Dinis, H. and Mendes, P.M. (2017). Recent advances on implantable wireless sensor networks. In: *Wireless Sensor Networks - Insights and Innovations* (ed. P. Sallis). InTech https://www.intechopen.com/books/wireless-sensor-networks-insights-and-innovations/recent-advances-on-implantable-wireless-sensor-networks.

15 M. Kiani, Uei-Ming Jow, and M. Ghovanloo, "Design and optimization of a 3-coil inductive link for efficient wireless power transmission," *IEEE Transactions on Biomedical Circuits and Systems*, vol. 5, no. 6, pp. 579–591, Dec. 2011.

16 Yilmaz, G. and Dehollain, C. (2017). *Wireless Power Transfer and Data Communication for Neural Implants*. Cham, Switzerland: Springer International Publishing AG.

17 Finkenzeller, K., Müller, D., and Handbook, R.F.I.D. (2010). *Fundamentals and Applications in Contactless Smart Cards, Radio Frequency Identification and Near-Field Communication*, 3e. New York, NY: Wiley.

18 Paret, D. (2016). *Design Constraints for NFC Devices*, 1e. Hoboken, NJ: Wiley.

19 B. Ozdenizci, V. Coskun, and K. Ok, "NFC internal: an indoor navigation system," *Sensors*, vol. 15, no. 4, pp. 7571–7595, Mar. 2015.

20 A. Lazaro, R. Villarino, and D. Girbau, "A survey of NFC sensors based on energy harvesting for IoT applications," *Sensors*, vol. 18, no. 11, p. 3746, Nov. 2018.

21 M. Boada, A. Lazaro, R. Villarino, and D. Girbau, "Battery-less soil moisture measurement system based on a NFC device with energy harvesting capability," *IEEE Sensors Journal*, vol. 18, no. 13, pp. 5541–5549, Jul. 2018.

22 Boada, M., Lazaro, A., Villarino, R., and Girbau, D. (2019). Battery-less NFC sensor for PH monitoring. *IEEE Access* 7: 33226–33239.

23 Lazaro, A., Boada, M., Villarino, R., and Girbau, D. (2019). Color measurement and analysis of fruit with a battery-less NFC sensor. *Sensors* 19 (7): 1741.

24 Agbinya, J.I. (2016). *Wireless Power Transfer*, 2e. Delft, The Netherlands: River Publishers.

25 A. Kurs, A. Karalis, R. Moffatt, J. D. Joannopoulos, P. Fisher, and M. Soljacic, "Wireless power transfer via strongly coupled magnetic resonances," *Science*, vol. 317, no. 5834, pp. 83–86, Jul. 2007.

26 B. L. Cannon, J. F. Hoburg, D. D. Stancil, and S. C. Goldstein, "Magnetic resonant coupling as a potential means for wireless power transfer to multiple small receivers," *IEEE Transactions on Power Electronics*, vol. 24, no. 7, pp. 1819–1825, Jul. 2009.

27 Gebhart, M., Bruckbauer, J., and Gossar, M. (July 2010). Chip impedance characterization for contactless proximity personal cards. *7th International Symposium on Communication Systems, Networks & Digital Signal Processing (CSNDSP 2010)*, Newcastle upon Tyne, 826–830

28 Gvozdenovic, N., Mayer, L.W., and Mecklenbrauker, C.F. (2014). Measurement of harmonic distortions and impedance of HF RFID chips. In: *in the 8th European Conference on Antennas and Propagation (EuCAP 2014), The Hague*, 2940–2944. Netherlands.

29 B. Couraud, T. Deleruyelle, E. Kussener, and R. Vauche, "Real-time impedance characterization method for RFID-type backscatter communication devices," *IEEE Transactions on Instrumentation and Measurement*, vol. 67, no. 2, pp. 288–295, Feb. 2018.

30 Jensen, H.T. and Lee, H. (2017). Chassis NFC antenna booster. U.S. Patent No. 9,729,210. 8 August 2017.

31 Lee, B., Kim, B., and Yang, S. (2014). Enhanced loop structure of NFC antenna for mobile handset applications. *International Journal of Antennas and Propagation* 2014: 1–6.

32 Zhu, J.-Q., Ban, Y.-L., Sim, C.-Y.-D., and Wu, G. (2017). NFC antenna with nonuniform meandering line and partial coverage ferrite sheet for metal cover smartphone applications. *IEEE Transactions on Antennas and Propagation* 65 (6): 2827–2835.

33 Liu, Q., Zhou, J., and Chen, D. (2018). NFC antenna assembly. U.S. Patent No. 9,941,572. 10 April 2018.

34 Lee, B. and Harackiewicz, F.J. (2017). Design of a simple structured NFC loop antenna for mobile phones applications. *Progress In Electromagnetics Research C* 76: 149–157.

35 Alibakhshi-Kenari, M., Naser-Moghadasi, M., Ali Sadeghzadeh, R. et al. (2016). Dual-band RFID tag antenna based on the Hilbert-curve fractal for HF and UHF applications. *IET Circuits, Devices and Systems* 10 (2): 140–146.

36 Jiang, B., Smith, J.R., Philipose, M. et al. (2007). Energy scavenging for inductively coupled passive RFID systems. *IEEE Transactions on Instrumentation and Measurement* 56 (1): 118–125.

37 Marechal, C. and Paret, D. (Oct. 2008). The loading effect of proximity contactless smart card. Incidences of impedance of the shunt regulator. *2008 4th International Conference on Wireless Communications, Networking and Mobile Computing*, Dalian, China, 1–4.

38 Gebhart, M. (June 2011). Analytical considerations for an ISO/IEC14443 compliant SmartCard transponder. *Proceedings of the 11th International Conference on Telecommunications*, Graz, 9–16.

39 Lu, C.-H., Li, J.-A., and Lin, T.-H. (April 2016) A 13.56-MHz passive NFC tag IC in 0.18-μm CMOS process for biomedical applications. *2016 International Symposium on VLSI Design, Automation and Test (VLSI-DAT)*, Hsinchu, 1–4.

40 Jamali, B., Ranasinghe, D.C., and Cole, P.H. (2005 2006). Analysis of UHF RFID CMOS rectifier structures and input impedance characteristics. *Proc. SPIE 6035, Microelectronics: Design, Technology, and Packaging II*, Brisbane, Australia, 60350D.

41 Teichmann, J., Burger, K., Hasche, W. et al. (Sept. 2003). One time programming (OTP) with Zener diodes in CMOS processes. *Electrical Performance of Electrical Packaging* (IEEE Cat. No. 03TH8710), Estoril, Portugal, 433–436.

42 Alcalde, F., Alpago, O., and Lipovetsky, J. (July 2014) CMOS design of an RFID interface compatible with ISO/IEC-14443 type a protocol. *2014 Argentine Conference on Micro-Nanoelectronics, Technology and Applications (EAMTA)*, Mendoza, Argentina, 41–47.

43 Hu, J., Tan, H., and Xu, Y. (Dec. 2007). A novel design of low power RFID tag interface circuit. *IET Conference on Wireless, Mobile and Sensor Networks 2007 (CCWMSN07)*, Shanghai, China, 478–481.

44 Mosis integrated circuit fabrication service. https://www.mosis.com (accessed 24 December 2019).

45 Tan, C.Y., Statham, B., Marks, R., and Payne, P.A. (1982). Skin thickness measurement by pulsed ultrasound; its reproducibility, validation and variability. *British Journal of Dermatology* 106: 657–667.

46 Huclova, S., Erni, D., and Fröhlich, J. (2012). Modelling and validation of dielectric properties of human skin in the MHz region focusing on skin layer morphology and material composition. *Journal of Physics D: Applied Physics* 45 (2): 025301.

47 Gabriel, C., Gabriel, S., and Corthout, E. (1996). The dielectric properties of biological tissues: I. Literature survey. *Physics in Medicine and Biology* 41 (11): 2231–2249.

48 Gabriel, S., Lau, R.W., and Gabriel, C. (1996). The dielectric properties of biological tissues: II. Measurements in the frequency range 10 Hz to 20 GHz. *Physics in Medicine and Biology* 41 (11): 2251–2269.

49 Gabriel, S., Lau, R.W., and Gabriel, C. (1996). The dielectric properties of biological tissues: III. Parametric models for the dielectric spectrum of tissues. *Physics in Medicine and Biology* 41 (11): 2271–2293.

50 Schormans, M., Valente, V., and Demosthenous, A. (2018). Practical inductive link design for biomedical wireless power transfer: a tutorial. *IEEE Transactions on Biomedical Circuits and Systems* 12 (5): 1112–1130.

51 Grover, F.W. (2009). *Inductance Calculations: Working Formulas and Tables.* Mineola, NY: Dover Publications.

52 Youbok, L. (2003). Antenna circuit design for RFID applications. Application Note AN710, Microchip Technology Inc. http://ww1.microchip.com/downloads/en/appnotes/00710c.pdf.

53 Donaldson, P.E.K. and Sayer, E. (1981). A technology for implantable hermetic packages. Part 2: An implementation. *Medical & Biological Engineering & Computing* 19 (4): 403–405.

54 H. A. Wheeler, "Formulas for the Skin Effect," *Proceedings of the IRE*, vol. 30, no. 9, pp. 412–424, Sep. 1942.

55 Dichtl, C., Sippel, P., and Krohns, S. (2017). Dielectric properties of 3D printed polylactic acid. *Advances in Materials Science and Engineering*, Article ID 6913835, 10 pages.

56 Fotopoulou, K. and Flynn, B. (2011). Wireless power transfer in loosely coupled links: coil misalignment model. *IEEE Transactions on Magnetics* 47 (2): 416–430.

57 Ishii, M. and Komiyama, K. (June-Jul. 2004). A measurement method for the antenna factor of small loop antenna by measuring the input impedance. *2004 Conference on Precision Electromagnetic Measurements*, London, UK, 80–81.

Index

Note: *Italic* page numbers refer to *figures* and **Bold** page numbers reference to **tables**.

Wireless Identification and Sensing Systems for Harsh and Severe Environments, First Edition. Edited by Smail Tedjini and Valentina Palazzi.

Published 2025 by John Wiley & Sons, Inc.

f

q

r

Printed and bound by CPI Group (UK) Ltd, Croydon, CR0 4YY
17/09/2026
14964308-0005